Handbook of Paper and Paperboard Packaging Technology

Handbook of Paper and Paperboard Packaging Technology

Second Edition

Edited by

Mark J. Kirwan
Paper and Paperboard Specialist,
Fellow of the Packaging Society,
London, UK

WILEY-BLACKWELL
A John Wiley & Sons, Ltd., Publication

This edition first published 2013 © 2013 by John Wiley & Sons, Ltd

Wiley-Blackwell is an imprint of John Wiley & Sons, formed by the merger of Wiley's global Scientific, Technical and Medical business with Blackwell Publishing.

Registered Office
John Wiley & Sons, Ltd, The Atrium, Southern Gate, Chichester, West Sussex, PO19 8SQ, UK

Editorial Offices
9600 Garsington Road, Oxford, OX4 2DQ, UK
The Atrium, Southern Gate, Chichester, West Sussex, PO19 8SQ, UK
2121 State Avenue, Ames, Iowa 50014-8300, USA

For details of our global editorial offices, for customer services and for information about how to apply for permission to reuse the copyright material in this book please see our website at www.wiley.com/wiley-blackwell.

The right of the authors to be identified as the authors of this work has been asserted in accordance with the UK Copyright, Designs and Patents Act 1988.

All rights reserved. No part of this publication may be reproduced, stored in a retrieval system, or transmitted, in any form or by any means, electronic, mechanical, photocopying, recording or otherwise, except as permitted by the UK Copyright, Designs and Patents Act 1988, without the prior permission of the publisher.

Designations used by companies to distinguish their products are often claimed as trademarks. All brand names and product names used in this book are trade names, service marks, trademarks or registered trademarks of their respective owners. The publisher is not associated with any product or vendor mentioned in this book. This publication is designed to provide accurate and authoritative information in regard to the subject matter covered. It is sold on the understanding that the publisher is not engaged in rendering professional services. If professional advice or other expert assistance is required, the services of a competent professional should be sought.

Library of Congress Cataloging-in-Publication Data

Handbook of paper and paperboard packaging technology / edited by Mark J. Kirwan. – 2nd ed.
　p.　cm.
　Earlier edition has title: Paper and paperboard packaging technology.
　Includes bibliographical references and index.
　ISBN 978-0-470-67066-8 (hardback : alk. paper) – ISBN 978-1-118-47091-6 (epdf/ebook) –
ISBN 978-1-118-47089-3 (emobi) – ISBN 978-1-118-47092-3 (epub) – ISBN 978-1-118-47093-0 (obook)
　1. Paper containers.　2. Paperboard.　3. Packaging.　I. Kirwan, Mark J.
　　TS198.3.P3P37 2013
　　658.7′85–dc23

2012024778

A catalogue record for this book is available from the British Library.

Wiley also publishes its books in a variety of electronic formats. Some content that appears in print may not be available in electronic books.

Front cover illustration courtesy of Smurfit Kappa Group plc.
Cover design by Meaden Creative

Set in 10/12pt Times by SPi Publisher Services, Pondicherry, India
Printed and bound in Malaysia by Vivar Printing Sdn Bhd

1　2013

Contents

Contributors	xv
Preface	xvii
Acknowledgements	xix

1 Paper and paperboard – raw materials, processing and properties — 1
Daven Chamberlain and Mark J. Kirwan

1.1	Introduction – quantities, pack types and uses		1
1.2	Choice of raw materials and manufacture of paper and paperboard		6
	1.2.1	Introduction to raw materials and processing	6
	1.2.2	Sources of fibre	7
	1.2.3	Fibre separation from wood (pulping)	8
	1.2.4	Whitening (bleaching)	10
	1.2.5	Recovered fibre	10
	1.2.6	Other raw materials	11
	1.2.7	Processing of fibre at the paper mill	12
	1.2.8	Manufacture on the paper or paperboard machine	13
	1.2.9	Finishing	19
1.3	Packaging papers and paperboards		20
	1.3.1	Introduction	20
	1.3.2	Tissues	20
	1.3.3	Greaseproof	20
	1.3.4	Glassine	21
	1.3.5	Vegetable parchment	21
	1.3.6	Label paper	21
	1.3.7	Bag papers	21
	1.3.8	Sack kraft	22
	1.3.9	Impregnated papers	22
	1.3.10	Laminating papers	22
	1.3.11	Solid bleached board (SBB)	22
	1.3.12	Solid unbleached board (SUB)	23
	1.3.13	Folding boxboard (FBB)	23
	1.3.14	White-lined chipboard (WLC)	24
1.4	Packaging requirements		25
1.5	Technical requirements of paper and paperboard for packaging		26
	1.5.1	Requirements of appearance and performance	26
	1.5.2	Appearance properties	26
	1.5.3	Performance properties	34
1.6	Specifications and quality standards		48
1.7	Conversion factors for substance (basis weight) and thickness measurements		48
	References		49

2 Environmental and resource management issues 51
Daven Chamberlain and Mark J. Kirwan

    2.1   Introduction   51
    2.2   Sustainable development   53
    2.3   Forestry   54
    2.4   Environmental impact of manufacture and use of paper and paperboard   61
           2.4.1   Issues giving rise to environmental concern   61
           2.4.2   Energy   62
           2.4.3   Water   66
           2.4.4   Chemicals   67
           2.4.5   Transport   68
           2.4.6   Manufacturing emissions to air, water and solid waste   68
    2.5   Used packaging in the environment   73
           2.5.1   Introduction   73
           2.5.2   Waste minimisation   74
           2.5.3   Waste management options   74
    2.6   Life cycle assessment   79
    2.7   Carbon footprint   81
           2.7.1   Carbon sequestration in forests   81
           2.7.2   Carbon stored in forest products   82
           2.7.3   Greenhouse gas emissions from forest product manufacturing facilities   82
           2.7.4   Greenhouse gas emissions associated with producing fibre   83
           2.7.5   Greenhouse gas emissions associated with producing other raw materials/fuels   83
           2.7.6   Greenhouse gas emissions associated with purchased electricity, steam and heat, and hot and cold water   83
           2.7.7   Transport-related greenhouse gas emissions   83
           2.7.8   Emissions associated with product use   83
           2.7.9   Emissions associated with product end of life   83
           2.7.10 Avoided emissions and offsets   83
    2.8   Conclusion   84
    References   86

3 Paper-based flexible packaging 91
Jonathan Fowle and Mark J. Kirwan

    3.1   Introduction   91
    3.2   Packaging needs which are met by paper-based flexible packaging   94
           3.2.1   Printing   94
           3.2.2   Provision of a sealing system   95
           3.2.3   Provision of barrier properties   95
    3.3   Manufacture of paper-based flexible packaging   99
           3.3.1   Printing and varnishing   99
           3.3.2   Coating   100
           3.3.3   Lamination   105

	3.4	Medical packaging		109
		3.4.1 Introduction to paper-based medical flexible packaging		109
		3.4.2 Sealing systems		112
		3.4.3 Typical paper-based medical packaging structures		113
	3.5	Packaging machinery used with paper-based flexible packaging		114
	3.6	Paper-based cap liners (wads) and diaphragms		118
		3.6.1 Pulpboard disc		119
		3.6.2 Induction-sealed disc		119
	3.7	Tea and coffee packaging		119
	3.8	Sealing tapes		121
	3.9	Paper cushioning		121
	References			123
	Websites			123
4	**Paper labels**			**125**
	Michael Fairley			
	4.1	Introduction		125
	4.2	Types of labels		128
		4.2.1 Glue-applied paper labels		128
		4.2.2 Pressure-sensitive labels		130
		4.2.3 In-mould labels		133
		4.2.4 Plastic shrink-sleeve labels		134
		4.2.5 Stretch-sleeve labels		135
		4.2.6 Wrap-around film labels		135
		4.2.7 Other labelling techniques		136
	4.3	Label adhesives		136
		4.3.1 Adhesive types		137
		4.3.2 Label adhesive performance		138
	4.4	Factors in the selection of labels		139
	4.5	Nature and function of labels		140
		4.5.1 Primary labels		140
		4.5.2 Secondary labels		141
		4.5.3 Logistics labels		141
		4.5.4 Special application or purpose labels		142
		4.5.5 Smart, smart-active and smart-intelligent labels		142
		4.5.6 Functional labels		144
		4.5.7 Recent developments		144
	4.6	Label printing and production		145
		4.6.1 Letterpress printing		146
		4.6.2 Flexography		148
		4.6.3 Lithography		149
		4.6.4 Gravure		150
		4.6.5 Screen process		151
		4.6.6 Hot-foil blocking/stamping process		152
		4.6.7 Variable information printing (VIP), electronically originated		153
		4.6.8 Digital printing		155

4.7	Print finishing techniques		156
	4.7.1	Lacquering	156
	4.7.2	Bronzing	156
	4.7.3	Embossing	156
4.8	Label finishing		156
	4.8.1	Introduction	156
	4.8.2	Straight cutting	157
	4.8.3	Die-cutting	157
	4.8.4	Handling and storage	159
4.9	Label application, labelling and overprinting		159
	4.9.1	Introduction	159
	4.9.2	Glue-applied label applicators	160
	4.9.3	Self-adhesive label applicators	160
	4.9.4	Shrink-sleeve label applicators	161
	4.9.5	Stretch-sleeve label applicators	162
	4.9.6	In-mould label applicators	162
	4.9.7	Modular label applicators	163
4.10	Label legislation, regulations and standards		163
	4.10.1	Acts of Parliament	163
	4.10.2	EC regulations and directives	163
	4.10.3	Standards	164
4.11	Specifications, quality control and testing		164
	4.11.1	Introduction	164
	4.11.2	Testing methods for self-adhesive labels	165
	4.11.3	Testing methods for wet-glue labels	165
4.12	Waste and environmental issues		167
Websites			168

5 Paper bags — 169
Smith Anderson Group Ltd, Fife, UK, and Welton Bibby & Baron Ltd, Radstock, Somerset, UK

5.1	Introduction		169
	5.1.1	Paper bags and the environment	170
5.2	Types of paper bags and their uses		170
	5.2.1	Types of paper bag	170
	5.2.2	Flat and satchel	170
	5.2.3	Strip window bags	172
	5.2.4	Self-opening satchel bags (SOS bags)	172
	5.2.5	SOS carrier bags with or without handles	174
5.3	Types of paper used		175
	5.3.1	Kraft paper – the basic grades	175
	5.3.2	Grease-resistant and greaseproof papers	176
	5.3.3	Vacuum dust bag papers	176
	5.3.4	Paper for medical use and sterilisation bags	176
	5.3.5	Wet-strength kraft	176
	5.3.6	Recycled kraft	176
	5.3.7	Coated papers	176

		5.3.8	Laminations	177
		5.3.9	Speciality papers	177
		5.3.10	Weights of paper	177
	5.4	Principles of manufacture		177
		5.4.1	Glue-seal bags	177
		5.4.2	Heat-seal bags	178
		5.4.3	Printing on bag-making machines	178
		5.4.4	Additional processes on bag-making machines	178
		5.4.5	Additional operations after bag making	179
	5.5	Performance testing		179
		5.5.1	Paper	179
		5.5.2	Paper bags	179
	5.6	Printing methods and inks		180
		5.6.1	Printing methods	180
		5.6.2	Inks	181
	5.7	Conclusion		181
		5.7.1	Development of the paper bag industry	181
		5.7.2	The future	181
	Reference			182
	Websites			182
6	**Composite cans**			**183**
	Catherine Romaine Henderson			
	6.1	Introduction		183
	6.2	Composite can (container)		185
		6.2.1	Definition	185
		6.2.2	Manufacturing methods	185
	6.3	Historical background		187
	6.4	Early applications		189
	6.5	Applications today by market segmentation		189
	6.6	Designs available		190
		6.6.1	Shape	190
		6.6.2	Size	190
		6.6.3	Consumer preferences	190
		6.6.4	Clubstore/institutional	190
		6.6.5	Other features	191
		6.6.6	Opening/closing systems	191
	6.7	Materials and methods of construction		194
		6.7.1	The liner	195
		6.7.2	The paperboard body	196
		6.7.3	Labels	197
		6.7.4	Nitrogen flushing	197
	6.8	Printing and labelling options		197
		6.8.1	Introduction	197
		6.8.2	Flexographic	197
		6.8.3	Rotogravure	198
		6.8.4	Lithography (litho/offset) printing	199
		6.8.5	Labelling options	199

	6.9	Environment and waste management issues		200
		6.9.1	Introduction	200
		6.9.2	Local recycling considerations	200
	6.10	Future trends in design and application		200
		6.10.1	Introduction	200
		6.10.2	Increase barrier performance of paper-bottom canisters	201
		6.10.3	Totally repulpable can	201
		6.10.4	Non-paper-backed liner	201
		6.10.5	Film label	201
		6.10.6	Killer paper	201
	6.11	Glossary of composite can-related terms		201
	References			203
	Websites			203
7	**Fibre drums**			**205**
	Fibrestar Drums Ltd., Cheshire, UK			
	7.1	Introduction		205
	7.2	Raw material		207
	7.3	Production		208
		7.3.1	Sidewall	208
		7.3.2	Drum base	210
		7.3.3	Lid	210
	7.4	Performance		212
	7.5	Decoration, stacking and handling		214
	7.6	Waste management		215
	7.7	Summary of the advantages of fibre drums		215
	7.8	Specifications and standards		216
	References			216
	Websites			216
8	**Multiwall paper sacks**			**217**
	Mondi Industrial Bags, Vienna, Austria			
	8.1	Introduction		217
	8.2	Sack designs		218
		8.2.1	Types of sacks	218
		8.2.2	Valve design	223
		8.2.3	Sewn closures	225
	8.3	Sack materials		226
		8.3.1	Sack body material	226
		8.3.2	Ancillary materials	230
	8.4	Testing and test methods		232
		8.4.1	Sack materials	232
		8.4.2	Sack testing	235
	8.5	Weighing, filling and closing systems		237
		8.5.1	Open mouth sacks	238
		8.5.2	Valve sacks	241

		8.5.3	Sack identification	245
		8.5.4	Sack flattening and shaping	247
		8.5.5	Baling systems	247
	8.6	Standards and manufacturing tolerances		248
		8.6.1	Standards	248
		8.6.2	Manufacturing tolerances	248
	8.7	Environmental position		250
	References			251
	Useful contacts			251
	Websites			251
9	**Rigid boxes**			**253**
	Michael Jukes			
	9.1	Overview		253
	9.2	Rigid box styles (design freedom)		254
	9.3	Markets for rigid boxes		256
	9.4	Materials		256
		9.4.1	Board and paper	256
		9.4.2	Adhesives	257
		9.4.3	Print	257
	9.5	Design principles		257
	9.6	Material preparation		258
	9.7	Construction		259
		9.7.1	Four-drawer box	261
	9.8	Conclusion		263
	References			263
	Websites			263
10	**Folding cartons**			**265**
	Mark J. Kirwan			
	10.1	Introduction		265
	10.2	Paperboard used to make folding cartons		267
	10.3	Carton design		268
		10.3.1	Surface design	268
		10.3.2	Structural design	269
	10.4	Manufacture of folding cartons		277
		10.4.1	Printing	277
		10.4.2	Cutting and creasing	280
		10.4.3	Creasing and folding	287
		10.4.4	Embossing	292
		10.4.5	Hot-foil stamping	293
		10.4.6	Gluing	294
		10.4.7	Specialist conversion operations	295
	10.5	Packaging operation		296
		10.5.1	Speed and efficiency	296
		10.5.2	Side seam-glued cartons	297
		10.5.3	Erection of flat carton blanks	298

		10.5.4	Carton storage	300
		10.5.5	Runnability and packaging line efficiency	300
	10.6	Distribution and storage		303
	10.7	Point of sale, dispensing, etc.		306
	10.8	Consumer use		307
	10.9	Conclusion		311
	References			311
	Suggested further reading			312
	Websites			312

11 Corrugated fibreboard packaging 313
Arnoud Dekker

	11.1	Introduction		313
		11.1.1	Overview	313
		11.1.2	Structure of corrugated fibreboard	313
		11.1.3	Types of corrugated fibreboard packaging	315
		11.1.4	History of corrugated fibreboard	317
	11.2	Functions		318
		11.2.1	Overview functions	318
		11.2.2	Corrugated fibreboard packaging production	318
		11.2.3	Packing lines	319
		11.2.4	Palletisation and logistic chain	319
		11.2.5	Communication	320
		11.2.6	Retail-ready	320
		11.2.7	Product safety	320
		11.2.8	Recycling and sustainability	321
	11.3	Board properties and test methods		321
		11.3.1	Overview of board properties and test methods	321
		11.3.2	Box tests	323
		11.3.3	Pallet tests	324
		11.3.4	Predictions	324
	11.4	Manufacturing		326
		11.4.1	Overview	326
		11.4.2	Paper production	326
		11.4.3	Corrugated board production	328
		11.4.4	Corrugated fibreboard converting	330
		11.4.5	Corrugated fibreboard printing	333
		11.4.6	Customer packing lines	335
		11.4.7	Good manufacturing practice	335
	11.5	Corrugated fibreboard and sustainability		335
		11.5.1	Sustainable sourcing of raw materials	336
		11.5.2	Sustainable production	337
		11.5.3	Sustainable packaging design	337
		11.5.4	Sustainable supply chain	338
	References			338
	Websites			338
	Suggested further reading			339

12 Solid board packaging 341
Mark J. Kirwan

- 12.1 Overview 341
- 12.2 Pack design 342
- 12.3 Applications 345
  - 12.3.1 Horticultural produce 345
  - 12.3.2 Meat and poultry 346
  - 12.3.3 Fish 346
  - 12.3.4 Beer (glass bottles and cans) 346
  - 12.3.5 Dairy products 346
  - 12.3.6 Footwear 346
  - 12.3.7 Laundry 346
  - 12.3.8 Engineering 346
  - 12.3.9 Export packaging 347
  - 12.3.10 Luxury packaging 347
  - 12.3.11 Slip sheets 347
  - 12.3.12 Partitions (divisions, fitments and pads) 348
  - 12.3.13 Recycling boxes 350
  - 12.3.14 Bag-in-box liquid containers 350
  - 12.3.15 Shelf-ready packaging 350
- 12.4 Materials 350
- 12.5 Water and water-vapour resistance 350
- 12.6 Printing and conversion 351
  - 12.6.1 Printing 351
  - 12.6.2 Cutting and creasing 352
- 12.7 Packaging operation 352
- 12.8 Waste management 352
- 12.9 Good manufacturing practice 352
- Reference 352
- Websites 352

13 Paperboard-based liquid packaging 353
Mark J. Kirwan

- 13.1 Introduction 353
- 13.2 Packaging materials 357
  - 13.2.1 Paperboard 357
  - 13.2.2 Barriers and heat-sealing layers 358
- 13.3 Printing and converting 360
  - 13.3.1 Reel-to-reel converting for reel-fed form, fill, seal packaging 360
  - 13.3.2 Reel-to-sheet converting for supplying printed carton blanks for packing 361
  - 13.3.3 Sheet-fed for bag-in-box 361
- 13.4 Carton designs 361
  - 13.4.1 Gable top 362
  - 13.4.2 Pyramid shape 362
  - 13.4.3 Brick shape 363

		13.4.4	Pouch	364
		13.4.5	Wedge	364
		13.4.6	Multifaceted design	365
		13.4.7	Bottle shapes	365
		13.4.8	Round cross section	365
		13.4.9	Bottom profile for gable top carton	367
		13.4.10	Bag-in-box	368
	13.5	Opening, reclosure and tamper evidence		369
	13.6	Aseptic processing		374
	13.7	Post-packaging sterilisation		375
	13.8	Transit packaging		376
	13.9	Applications for paperboard-based liquid packaging		378
	13.10	Environmental issues		378
		13.10.1	Resource reduction	379
		13.10.2	Life-cycle assessment	379
		13.10.3	Recovery and recycling	381
	13.11	Systems approach		382
	References			382
	Suggested further reading			383
	Websites			383
14	**Moulded pulp packaging**			**385**
	Cullen Packaging Ltd, Glasgow, UK			
	14.1	Introduction		385
	14.2	Applications		385
	14.3	Raw materials		388
	14.4	Production		389
	14.5	Product drying		391
	14.6	Printing/decoration		392
	14.7	Conclusion		392
	Website			392
	Appendix: Checklist for a packaging development brief			**393**
		Reference		398
		Further reading		398
	Index			399

Contributors

Smith Anderson Group
Anderson Group Ltd
Fife, UK

Welton Bibby & Baron Ltd
Radstock, Somerset, UK

Cullen Packaging Ltd
Glasgow, UK

Daven Chamberlain
Editor
Paper Technology
Bury, Lancashire, UK

Arnoud Dekker
InnoTools Manager
Smurfit Kappa Development Centre
Hoogeveen, The Netherlands

Michael Fairley
Labels & Labelling Consultancy
Hertfordshire, UK

Fibrestar Drums Ltd
Cheshire, UK

Jonathan Fowle
Innovati Partners, Shepton Mallet,
Somerset, UK

Catherine Romaine Henderson
Integrated Communication Consultants
Greer, SC, USA

Michael Jukes
London Fancy Box Company
Kent, UK

Mark J. Kirwan
Paper and Paperboard Specialist
Fellow of the Packaging Society
London, UK

Mondi Industrial Bags
Vienna, Austria

Preface

This book discusses all the main types of packaging based on paper and paperboard. It considers the raw materials and manufacture of paper and paperboard, and the basic properties and features on which packaging made from these materials depends for its appearance and performance. The manufacture of 12 of the main types of paper- and paperboard-based packaging is described, together with their end-use applications and the packaging machinery involved. The importance of pack design is stressed, including how these materials offer packaging designers opportunities for imaginative and innovative design solutions.

Authors have been drawn from major manufacturers of paper- and paperboard-based packaging in the UK, the Netherlands, Austria and the USA, and companies over a much wider area have helped with information and illustrations. The editor has wide experience in industry having spent his career in technical roles in the manufacture, printing, conversion and use of paper, paperboard and packaging.

Packaging represents the largest usage of paper and paperboard and therefore both influences and is influenced by the worldwide paper industry. Paper is based mainly on cellulose fibres derived from wood, which in turn is obtained from forestry. The paper industry is a major user of energy and other resources. The industry is therefore in the forefront of current environmental debates. This book discusses these issues and indicates how the industry stands in relation to the current requirement to be environmentally sound and the need to be sustainable in the long term. Other related issues discussed are packaging reduction, life-cycle analysis and assessment, and the options for waste management.

The book is directed at those joining companies which manufacture packaging grades of paper and paperboard, companies involved in the design, printing and production of packaging and companies which manufacture inks, coatings, adhesives and packaging machinery. It will be essential reading for students of packaging technology in the design and use of paper- and paperboard-based packaging as well as those working in the associated media.

The 'packaging chain' mainly comprises:

- Those responsible for sourcing and manufacturing packaging raw materials.
- Printers and manufacturers of packaging, including manufacturers of inks, adhesives, coatings of all kinds and the equipment required for printing and conversion.
- Packers of goods, for example within the food industry, including manufacturers of packaging machinery and those involved in distribution.
- The retail sector, supermarkets, high street shops, etc., together with the service sector, hospitals, catering, education, etc.

The packaging chain creates a large number of supplier/customer interfaces, both between and within companies, which require knowledge and understanding. The papermaker needs to understand the requirements of printing, conversion and use. Equally, those involved in printing conversion and use need to understand the technology and logistics of papermaking together with the packaging needs of their customers and society. Whatever your position within the packaging chain, it is important to be knowledgeable about the technologies both upstream and downstream from your position.

Packaging technologists play a pivotal role in defining packaging needs and cooperating with other specialists to meet those needs in cost-effective and environmentally sound ways. They work with suppliers to keep abreast of innovations in the manufacture of materials and innovations in printing, conversion and use. They need to be aware of trends in distribution, retailing, point-of-sale/dispensing, consumer use, disposal options and all the societal and environmental issues relevant to packaging in general.

Acknowledgements

My thanks go to the contributing authors and their companies. It is not easy these days to find time for such additional work, and their contributions are much appreciated.

The text has been greatly enhanced by the diagrams kindly provided by a large number of organisations and by the advice and information that I have received from many individuals in packaging companies and organisations involved in the paper, paperboard, packaging and allied industries.

In particular, I would like to acknowledge the help that I have received from the following:

The Confederation of European Paper Industries (CEPI), Pira International (now Smithers Pira), The Packaging Society (IOM3, Institute of Materials, Minerals & Mining), Pro Carton, British Carton Association (BPIF Cartons), Swedish Forest Industries Federation, PITA, Confederation of Paper Industries, Sonoco, Fibrestar Drums, London Fancy Box, INCPEN, Iggesund Paperboard, M-Real (now Metsä Board Corporation), Stora Enso, Bobst SA, AMCOR Flexibles, Billerud Beetham (manufacturer of medical packaging paper, formerly Henry Cooke), Bill Inman (former Technical Manager at Henry Cooke), Alexir Packaging (folding cartons), Smurfit Kappa Packaging (for corrugated fibreboard and solid board), Smurfit Kappa Lokfast, Tetra Pak, Elopak, SIG Combibloc, Rapak, Lamican, HayssenSandiacre Europe Ltd (for Rose Forgrove), Marden Edwards, Robert Bosch GmbH, Rovema Packaging Machines, IMA (tea packaging machinery), Easypack Ltd, DieInfo, Bernal, Atlas, Michael Pfaff (re. rotary cutting and creasing), Diana Twede (School of Packaging, Michigan State University), Neil Robson (re. packaging issues in relation to the developing world), Mondi Industrial Bags, Cullen Packaging, National Starch and Chemical (adhesives), Sun Chemical (inks), Smith Anderson Group, Welton, Bibby & Baron, Interflex Group (wax/paper flexible packaging) and John Wiley & Sons.

This book would not have been attempted without the experience gained in my packaging career, for which I thank former colleagues at Reed Medway Sacks, Bowater Packaging (carton, paper bag and flexible packaging manufacture), Cadbury Schweppes (foods packaging), Glaxo (ethical and proprietary pharmaceuticals packaging), Thames Group (paperboard manufacture) and, in particular, Iggesund Paperboard, who encouraged me to become involved in technical writing.

In helping me to complete this second edition, I acknowledge the consistent help I have received from Richard Coles (Open University and Greenwich University), who involved me in lecturing on packaging technology at B.Sc. and Institute of Packaging Diploma level at West Herts College, Watford, UK, and Daven Chamberlain, Editor of *Paper Technology*, whom I have also known for many years and who has a wide technical background in the paper industry.

I am indebted to Professor Frank Paine, who stimulated my interest in producing this book in the first place, for his cheerful support and encouragement over many years. I first met Frank as a colleague in Bowater in the 1960s. He has 60 years of international experience in packaging technology research and practice and substantial experience in authorship and editing.

I would like to dedicate this book to the memory of my former colleague, Richard Slade (Findus and Cadbury Schweppes), who involved me in packaging development from the 1960s onwards, and to that of Dennis Hine, who led, lectured and wrote about much of the investigative work on carton performance and packaging machine/packaging materials interactions at Pira International over many years.

Mark J. Kirwan

1 Paper and paperboard – raw materials, processing and properties

Daven Chamberlain[1] and Mark J. Kirwan[2]

[1] Paper Technology, Bury, Lancashire, UK
[2] Paper and Paperboard Specialist, Fellow of the Packaging Society, London, UK

1.1 Introduction – quantities, pack types and uses

Paper and paperboard are manufactured worldwide. The world output for the years quoted is shown in Table 1.1. The trend has been upward for many years; indeed, worldwide production has more than doubled in just three decades. Both materials are produced in all regions of the world. The proportions produced per region in 2010 are shown in Table 1.2.

Paper and paperboard have many applications. These include newsprint, books, tissues, stationery, photography, money, stamps, general printing, etc. The remainder comprises packaging and many industrial applications, such as plasterboard base and printed impregnated papers for furniture. In 2010, paper and paperboard produced for packaging applications accounted for 51% of total paper and paperboard production (BIR, 2011).

A single set of figures for world production of paper and paperboard hides a very significant change that has taken place in the last decade. A large amount of investment has poured into Asia, resulting in the creation of many new mills with large and fast machines. Consequently, the proportion of world production originating from Asia has increased by 10% since 2003; Europe and North America have been the casualties, and both regions have experienced significant numbers of mill closures during this period.

As a result of the widespread uses of paper and paperboard, the apparent consumption of paper and paperboard per capita can be used as an economic barometer, i.e. indication, of the standard of economic life. The apparent consumption per capita in the various regions of the world in 2010 is shown in Table 1.3.

The manufacture of paper and paperboard is therefore of worldwide significance and that significance is increasing. A large proportion of paper and paperboard is used for packaging purposes. About 30% of the total output is used for corrugated and solid fibreboard, and the overall packaging usage is significant. Amongst the membership of CEPI (Confederation of European Paper Industries), 43% of all paper and paperboard output during 2011 was used in packaging, (CEPI, 2011).

Not only is paper and paperboard packaging a significant part of the total paper and paperboard market, it also provides a significant proportion of world *packaging* consumption.

Handbook of Paper and Paperboard Packaging Technology, Second Edition. Edited by Mark J. Kirwan.
© 2013 John Wiley & Sons, Ltd. Published 2013 by John Wiley & Sons, Ltd.

Table 1.1 World production of paper and paperboard

Year	Total tonnage (million tonnes)
1980	171
1990	238
2000	324
2005	367
2006	382
2007	394
2008	391
2009	371
2010	394

Source: BIR (2010).

Table 1.2 World production % of paper and paperboard by region for 2010

Region	% of world production
Europe	27.1
Latin America	5.2
North America	22.5
Africa	1.1
Asia	43.1
Australasia	1.0

Source: BIR (2010).

Table 1.3 Apparent per capita consumption of all types of paper and paperboard in 2010

Location	Apparent consumption (kg)
North America	234.8
Europe	142.0
Australasia	135.0
Latin America	45.5
Asia	40.0
Africa	7.8

Source: BIR (2010).

Up to 40% of all *packaging* is based on paper and paperboard, making it the largest packaging material used, by weight. Paper and paperboard packaging is found wherever goods are produced, distributed, marketed and used.

Many of the features of paper and paperboard used for packaging, such as raw material sourcing, principles of manufacture, environmental and waste management issues, are identical to those applying to all the main types of paper and paperboard. It is therefore important to view the packaging applications of paper and paperboard within the context of the worldwide paper and paperboard industry.

According to Robert Opie (2002), paper was used for wrapping reams of printing paper by a papermaker around 1550; the earliest printed paper labels were used to identify bales of cloth in the sixteenth century; printed paper labels for medicines were in use by 1700 and paper labels for bottles of wine exist from the mid-1700s. One of the earliest references to the use of paper for packaging is in a patent taken out by Charles Hildeyerd on 16 February 1665 for 'The way and art of making blew paper used by sugar-bakers and others' (Hills, 1988). For an extensive summary of packaging from the 1400s using paper bags, labels, wrappers and cartons, see Davis (1967).

The use of paper and paperboard packaging accelerated during the latter part of the nineteenth century to meet the developing needs of manufacturing industry. The manufacture of paper had progressed from a laborious manual operation, one sheet at a time, to continuous high-speed production with wood pulp replacing rags as the main raw material. There were also developments in the techniques for printing and converting these materials into packaging containers and components and in mechanising the packaging operation.

Today, examples of the use of paper and paperboard packaging are found in many places, such as supermarkets, traditional street markets, shops and departmental stores, as well as for mail order, fast food, dispensing machines, pharmacies, and in hospital, catering, military, educational, sport and leisure situations. For example, uses can be found for the packaging of:

- dry food products – for example cereals, biscuits, bread and baked products, tea, coffee, sugar, flour and dry food mixes
- frozen foods, chilled foods and ice cream
- liquid foods and beverages – milk, wines and spirits
- chocolate and sugar confectionery
- fast foods
- fresh produce – fruits, vegetables, meat and fish
- personal care and hygiene – perfumes, cosmetics and toiletries
- pharmaceuticals and health care
- sport and leisure
- engineering, electrical and DIY
- agriculture, horticulture and gardening
- military stores.

Papers and paperboards are sheet materials comprising an overlapping network of cellulose fibres that self-bond to form a compact mat. They are printable and have physical properties which enable them to be made into various types of flexible, semi-rigid and rigid packaging.

There are many different types of paper and paperboard. Appearance, strength and many other properties can be varied depending on the type(s) and amount of fibre used, and how the fibres are processed in fibre separation (pulping), fibre treatment and in paper and paperboard manufacture.

In addition to the type of paper or paperboard, the material is also characterised by its weight per unit area and thickness. Indeed, the papermaking industry has many specific terms, and a good example is the terminology used to describe weight per unit area and thickness.

Weight per unit area may be described as 'grammage' because it is measured in grammes per square metre (g m^{-2}). Other area/weight-related terms are 'basis weight' and 'substance', which are usually based on the weight in pounds of a stated number of sheets of specified dimensions, also known as a 'ream', for example 500 sheets of 24 in. × 36 in., which equates to total ream area of 3000 ft^2. The American organisation TAPPI (Technical Association of the Pulp & Paper Industry, 2002–2003) issues a standard that describes basis weight in great detail; currently there are 14 different areas used for measurement, depending upon the grade being measured. It is therefore important when discussing weight per unit area, as with all properties, to be clear as to the methods and units of measurement.

Thickness, also described as 'caliper', is measured either in microns (μm), 0.001 mm or in thou. (0.001 in.), also referred to as *points*.

Appearance is characterised by the colour and surface characteristics, such as whether it is smooth or rough and has a high gloss, satin or matte finish.

Paperboard is thicker than paper and has a higher weight per unit area, although the dividing line between the two is somewhat blurred. Paper over $225\,g\,m^{-2}$ is defined by ISO (International Organization for Standardization) as paperboard, board or cardboard. Some products are, however, known as paperboard even though they are manufactured at lower grammages; for example, many producers and merchants now class products of $180–190\,g\,m^{-2}$ upwards as paperboard, because improvements in manufacturing techniques mean these lightweight materials can now be produced with similar strength properties to older heavyweight grades.

The main types of paper and paperboard-based packaging are:

- bags, wrappings and infusible tissues, for example tea and coffee bags, sachets, pouches, overwraps, sugar and flour bags, and carrier bags
- multiwall paper sacks
- folding cartons and rigid boxes
- corrugated and solid fibreboard boxes (transit or shipping cases)
- paper-based tubes, tubs and composite containers
- fibre drums
- liquid packaging
- moulded pulp containers
- labels
- sealing tapes
- cushioning materials
- cap liners (sealing wads) and diaphragms (membranes).

Paper and paperboard-based packaging is widely used because it meets the criteria for successful packing, namely to:

- contain the product
- protect goods from mechanical damage
- preserve products from deterioration
- inform the customer/consumer
- provide visual impact through graphical and structural designs.

These needs are met at all three levels of packaging, namely:

- primary – product in single units at the point of sale or use, for example cartons
- secondary – collections of primary packs grouped for storage and distribution, wholesaling and 'cash and carry', for example transit trays and cases
- tertiary – unit loads for distribution in bulk, for example heavy-duty fibreboard packaging.

Paper and paperboard, in many packaging forms, meet these needs because they have appearance and performance properties which enable them to be made into a wide range of packaging structures cost-effectively. They are printable, varnishable and can be laminated to other materials. They have physical properties which enable them to be made into flexible, semi-rigid and rigid packages by cutting, creasing, folding, forming, winding, gluing, etc.

Paper and paperboard packaging is used over a wide temperature range, from frozen-food storage to the temperatures of boiling water and heating in microwave and conventional ovens.

Whilst it is approved for direct contact with many food products, packaging made solely from paper and paperboard is permeable to water, water vapour, aqueous solutions and emulsions, organic solvents, fatty substances (except grease-resistant papers), gases such as oxygen, carbon dioxide and nitrogen, aggressive chemicals, and volatile vapours and aromas. Whilst paper and paperboard can be sealed with several types of adhesive, with certain special exceptions, such as tea-bag grades, it is not itself heat sealable.

Paper and paperboard can acquire barrier properties and extended functional performance, such as heat sealability, heat resistance, grease resistance, product release, etc., by coating, lamination and impregnation. Traditional materials used for these purposes include:

- extrusion coating with polyethylene (PE), polypropylene (PP), polyethylene terephthalate (PET or PETE), ethylene vinyl alcohol (EVOH) and polymethyl pentene (PMP)
- lamination with plastic films or aluminium foil
- treatment with wax, silicone or fluorocarbon
- impregnated with a vapour-phase metal-corrosion inhibitor, mould inhibitor or coated with an insect repellent.

Recently, the use of various biopolymers has gained predominance because their use does not impede biodegradation of treated paper or paperboard. Biopolymers based upon proteins (casein and caseinates, whey, soy, wheat gluten or corn zein), polysaccharides (chitosan, alginate or starch) and lipids (long chain fatty acids and waxes) have all been used, singularly or in combination, to form barriers against gases, water vapour or grease. Furthermore, these coatings can be rendered bioactive by addition of natural antimicrobial agents, such as lactic acid, nisin, carvacrol or cinnamaldehyde (Khwaldia et al., 2010).

Packaging made solely from paperboard can also provide a wide range of barrier properties by being *overwrapped* with a heat-sealable plastic film, such as polyvinylidene chloride (PVdC), coated oriented polypropylene (OPP or, as it is sometimes referred to, BOPP) or regenerated cellulose films, such as Cellophane™.

Several types of paper and paperboard-based packaging may incorporate metal or plastic components, examples being as closures in liquid-packaging cartons and as lids, dispensers and bases in composite cans.

In an age where environmental and waste management issues have a high profile, packaging based on paper and paperboard has important advantages:

- The majority of paper-based packaging grades are now produced using recovered fibre. As such, paper and paperboard packaging forms a very important end product for the recovered paper sector.
- The main raw material (wood or other suitable vegetation) is based on a naturally renewable resource. In most cases it is sustainably sourced from certified plantations.
- The growth of these raw materials removes carbon dioxide from the atmosphere, thereby reducing the greenhouse effect. As such they have a smaller carbon footprint than materials made from non-renewable resources, such as petrochemical derivatives.

- When the use of the package is completed, most types of paper and paperboard packaging can be recovered and recycled. Furthermore, they can all be incinerated with energy recovery, and if none of these options is possible, most are biodegradable in landfill.

1.2 Choice of raw materials and manufacture of paper and paperboard

1.2.1 Introduction to raw materials and processing

So far we have indicated that paper and paperboard-based packaging provides a well-established choice for meeting the packaging needs of a wide range of products. We have defined paper and paperboard and summarised the reasons why this type of packaging is used. We now need to discuss the underlying reasons why paper and paperboard packaging is able to meet these needs.

This discussion falls into four distinct sections:

- choice and processing of raw materials
- manufacture of paper and paperboard
- additional processes which enhance the appearance and performance of paper and paperboard by coating and lamination
- use of paper and paperboard in the printing, conversion and construction of particular types of packaging.

Cotton, wool and flax are examples of fibres, and we know that they can be spun into a thread and that thread can be woven into a sheet of cloth material. Papers and paperboards are also based on fibre, but the sheet is a three-dimensional self-bonded structure formed by random overlapping of fibres. The resulting structure, which is known as a *sheet* or *web*, is sometimes described as being 'non-woven'. The fibres are prepared by mixing them with water to form a very dilute suspension, which is poured onto a porous mesh. The paper structure forms as an even layer on this mesh, which is known as a wire and which acts as a sieve. Most of the water is then removed successively by drainage, pressure and heat.

So why does this structure have the strength and toughness which makes it suitable for printing and conversion for use in many applications, including packaging? To answer this question we need to examine the choices which are available in the raw materials used and how they are processed.

According to tradition, paper was first made in China around the year AD 105 using fibres such as cotton and flax. Such fibres are of vegetable origin, based on cellulose, which is a natural polymer, formed in green plants and some algae from carbon dioxide and water by the action of sunlight. The process initially results in natural sugars based on a multiple-glucose-type structure comprising carbon, hydrogen and oxygen in long chains of hexagonally linked carbon atoms, to which hydrogen atoms and hydroxyl (OH) groups are attached. This process is known as photosynthesis; oxygen is the by-product and the result is that carbon is removed (fixed) from the atmosphere. Large numbers of cellulose molecules form fibres – the length, shape and thickness of which vary depending on the plant species concerned. Pure cellulose is non-toxic, tasteless and odourless.

The fibres can bond at points of interfibre contact as the fibre structure dries during water removal. It is thought that bonds are formed between hydrogen (H) and OH units in adjacent cellulose molecules causing a consolidation of the three-dimensional sheet structure. The degree of bonding, which prevents the sheet from fragmenting, depends on a number of factors which can be controlled by the choice and treatment of the fibre prior to forming the sheet.

The resulting non-woven structure which we know as paper ultimately depends on a three-dimensional overlapped fibre network and the degree of interfibre bonding. Its thickness, weight per unit area and strength can be controlled, and in this context paperboard is a uniform thicker paper-based sheet. It is flat, printable, creasable, foldable, gluable and can be made into many two- and three-dimensional shapes. These features make paper and paperboard ideal wrapping and packaging materials.

Over the centuries, different cellulose-based raw materials, particularly rags incorporating cotton, flax and hemp, were used to make paper, providing good examples of recycling. During the nineteenth century, the demand for paper and paperboard increased, as wider education for the increasing population created a rising demand for written material. This in turn led to the search for alternative sources of fibre. Esparto grass was widely used but eventually processes for the separation of the fibres from wood became technically and commercially successful and from that time (1880 onwards) wood has become the main source of fibre. The process of fibre separation is known as pulping.

Today there are choices in:

- source of fibre
- method of fibre separation (pulping)
- whether the fibre is whitened (bleached) or not
- preparation of the fibre (stock) prior to use on the paper or paperboard machine.

1.2.2 Sources of fibre

Basically, the choice is between *virgin*, or primary, fibre derived from vegetation, of which wood is the principal source, and *recovered*, or secondary, fibre derived from waste paper and paperboard. Until 2005, virgin pulp formed the main fibre source for paper manufacture. Since that date, recovered paper has become the principal fibre used worldwide (BIR, 2006). In 2010, about 45% of the fibre used worldwide was virgin fibre and the rest, 55%, was from recovered paper. It must be appreciated at the outset that:

- fibres from all sources, virgin and recovered, are not universally interchangeable with respect to the paper and paperboard products which can be made from them
- some fibres by nature of their use are not recoverable and some that are recovered are not suitable for recycling on grounds of hygiene and contamination
- fibres cannot be recycled indefinitely.

The properties of virgin fibre depend on the species of tree from which the fibre is derived. The flexibility, shape and dimensional features of the fibres influence their ability to form a uniform overlapped network. Some specialised paper products incorporate other cellulose fibres such as cotton, hemp, or bagasse (from sugar cane), and there is also some use of synthetic fibre.

The paper or paperboard maker has a choice between trees which have relatively *long* fibres, such as spruce, fir and pine (*coniferous species*), which provide strength, toughness

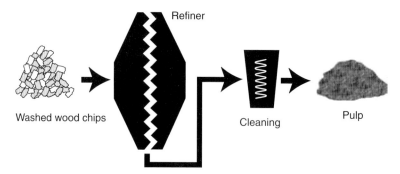

Figure 1.1 Production of mechanically separated pulp. (Courtesy of Pro Carton.)

and structure, and *shorter* fibres, such as those from birch, eucalyptus, poplar (aspen), acacia and chestnut (*deciduous species*), which give high bulk (low density), closeness of texture and smoothness of surface.

The long, wood-derived fibres used by the paper and paperboard industry are around 3–4 mm in length and the short fibres are 1–1.5 mm. The fibre tends to be ribbon shaped, about 30 microns across and therefore visible to the naked eye.

The terms 'long' and 'short' are relative to the lengths of fibres from wood as, by contrast, cotton and hemp fibres may be as long as 20–30 mm.

1.2.3 Fibre separation from wood (pulping)

In trees, the cellulose fibres are cemented together by a hard, brittle material known as lignin, another complex polymer, which forms up to 30% of the tree. The separation of fibre from wood is known as pulping. The process may be based on either mechanical or chemical methods.

Mechanical pulping applies mechanical force to wood in a crushing or grinding action, which generates heat and softens the lignin thereby separating the individual fibres. As it does not remove lignin, the yield of pulp from wood is very high. The presence of lignin on the surface and within the fibres makes them hard and stiff. They are also described as being dimensionally more stable. This is related to the fact that cellulose fibre absorbs moisture from the atmosphere when the relative humidity (RH) is high and loses moisture when the RH is low – a process that is accompanied by dimensional changes, the magnitude of which is reduced if the fibre is coated with a material such as lignin. The degree of interfibre bonding with such fibres is not high, so sheets tend to be weak. Products made from mechanically separated fibre have a 'high bulk' or low density, i.e. a relatively low weight per unit area for a given thickness. This, as will be discussed later, has technical and commercial implications. Figure 1.1 illustrates the production of mechanically separated pulp.

The most basic form of mechanical pulping, which is still practised in some mills today, involves forcing a debarked tree trunk against a rotating grinding surface. This process uses a large amount of energy and results in a very high-yield product known as stone groundwood (SGW) pulp. Alternatively, lignin can be softened using heat or by the action of certain chemicals; this reduces the mechanical energy needed to separate fibres during pulping and reduces fibre damage, leading to higher quality pulp. Wood in chip form may be heated prior

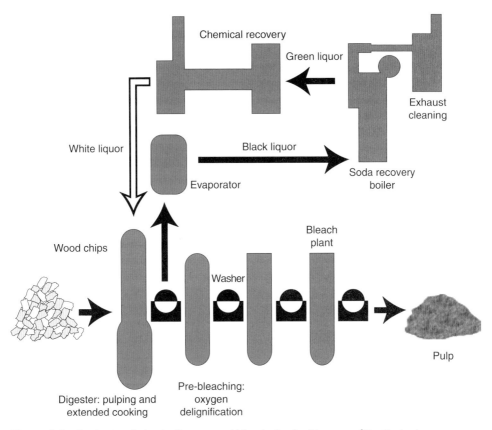

Figure 1.2 Production of chemically separated bleached pulp. (Courtesy of Pro Carton.)

to or during pulping, in which case the pulp is known as thermomechanical pulp (TMP); application of chemicals such as sodium sulphite, sodium hydroxide or oxalic acid yields chemimechanical pulp (CMP); and when the two processes are combined, the resulting pulp is called chemi-thermomechanical pulp (CTMP). Mechanical pulp retains the colour of the original wood although CTMP is lighter in colour because more lignin is removed.

Chemical pulping uses chemicals to separate the fibre by dissolving the non-cellulose and non-fibrous components of the wood (Fig. 1.2). There are two main processes characterised by the names of the types of chemicals used. The sulphate, or Kraft, process uses strong alkali; it is most widely used today because it can process all the main types of wood, and the chemicals can be recovered and reused. The other main process is known as the sulphite process, which uses strong acid. In both processes, the non-cellulose and non-fibrous material extracted from the wood is used as the main energy source in the pulp mill and in what are referred to as 'integrated' mills, which manufacture both pulp and paper/paperboard on the same site.

Chemically separated pulp comprises 74% of virgin wood fibre production. It has a lower yield than the mechanically separated pulp due to the fact that the non-cellulose constituents of the wood have been removed. This results in pulp which can undergo a high degree of interfibre bonding. Furthermore, the average fibre length of wood from

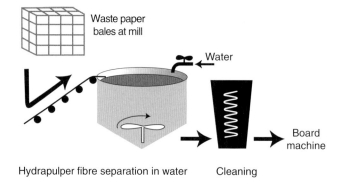

Figure 1.3 Production of pulp from recovered paper/board (recycling). (Courtesy of Pro Carton.)

the same species is longer than for mechanically separated fibre. It is also more flexible. These factors result in a stronger and more flexible sheet. The colour is brown.

1.2.4 Whitening (bleaching)

Chemically separated pulp can be whitened or bleached by processes which remove residual lignin and traces of any other wood-based material. Bleached pulp is white in colour even though individual cellulose fibres are colourless and translucent. Chemically separated and bleached fibre is pure cellulose, and this has particular relevance in packaging products where there is a need to prevent materials originating from the packaging affecting the flavour, odour or aroma of the product. Examples of such sensitive products are chocolate, butter, tea and tobacco.

Bleaching has been subject to criticism on environmental grounds. This was due to chloro-organic by-products in the effluent from mills where chlorine gas was used to treat the pulp. The criticism is no longer valid, as today the main bleaching process is elemental chlorine free (ECF) which uses oxygen, hydrogen peroxide and chlorine dioxide. The by-products of this process are simple and harmless. Another process called totally chlorine free (TCF) is based on oxygen and hydrogen peroxide. Ozone is also becoming an increasingly common and powerful bleaching agent that is being incorporated into both TCF and ECF processes (Métais et al., 2011).

Bleached cellulose fibre has high light stability, i.e. the tendency for fading or yellowing in sunlight is much reduced.

1.2.5 Recovered fibre

Waste paper and paperboard are also collected, sorted and repulped by mechanical agitation in water (Fig. 1.3). There are many different qualities of repulped fibre depending on the nature of the original fibre, how it was processed and how the paper or paperboard product was converted and used. Each time paper or paperboard is repulped, the average fibre length and the degree of interfibre bonding is reduced. This, together with the fact that some types of paper and paperboard cannot be recovered by nature of their use, means that new or virgin fibre, made directly from primary vegetation such as wood, must be introduced into the market on a regular basis to maintain *quantity* and *quality*.

There are many classifications, based on type and source, of recovered paper and paperboard which reflect their value for reuse. Classifications range from 'white shavings' (highly priced), newspapers (medium priced) to 'mixed recovered paper and board' (lowest priced). Generally, where recovered paper and paperboard which are printed are used in the manufacture of packaging grades, they are not de-inked as part of the process. Industry-agreed classification lists have been developed in Europe, where EN643 (2001) categorises 67 defined grades; the USA and Japan both have similar grading systems.

Some paper and paperboard products are either made exclusively from recycled pulp or contain a high proportion of recycled fibre. Others are made exclusively from either chemical pulp or a mixture of chemical and mechanical pulp.

1.2.6 Other raw materials

In addition to fibre, which constitutes around 89% of the raw material used for paper and paperboard, there are a number of non-fibrous additives (Zellcheming, 2008). These comprise:

- mineral pigments used for surface coatings or as fillers
- internal sizing additives
- strength additives
- surface sizing additives
- chemicals used to assist the process of paper manufacture.

They all assist in one way or another in improving either the appearance or performance of the product or the productivity of the process.

Coating the paper or paperboard involves the application of one or more layers of mineral pigment to one or both surfaces. Coatings control surface appearance, smoothness, gloss, colour (usually whiteness) and printability. Coatings comprise a pigment, generally of china clay, calcium carbonate (chalk) or titanium dioxide; an adhesive or binder, which ensures the particles both adhere together and to the surface being coated; and water (the vehicle), which facilitates the application and smoothing of the wet coating. Additional components could be optical brightening agents (OBA), also known as fluorescent whitening agents (FWA), dyes and processing aids such as anti-foaming agents.

Fillers are white inorganic materials, similar to pigments used in coatings but of larger particle size, added to the stock during paper manufacture. They fill voids in the fibre structure and increase light scattering, improving the smoothness and printability of the surface and also the opacity and shade of uncoated papers. In general, fillers are not widely used in the manufacture of packaging grades of paper and paperboard.

Mineral pigment used for coating and as fillers account for 8% of the raw materials used by the paper industry.

Internal sizing is the technique whereby the surfaces of cellulose fibres which are naturally absorbent to water are treated to render them water repellent. Traditionally, this has been carried out by what is known as alum sizing. The material used comprises a rosin size derived from pine tree gum which, after treatment to make it water soluble, is added together with aluminium sulphate at the stock-preparation stage – hence the name 'alum'. The aluminium sulphate reacts with rosin to produce a modified resin soap which is deposited onto the fibre surface. The process has been progressively developed using both rosin and chemically unrelated synthetic resins.

Alum sizing still accounted for around 30% of the world market in 2005. However, most internal sizing is performed using alkyl ketene dimer (AKD), a waxy substance, alkenyl succinic anhydride (ASA) or other synthetic sizing agents.

Resins, such as urea-formaldehyde (UF), melamine-formaldehyde (MF) and polyamide-epichlorohydrin (PAE) can be added to ensure that a high proportion of the dry strength of paper is retained when it is saturated with water, as would be necessary for multiwall paper sacks which may be exposed to rain, or carriers for cans or bottles of beer in the wet environment of a brewery. Starch is used to increase strength by increasing interfibre bonding within the sheet and interply bonding in the case of multi-ply paperboard.

Starch is also applied as a *surface size* in the drying section of a paper or paperboard machine to one or both surfaces. The purpose is to increase the strength of the sheet and in particular the surface strength which is important during printing. It helps to bind the surface fibres into the surface, thereby preventing fibre shedding, which would lead to poor printing results. It also prepares the surface for mineral pigment coating. Other additives used to modify performance include wax (resistance to moisture, permeation of taint and odour, heat sealability and gloss), acrylic resins (moderate moisture resistance) and fluorocarbons (grease resistance).

Various other chemicals are used to assist the process of manufacture. Examples are anti-foaming agents, flocculating agents to improve drainage during the forming of the wet sheet, biocides to restrict microbiological activity in the mill and pitch-control chemicals which prevent pitch (wood resins) from being deposited on the paper machine where it can build up and break away causing machine web breaks and subsequent problems with particles (fragments) in printing. Some of the newer chemicals are based upon the use of renewable feedstock in place of petrochemicals (Greenall and Bloembergen, 2011), and the use of nanotechnology is also becoming increasingly common (Patel, 2009).

1.2.7 Processing of fibre at the paper mill

Preparing fibres for paper manufacture is known as 'stock preparation'. The properties of fibres can be modified by processing and the use of additives at the stock-preparation stage prior to paper or paperboard manufacture. In this way, the papermaker can in theory start with, for example, a suspension of bleached chemically separated fibre in water, and by the use of different treatments produce modified pulps which can be used to make grades as diverse as blotting paper, bag paper or greaseproof paper.

The surface structure of the fibre can be modified in a controlled way by mechanical treatment in water. This was originally performed in a beater. Beating is a *batch* process in which the pulp suspension is drawn between moving and stationary bars. The moving bars are mounted on the surface of a beater roll which rotates at a fixed, adjustable distance above a bedplate, which also carries bars. The motion of the beater roll draws the suspension between the bars causing fibrillation of the fibre surface and swelling of the fibre. The suspension is thrown over the backfall and around the midfeather to the front of the roll for further treatment as shown in the illustration from the Paper Industry Technical Association, (PITA), (Fig. 1.4). Today, unless the grade of paper being produced requires beating in this way, for example greaseproof paper, where the pulp is highly beaten to an almost gelatinous consistency (Fig. 1.5), this treatment is carried out as a continuous, *in-line*, process through a refiner. Refiners also have stationary and moving bars, mounted either conically or on parallel discs (Grant et al., 1978).

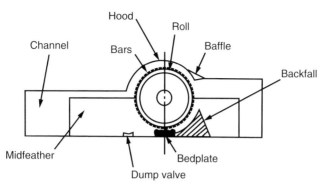

Figure 1.4 Beater. (Courtesy of PITA.)

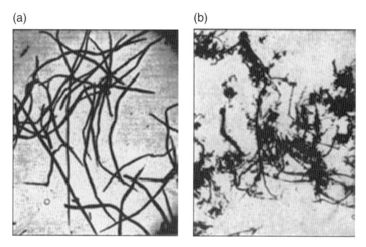

Figure 1.5 Fibres on (a) before beating and (b) after being well beaten for greaseproof paper manufacture. (Courtesy of PITA.)

1.2.8 Manufacture on the paper or paperboard machine

The basic principles of papermaking today are the same as they have always been:

- prepare a dilute suspension of fibres in water
- form a sheet consisting of an overlapping network of fibres
- remove most of the water progressively by drainage, pressure and evaporation (drying).

Traditionally, forming was achieved manually by dipping a finely woven flat wire mesh set in a wooden frame, called a mould, into a vat of fibres suspended in water and allowing excess suspension to flow over a separate wooden frame, or deckle, fitted around the edge of the mould. Water was drained through the wire mesh. The deckle was removed when the layer of fibres had consolidated. This resulted in a sheet where the fibres were randomly and evenly distributed (Fig. 1.6).

The mould was then inverted and the sheet transferred (couched) to a wetted felt blanket (Fig. 1.7). The wet sheet was covered by another felt. This process was repeated several

Figure 1.6 Vatman hand-forming paper sheet using mould. (Courtesy of PITA.)

Figure 1.7 Coucher removing wet sheet from mould. (Courtesy of PITA.)

times to form a pile of alternate layers of wet sheets and felts, known as a post, which was then subjected to pressure in a mechanical or hydraulic press to squeeze water from the sheets. After this process, the sheets were strong enough to be handled and separated from the felts.

Further pressing without felts removed more water, after which the sheets were dried in air. Sheets intended for printing would then be tub-sized by immersion in a solution of gelatine, pressed and re-dried in air (Fig. 1.8).

Figure 1.8 Loft-drying handmade paper. (Courtesy of PITA.)

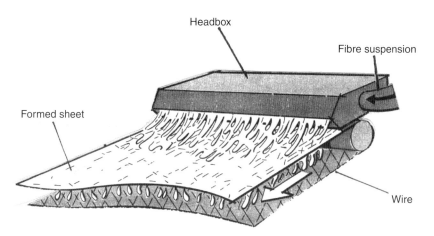

Figure 1.9 Simplified diagram of the forming process (Fourdrinier). (Courtesy of Iggesund Paperboard.)

The Industrial Revolution facilitated progress from laborious manual operations, one sheet at a time, to today's continuous high-speed production, using computerised process control.

Whilst the principles of sheet forming, pressing and drying are common to all paper and paperboard, the way in which it is carried out depends on the specific requirements and the most cost-effective method available.

Since the early 1800s, pulp has been mechanically applied to a wire mesh on a paper or paperboard machine. Mechanical forming (Fig. 1.9) induces a degree of directionality in the way the fibres are arranged in the sheet. As the fibres are relatively long in relation to their width, they tend to line up, as the sheet is formed, in the direction of motion along the machine. This direction is known as the machine direction (MD), and whilst fibres line up

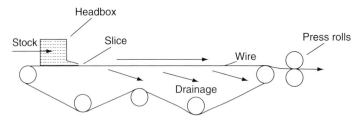

Figure 1.10 Sheet forming on moving wire – Fourdrinier process.

in all directions the least number of fibres will line up in the direction at right angles to the MD, which is known as the cross direction (CD).

The techniques used prior to and during forming are very important for the performance of the product. Strength properties and other features show variations characterised by these two directions of property measurement. The significance of this feature for the printing, conversion and use of paper and paperboard packaging varies depending on the application, and users should be aware of the implications in specific situations. Examples are discussed in the chapters on the various types of packaging.

It is important that the weight per unit area and the distribution of fibre orientation are functionally adequate and consistent, both within makings and from making to making, for the intended use.

There are two main methods of forming. Wire forming – where the pulp suspension in water, with a consistency of around 1% fibre and 99% water, flows from a headbox out of a narrow horizontal slot, known as the slice, onto a moving wire – was originated by Nicolas Robert in France in 1798. The initial method required further development by others, particularly Bryan Donkin, before the first continuous paper machine, financed by the Fourdrinier brothers, was installed at Frogmore Mill, Herts, England, in 1803. The name 'Fourdrinier' has been adopted to describe this method of forming (Fig. 1.10).

The wire, which is usually a plastic mesh today, may have a transverse 'shake' from side to side to produce a more random orientation of fibre. Water is drained from the underside of the wire using several techniques, including vacuum. The wet web is removed from the wire when it can support its own weight, at which point its solids content will be around 20%. This section of the machine is called the wet end, and the wire, which is a continuous band, carries on around to receive more pulp suspension.

An alternative method of mechanical forming using a wire-covered cylinder was also being developed at the same time. The patent which led to a successful process was taken out by John Dickinson in 1809 and he was making paper in quantity at Apsley and Nash mills also in Herts, England, by 1812. In this process, the wire-covered cylinder revolves in a vat of pulp which forms a sheet on the surface of the cylinder as a result of the maintenance of a differential pressure between the outside and inside of the cylinder. Figure 1.11 shows a 'uniflow' vat where the pulp suspension flows through the vat in the same direction as the wire mould is rotating. This arrangement results in good sheet formation, whereas if the pulp suspension flows in the opposite direction, in what is known as a contraflow vat, higher weights of pulp are formed on the mould.

Multi-web, or *multi-ply*, sheet forming can be achieved by using several wires or several vats. A modification of the wire-forming method is the Inverform process where a second and subsequent headboxes add additional layers of pulp. As each layer is added, a top wire contacts the additional layer and drains water upwards as a result of the

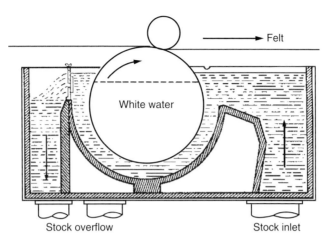

Figure 1.11 Uniflow vat cylinder mould forming. (Courtesy of PITA.)

mechanical design which is assisted with vacuum. This resulted in a significant increase in productivity without loss of quality.

Multilayering enables the manufacturer to make heavier weight per unit area products and use different pulps in the various layers to achieve specific functional needs cost-effectively. Multilayering in the case of thicker grades of paperboard also facilitates weight control and good creasing properties.

Following the forming process, the next water removal operation occurs in the press section. More water is removed by pressing the sheet, sandwiched between supporting blankets or felts, often accompanied by vacuum-induced suction. This increases the solids content to around 45–55%. Thereafter the sheet is dried in contact with steam-heated steel cylinders, producing a typical final solids content of 92–95%.

Some products are made on machines with a large diameter machine glazing (MG) or 'Yankee' cylinder. The web is applied to the cylinder whilst the moisture content is still high enough for it to adhere to the polished hot surface. This process not only dries the web but also promotes a polished or glazed smooth surface. Papers produced in this way have the prefix 'MG'. An important aspect of this for some products is that a smooth surface is achieved without loss of thickness – a feature which preserves stiffness, the importance of which will be discussed later.

Surface sizing can be applied to one or both surfaces during drying. Starch may be used to improve strength and prevent any tendency for fibre shedding during printing. Grease-resistant additives may also be applied in this way. A wax size may be applied as an emulsion in water, and with the heat from the drying cylinders the wax impregnates the paper or paperboard. However, the majority of wax treatments are applied as secondary conversion processes, i.e. off-machine. Sometimes, pigment is mixed with the starch added at this point to produce a lightweight coating to improve printability.

The final process before the web is wound into a reel involves calendering, where the dry paper is passed between smooth cylinders; this both enhances smoothness or finish and controls thickness and thickness profile. Calendering can be applied in several ways depending on the product and the surface finish required. Cylinders may be hard or soft, heated or chilled, and in some cases water is applied to the surface of the material during the process. At its simplest, calendering comprises two steel rolls though more could be used – this is the

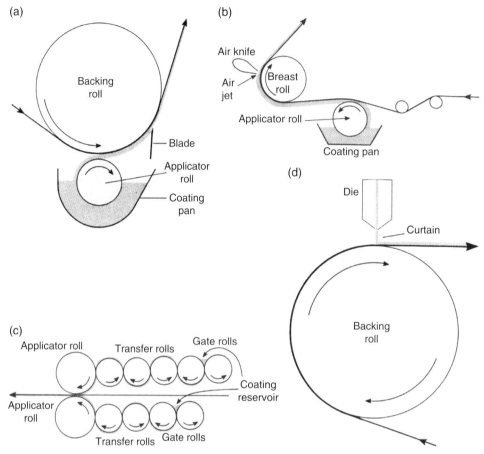

Figure 1.12 (a) Blade coating, (b) air knife coating, (c) roll coating and (d) curtain coating. (Courtesy of PITA.)

oldest and most basic method, which can result in significant reduction in thickness. Paperboard in general requires light calendering to control thickness without compressing the material excessively, which would reduce stiffness, as will be discussed later; this is best achieved by using soft calender bowls, rather than steel rolls. With soft bowls, high smoothness can be achieved without crushing the material, and reduction of caliper is minimised, so stiffness is maintained.

There are paper machines with up to seven calender rolls where steel and composite rolls are used alternatively to provide smoother and glossier finishes. An off-machine 'supercalender' (SC) produces a much smoother and glossier finish. In the case of glassine, as many as 14 rolls are used in SC.

White pigmented mineral-based coatings are applied, to one or both sides of the sheet, and smoothed and dried, to improve appearance in respect of colour, smoothness, printing and varnishing. One, two or three coating layers may be applied, depending on the needs of the product. In general, if a single layer is applied, it may be done in-line, after surface sizing and before calendering. Subsequent layers are more often applied off-line, in a separate process.

The method of coating (Fig. 1.12) has been adopted to describe the types of coating, such as blade, air knife, roll, bar or curtain. For packaging applications, blade has traditionally been

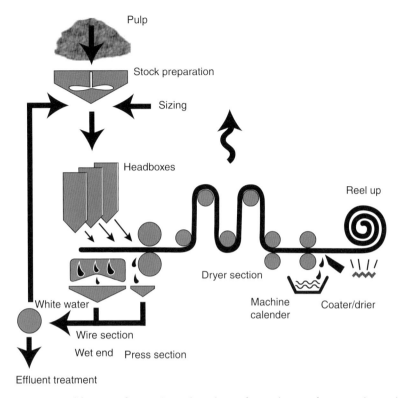

Figure 1.13 Principal features of paper/paperboard manufacture by wire forming – the number of headboxes will vary depending on the product and the machine design. (Courtesy of Pro Carton.)

the most popular technique, although curtain is just starting to gain in popularity. The advantage of curtain coating over the other methods is that there is no physical contact between the coater and paper web; this reduces the occurrence of web breaks, so increasing productivity.

The coating process described produces a surface with a matt finish. A more light-reflective gloss finish can be achieved by brushing, friction glazing or calendering.

A specific type of mineral pigment coating is known as cast coating, which is applied off-machine as a separate process. This is a reel-to-reel process in which the coating mix is applied to the paper or paperboard surface, smoothed, and whilst still wet is cast against the surface of a highly polished heated cylinder, similar to a Yankee or MG cylinder. When dry, the coated surface peels away from the cylinder leaving a coating on the paper or paperboard with a very smooth and extremely high gloss finish.

Paper and paperboard machines vary in width from around 1 m to as much as 10 m. The size is geared to output and output is geared to market size. The output-limiting factors for a given width are the amount of pulp used per unit area and the linear speed – both of which relate to the amount of water that has to be removed by a combination of drainage, vacuum, pressing and drying. Principal features of paper/paperboard manufacture by wire forming are shown in Fig. 1.13.

1.2.9 Finishing

This is the name given to the processes which are carried out after the paper and paperboard have left the papermaking machine. There are a number of options depending on customer requirements. Large reels ex-machine are slit to narrower widths and smaller diameters; subsequently, such reels may be slit, sheeted, counted, palletised, wrapped and labelled. The product is normally wrapped in moisture-resistant material such as PE, film and stretch or shrink wrap.

Papers and paperboards produced in the way described may be given secondary treatments by way of coating, lamination and impregnation with other materials to achieve specific functional properties. These are known as 'conversion' processes. They are performed after the paper and paperboard have left the mill either by specialist converters, such as laminators or plastic-extrusion coaters, or they may be integrated within the plants making packaging materials and containers. These processes are discussed in the packaging-specific chapters of this book.

We have now identified the nature of paper and paperboard, the raw materials and the processing which can be undertaken to make a wide range of substrates. We now need to review the various paper and paperboard products which are used to manufacture packaging materials and containers.

1.3 Packaging papers and paperboards

1.3.1 Introduction

A wide range of papers and paperboards are commercially available to meet market needs based on the choice of fibre (bleached or unbleached, chemically or mechanically separated, virgin or recovered fibre), the treatment and additives used at the stock-preparation stage.

We have noted that paper- and paperboard-based products can be made in a wide range of grammages and thicknesses. The surface finish (appearance) can be varied mechanically. Additives introduced at the stock-preparation stage provide special properties. Coatings applied to either one or both surfaces, smoothed and dried, offer a variety of appearance and performance features which are enhanced by subsequent printing and conversion, thereby resulting in various types of packaging material. To illustrate these features of paper and paperboard, some typical product examples are described later.

1.3.2 Tissues

These are lightweight papers with grammages from 12 to $30\,\mathrm{g\,m^{-2}}$. The lightest tissues for tea and coffee bags which require a strong porous sheet are based on long fibres such as those derived from Manila hemp. To maintain strength during immersion in boiling water, wet strength additives are used. The Constanta-type bag with the lowest grammage is folded and stapled. Heat-sealed tea and coffee bags require the inclusion of a heat-sealing fibre, such as PP. Single-portion tea bags have grammages in the range $12–17\,\mathrm{g\,m^{-2}}$, but larger bags would require higher grammages.

Neutral pH tissue grades with low chloride and sulphate residues are used as wrapping materials for archival purposes; such specialist tissue may also be laminated to aluminium foil and used to wrap silverware, jewellery, clothing, etc. (Packaging tissues should not be confused with absorbent tissues used for hygienic purposes, which are made on a different type of paper machine using different types of pulp.)

1.3.3 Greaseproof

The hydration (refining) of fibres at the stock-preparation stage, already described, is taken much further than normal. Hydration can also be carried out as a batch process in a beater. The fibres are treated (hydrated) so that they become almost gelatinous. Grammage range is 30–70 g m^{-2}.

1.3.4 Glassine

This is a SC greaseproof paper. The calendering produces a very dense sheet with a high (smooth and glossy) finish. It is non-porous, greaseproof and can be laminated to paperboard. It may be plasticised with glycerine. It may be embossed, PE coated, aluminium foil laminated, metallised or release-treated with silicone to facilitate product release. It is produced in plain and coloured versions, for example chocolate brown. Grammage range is 30–80 g m^{-2}.

1.3.5 Vegetable parchment

Bleached chemical pulp is made into paper conventionally and then passed through a bath of sulphuric acid, which produces partial hydrolysis of the cellulose surface of the fibres. Some of the surface cellulose is gelatinised and redeposited between the surface fibres forming an impervious layer closing the pores in the paper structure. The process is stopped by chemical neutralisation and the web is thoroughly washed in water. This paper has high grease resistance and wet strength. It can be used in the deep freeze (i.e. −20°C storage environment) and in both conventional and microwave ovens. It can be silicone treated for product release. Grammage range is 30–230 g m^{-2}.

1.3.6 Label paper

These may be coated, MG (machine glazed) or MF (machine finished – calendered) kraft papers (100% sulphate chemical pulp) in the grammage range 70–90 g m^{-2}. The paper may be coated on-machine or cast coated for the highest gloss in an off-machine or secondary process.

The term 'finish' in the paper industry refers to the surface appearance. This may be:

- machine finish (MF) – smooth but not glazed
- water finish (WF) – where one or both sides are dampened and calendered to be smoother and glossier than MF
- machine glazed (MG) – with high gloss on one side only
- supercalendered (SC) – which is dampened and polished off-machine to produce high gloss on both sides.

Depending upon the environment in which the label is to be used, various functional chemicals may need to be added, for example for labelling packages containing fatty products, grease-resistant chemicals, such as fluorocarbons, may be included.

1.3.7 Bag papers

'Imitation kraft' is a term on which there is no universally agreed definition; it can be either a blend of kraft virgin fibre with recycled fibre or 100% recycled. It is usually dyed brown.

It has many uses for wrapping and for bags where it may have an MG and a ribbed finish. Thinner grades may be used for lamination with aluminium foil and PE for use on form, fill, seal machines. For sugar or flour bags, coated or uncoated bleached kraft in the range 90–100 g m^{-2} is used.

1.3.8 Sack kraft

Usually this is unbleached kraft (90–100% sulphate chemical) pulp, though there is some use of bleached kraft. The grammage range is 70–100 g m^{-2}.

Paper used in wet conditions needs to retain considerable strength, at least 30%, when saturated with water. To achieve this, resins such as UF and MF are added to the stock. These chemicals cross-link during drying and are deposited on the surface of the cellulose fibres making them more resistant to water absorption.

Microcreping, as achieved for example by the Clupak process, builds an almost invisible crimp into paper during drying, enabling it to stretch up to 7% in the MD compared to a more normal 2%. When used in paper sacks, this feature improves the ability of the paper to withstand dynamic stresses, such as occur when sacks are dropped.

1.3.9 Impregnated papers

Papers are made for subsequent impregnation off-machine. Such treatment can, for example, be with wax, vapour-phase inhibitor for metal packaging and mould inhibitors for soap wrapping. (Mills have ceased to impregnate these products on-machine for technical and commercial reasons.)

1.3.10 Laminating papers

Coated and uncoated papers based on both kraft (sulphate) and sulphite pulps can be laminated to aluminium foil and extrusion coated with PE. The heavier weights can be PE laminated to plastic films and wax or adhesive laminated to unlined chipboard. The grammage range is 40–80 g m^{-2}.

1.3.11 Solid bleached board (SBB)

This board (Fig. 1.14) is made exclusively from bleached chemical pulp. It usually has a mineral pigment-coated top surface, and some grades are also coated on the back. The term 'solid bleached sulphate' (SBS), derived from the method of pulp production, is sometimes used to describe this product.

This paperboard has excellent surface and printing characteristics. It gives wide scope for innovative structural designs and can be embossed, cut, creased, folded and glued with ease. This is a pure cellulose primary (virgin) paperboard with consistent purity for food product safety, making it the best choice for the packaging of aroma and flavour-sensitive products. Examples of use include chocolate packaging, frozen, chilled and reheatable products, tea, coffee, liquid packaging and non-foods such as cigarettes, cosmetics and pharmaceuticals.

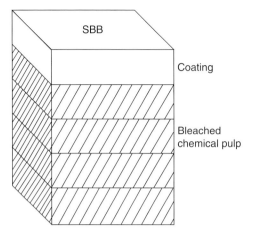

Figure 1.14 Solid bleached board. (Courtesy of Iggesund Paperboard.)

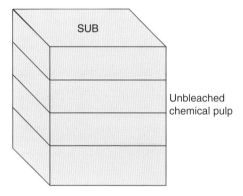

Figure 1.15 Solid unbleached board (SUB). (Courtesy of Iggesund Paperboard.)

1.3.12 Solid unbleached board (SUB)

This board (Fig. 1.15) is made exclusively from unbleached chemical pulp. The base board is brown in colour. This product is also known as 'solid unbleached sulphate' (SUS). To achieve a white surface, it can be coated with a white mineral pigment coating, sometimes in combination with a layer of bleached white fibres under the coating.

SUB is used where there is a high strength requirement in terms of puncture and tear resistance and/or good wet strength is required such as for bottle or can multipacks and as a base for liquid packaging.

1.3.13 Folding boxboard (FBB)

This board (Fig. 1.16) comprises middle layers of mechanical pulp sandwiched between layers of bleached chemical pulp. The top layer of bleached chemical pulp is usually coated with a white mineral pigment coating. The back is cream (manila) in colour. This is because the back layer of bleached chemical pulp is translucent, allowing the colour of the middle layers to show through. However, if the mechanical pulp in the middle layers is given a mild chemical treatment, it becomes lighter in colour and this makes the reverse side colour lighter in shade.

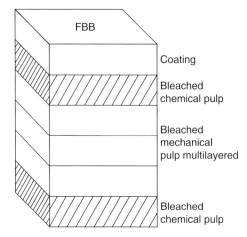

Figure 1.16 Folding boxboard. (Courtesy of Iggesund Paperboard.)

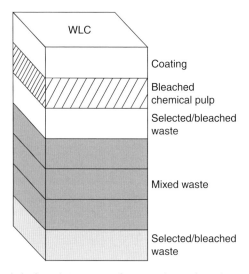

Figure 1.17 White-lined chipboard. (Courtesy of Iggesund Paperboard.)

The back layer may, however, be thicker or coated with a white mineral pigment coating, thus becoming a white back-folding boxboard. The combination of inner layers of mechanical pulp and outer layers of bleached chemical pulp creates a paperboard with high stiffness.

Fully coated grades have a smooth surface and excellent printing characteristics. This paperboard is a primary (virgin fibre) product with consistent purity for food product safety and suitable for the packing of aroma- and flavour-sensitive products. It is used for packing confectionery, frozen, chilled and dry foods, healthcare products, cigarettes, cosmetics, toys, games and photographic products.

1.3.14 White-lined chipboard (WLC)

WLC (Fig. 1.17) comprises middle plies of recycled pulp recovered from mixed papers or carton waste. The middle layers are grey in colour. The top layer, or liner, of bleached chemical pulp is usually white mineral pigment coated. The second layer, or under liner, may also comprise bleached chemical pulp or mechanical pulp. This product is also known as newsboard or chipboard, though the latter name is more likely to be associated with unlined grades, i.e. no white, or other colour, liner.

The reverse side outer layer usually comprises specially selected recycled pulp and is grey in colour. The external appearance may be white due to the use of bleached chemical pulp and, possibly, a white mineral pigment coating (white PE has also been used).

There are additional grades of unlined chipboard and grades with specially coloured (dyed) liner plies. (WLC with a blue inner liner was used for the packing of cube sugar.)

The overall content of WLC varies from about 80 to 100% recovered fibre depending on the choice of fibre used in the various layers. WLC is widely used for dry foods, frozen and chilled foods, toys, games, household products and DIY.

1.4 Packaging requirements

Packaging protects and identifies the product for customers/consumers. When we think about packaging requirements, we may initially think of the needs of the customer who is the purchaser of a branded product in a supermarket (supplier). However, the purchaser is not always the consumer or user of the product, and closer investigation also brings the realisation that there is a wide range of important needs which must be met at several supplier/customer interfaces in a chain, which links the:

- supply of raw materials, for example pulp, coatings, etc.
- manufacture of packaging material, for example papermaker
- manufacture of the pack or package components, for example printer, laminator, converter
- packing/filling of the product, for example food manufacturer
- storage and distribution, for example regional depot, wholesaler or 'cash and carry'
- point of sale, provision or dispensing to customer, for example retailer, pharmacy, etc.

At every stage, there are functional requirements which must be met. These requirements must be identified and built into the specification of the pack and the materials used for its construction. The specification for a primary consumer-use pack must also be compatible with the specification of the secondary distribution pack and the tertiary palletisation or other form of unit load.

Packaging requirements can be identified with respect to:

- *protection, preservation and containment* of the product to meet the needs of the packaging operation and the proposed distribution and use within the required shelf life
- *efficient production* of the packaging material, the pack in printing and converting, in packing, handling, distribution and storage, taking account of all associated hazards
- *promotion* requirements of visual impact, display and information throughout the packaging, sale and use of the product.

Checklists can be used to carry out these tasks logically to ensure that all important potential needs are examined (see Appendix, "Checklist for a packaging development brief").

Whilst the overall needs are defined by marketing and those responsible for the product itself, these needs have to be interpreted by packaging technologists and packaging buyers in both end-user and retailer and by production, purchasing and technical departments in printers, converters and manufacturers of the packaging.

The next step is to match these requirements with the knowledge about the ability of the proposed material, and the package which can be manufactured from that material, in order to achieve the requirements effectively. This implies making *choices*. A technologist, using this term in a general sense, assists this process using his/her knowledge about materials and the packages which can be made from these materials, methods of packing and the general logistical environment within which the business concerned operates. Ultimately, packaging must meet the needs of society in a sustainable way by:

- minimising product waste
- improving the quality of life
- protecting the environment
- managing packaging waste through recovery and recycling.

All these requirements have technical implications, and in order to meet the requirements at every stage in the manufacture and use of the packaging, paper and paperboard must be carefully selected on the basis of their properties and other relevant features.

We will now examine those physical properties and other features of paper and paperboard with technical implications which relate to their performance in printing, conversion and use.

1.5 Technical requirements of paper and paperboard for packaging

1.5.1 Requirements of appearance and performance

The properties of paper and paperboard correlating with the needs of printing, its conversion into packages and their use in packing, distribution, storage, product protection and consumer use can be identified and measured.

All paper and paperboard properties depend on the ingredients used, for example type and amount of fibre and other materials, together with the manufacturing processes. These properties of the paper and paperboard are related to the visual appearance and technical performance of packaging incorporating such materials:

- appearance that relates to colour, visual impression and the needs of any processes, such as printing, which have a major impact on the appearance of the packaging
- performance that relates to strength, product/consumer protection and the efficiency of all the production operations involved in making and using the packaging.

1.5.2 Appearance properties

1.5.2.1 *Colour*

Colour is a perceived sensation of the human eye and brain, which depends on the viewing light source and the ability of the illuminated surface to absorb, reflect and scatter that illumination. The lighting conditions under which the colour of paper and paperboard is viewed have been standardised so that different observers can make judgements about colour. Colour measurement has been standardised so that specifications can be defined.

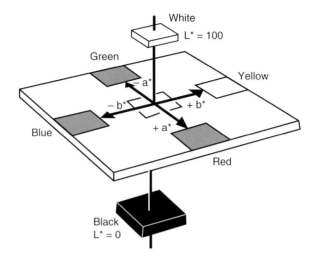

Figure 1.18 CIELAB colour coordinates. (Courtesy of Iggesund Paperboard.)

The colour of paper and paperboard is usually white or brown depending on whether the fibre is bleached or unbleached (brown). The outer surface, and sometimes the reverse side, may be pigment coated. Coating is usually white though other colours are possible. Other colours are also possible in uncoated products through the use of dyes added at the stock-preparation stage. Recycled mixed fibres which are not de-inked have a grey colour commonly seen on the reverse side of WLC.

Colour is assessed by eye under specified conditions of lighting. It is measured using reflected light from a standard light source in a reflectance spectrophotometer and calculating the colour values (Commission International de l'Éclairage (CIE) coordinates).

Natural daylight, or a simulated equivalent, is used as the source for viewing. CIE is recognised as the scientific authority with respect to colour in the paper, printing and packaging industries. The CIE colour coordinates (Fig. 1.18), L^*, a^* and b^*, are used to express lightness and colour using a standard D65 light source which simulates natural daylight.

Positive figures for a^* indicate redness, negative figures greenness; positive b^* indicates yellowness, negative figures blueness; and L^* is the percentage for luminance (intensity of light) on a scale where black is zero and pure white 100%. (A top of the range specification for a white paperboard-coated surface would be around a^* +2, b^* −5 and L^* 97.)

CIE whiteness is another measurement that is often quoted for paper and paperboard. CIE whiteness values are calculated from the same reflectance data used to produce CIE colour coordinates. However, instead of producing three values, the fundamental reflectance data are manipulated to produce just two numbers: one for whiteness and the other for tint. The former ranges from around 50, which has a noticeably yellow shade, to around 180 – a value that is only possible by use of OBA or FWA. Tint refers to the relative deviation from true white towards green or red; for a paper to be classed as white, the tint value must lie between −3 and +3. Whiteness is the preferred measurement over L^* or brightness for 'white' papers because it highlights reflectance data from the wavelength region where the human eye is most sensitive and it expands the scale compared to the other two similar variables, making differentiation of substrates easier.

There are many different examples of whiteness, or hue, found in packaging papers and paperboards. It is relatively easy to develop a specific whiteness/hue when using a mineral-based coating, and the perceived colour can be modified with additional components such as dyes and OBA or FWA. However, the base sheet colour, which depends on its composition, influences the colour of a pigment-coated surface. In packaging today, a whiteness with a bluish hue is preferred for primary retail packaging as this is felt to give the best appearance for food products, suggesting freshness, hygiene and high quality under shop (retail) lighting. Meanwhile, many secondary packaging grades of paper and paperboard, such as cardboard, are brown as they do not incorporate bleached fibre.

Brightness is often mentioned in the same context as colour and whiteness, but it is *not* comparable. Brightness is the percentage of blue light of nominal wavelength 457 nm reflected from the surface. The human eye perceives a range of wavelengths from under 400 to over 700 nm but has its greatest sensitivity in the range 470–650 nm, so the wavelength chosen for brightness measurement is below the optimum perceptual range for humans. This means values for brightness do not correlate well with human perception. Normally brightness is only measured on pulp. As it measures reflectance at a narrow band of wavelengths of blue light, it is of little value to a printer or end-user of packaging.

1.5.2.2 Surface smoothness

Surface smoothness is an important aesthetic feature and is also functionally important with respect to printing and varnishing. With some print processes, a rough paper would not faithfully reproduce the printing image as a result of 'dot skip', where ink has not been transferred from the plate, for example in the gravure printing process, to the surface being printed.

Surface smoothness is usually measured as surface roughness by air leak methods (Fig. 1.19) – the rougher the surface, the greater the rate of air leakage, at a specified air pressure, from under a cylindrical knife edge placed on the surface. Hence, the rougher the surface, the higher the value. As papers and paperboards are compressible, the pressure exerted by the knife edge is specified. By measuring roughness at two specified knife-edge pressures, an indication of compressibility is also achieved. (Compressibility is also important in printing.) The most common roughness measurements are based on the Parker Print Surf (PPS), Bendtsen and Sheffield instruments.

Occasionally, mainly for historical reasons, some markets measure surface smoothness using alternative air leak instruments, such as the Bekk smoothness tester. In this apparatus, a sample is placed against an optically smooth glass plate with a hole in the centre, and a chamber under the plate is evacuated to produce an under-pressure using a vacuum pump. When the pump is stopped, air bleeds between the paper surface and glass plate, enters the chamber and reduces the under-pressure. The time taken for a known change in pressure, corresponding to the passage of a known volume of air between paper and plate, is measured; high values indicate smoother paper surfaces. The clamping pressure and degree of paper or paperboard compression is different from that used in roughness measurement, and experience shows Bekk smoothness is not as reliable at predicting surface quality as roughness measurement techniques.

1.5.2.3 Surface structure

Surface structure is assessed visually by observing the surface under low-angle illumination (Fig. 1.20) which highlights any irregularities in the surface. The appearance varies depending on the direction, i.e. MD or CD, of observation and illumination. The surface

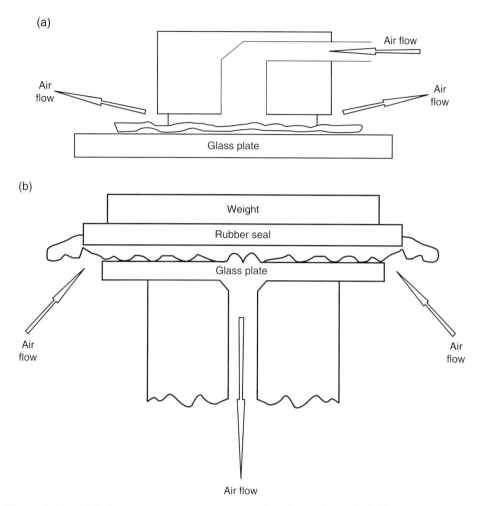

Figure 1.19 (a) Surface roughness, measuring principle of the Bendtsen, Sheffield and Parker Print Surf roughness testers and (b) surface roughness, measuring principle of the Bekk smoothness tester.

Figure 1.20 Low-angle illumination to examine surface irregularities. (Courtesy of Iggesund Paperboard.)

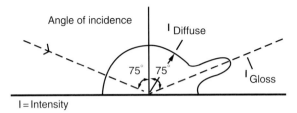

Figure 1.21 Principle of gloss measurement for paper and paperboard. (Courtesy of Iggesund Paperboard.)

structure is usually not fully apparent until the surface is varnished or laminated with either a transparent film or aluminium foil.

1.5.2.4 Gloss

Gloss is defined as the percentage of light reflected from the surface at the same angle as the angle of incidence. For better discrimination, the gloss of paper and paperboard is measured at an angle of 75° (Fig. 1.21), and printed and varnished surfaces are measured at 60°. Glossy surfaces are usually achieved with mineral pigment-coated surfaces which have been calendered, brush burnished, friction glazed or cast coated.

With uncoated papers, gloss is achieved by drying the paper or paperboard on an MG cylinder with a polished surface, as, for example, with MG bleached or unbleached kraft papers. The high gloss of a glassine is developed on an SC, where it is passed through several nips, i.e. the gaps between alternate hard metal and soft rolls made from compressed fibrous material.

1.5.2.5 Opacity

Opacity relates to the capacity of a sheet to obscure print on an underlying sheet or on the reverse of the sheet itself. This is required in a packaging context where paper is used as an overwrap on top of a printed surface. Opacity is measured by comparing light reflectance, using a spectrophotometer, from the surface of a single sheet over a black background with the reflectance from a pile of 100 sheets of the same grade.

1.5.2.6 Printability and varnishability

Packaging is usually printed to provide information, illustrations and to enhance visual display. Print may be varnished or sealed to give improved gloss and to protect print. The colour of the surface and the print design, text, solid colour and half-tone illustrations, and whether the print is varnished or sealed, all have a major impact on the appearance.

There is a wide range of print design in packaging as evidenced by the needs of, for example, multiwall paper sacks for cement, sugar bags, labels for beer bottles, cartons for breakfast cereals and the packaging used for brand leaders in chocolate assortments or expensive cosmetics. These examples will also be different compared with the printing on transit or shipping cases, used in distribution, or the labels on hazardous chemicals packaging.

Several printing processes are used commercially today. They are discussed in the package-specific chapters which follow. These processes include offset lithography, flexography, letterpress, gravure, silkscreen and digital printing. They vary in several important characteristics, chiefly in relation to the ink and varnish formulations, the techniques by which they are applied to the paper or paperboard substrate and the processes by which they dry and become permanent and durable.

Despite the variability, there are common features relating to printability which apply to all papers and paperboards. They comprise surface smoothness, surface structure, gloss level, opacity, surface strength, ink and varnish absorption, drying, rub resistance together with edge and surface cleanliness. In specific cases, surface pH and surface tension or wettability are also relevant.

Print colour can be measured using a spectrophotometer or with a densitometer. It can also be compared visually, under standardised lighting, with pre-set colour standards, to ensure that colours remain within acceptable light, standard and dark limits.

1.5.2.7 Surface strength

Adequate surface strength is necessary to ensure good appearance in printing and post embossing. The offset lithographic printing process uses tacky inks, and it places a high requirement for surface strength at the point of separation between the ink left on the sheet and the ink left on the offset rubber blanket. The IGT pick and printability test simulates this process by increasing tack on a printed strip of paper or paperboard to the point of failure, which is either a surface pick or blister. Measurement of the point of failure against a specification which is known to be satisfactory provides a means of predicting a satisfactory result.

IGT pick testing is usually performed on dry paper and paperboard surfaces. However, in some circumstances it can be useful to apply a controlled quantity of water to the surface and then perform the test with a tacky ink. This is especially so for some coated substrates, where the binder system may give adequate strength when dry, but very poor strength when wetted, which has obvious implications for wet offset lithographic printing.

Embossing is a means of producing a defined design feature, pattern or texture on the surface of paper and paperboard in relief. Surface strength in the fibre, interfibre bonding and, where present, the coating are necessary to achieve the required result depending on the depth, sharpness of detail and area of the surface which is distorted.

An alternative approach to the measurement of surface strength is to apply a number of wax sticks, which are tack graded, to the surface. Tack grading relates to their ability, when molten, to stick to a flat surface. The result, Dennison Wax Number, is the highest number wax which does not disrupt the surface, when removed in the specified manner. High wax numbers indicate high surface strength. This test is only suitable for uncoated surfaces, as when a coating is present the melted wax fuses with the binder in the coating giving an unrealistic result. It can also be unsuitable for very hard sized uncoated sheets, because the molten wax is unable to stick to the surface and so gives a false indication of high strength. Nevertheless, for most uncoated surfaces, this test is relevant for both printing and adhesion.

With adhesion, it is necessary for an adhesive to pull fibre at a reasonable level of surface strength – if, however, the strength is too high it could cause poor adhesion in practice. This is relevant to adhesion with water-based adhesives, hot melts and to the adhesion of heat-sealed blister packs on heat seal-varnished and printed paperboard cards.

1.5.2.8 Ink and varnish absorption and drying

Inks comprise a vehicle, usually an oil, organic solvent or water, a pigment to give colour, or dye in some cases, and a resin which binds the pigment to the substrate. A varnish is similar, though without the addition of colour. The vehicle, which depends on the ink and the printing process, is necessary to transfer the ink from the ink duct or reservoir of the printing press, via the plate, to the substrate. Once printed, the vehicle has to be removed by evaporation, absorption or by being changed chemically to a solid state, so that the ink dries by oxidation or cross-linking as a result of ultraviolet (UV) or electron beam (EB) curing. Inks which are not fully dry as they leave the printing press, such as conventional oil-based litho and letterpress inks, must at least 'set', by a degree of adsorbency, to an extent where they do not set-off against the reverse side of adjacent sheets as they are stacked off the press.

As with most paper and paperboard properties, the key to satisfactory performance is uniformity. Lack of uniformity can lead to set-off, mottle and strike through. Tests based on the absorption of a standard ink or ink vehicle or solvent are used to check uniformity and achievement of a satisfactory specification.

Additionally, in the conventional offset lithographic process, the second colour printed in-line is transferred to a substrate which has been wetted. Under certain circumstances, this can result in print mottle, and hence a test has been devised to check ink repellency on a water-dampened surface.

Water-based sealers are increasingly displacing varnish as an after-treatment following printing. These consist of emulsified film-forming chemicals in water. They are applied by a coating unit at the end of the printing press, occasionally with an infrared drying unit directly afterwards to aid removal of the water vehicle. The emulsified chemicals film-form during evaporation of the vehicle, forming a protective layer over the ink film, which in particular virtually eliminates set-off. Various functional chemicals, such as grease-resistant materials, can be added to impart extra functionality at this stage.

1.5.2.9 Surface pH

For oil-based inks which dry by oxidation, a surface pH of around 6–8 is preferred. A surface pH of 5 or less is unsatisfactory as it can lead to poor ink drying with some types of ink, for example oil-based litho inks. The test is carried out by measuring the pH of a drop of distilled water using a pH electrode. This range is also important for papers or paperboards which are printed with metal pigments such as bronze and those required for laminating with aluminium foil.

1.5.2.10 Surface tension or surface energy

This property is important in the printing and adhesion of non-absorbent surfaces such as plastic extrusion-coated papers and paperboards. These plastic surfaces require treatment to prevent inks from reticulating. This is done by surface oxidation using an electric corona discharge or a gas flame. The effect of the treatment can be measured by checking the surface tension using Dyne measuring pens. It should be noted that the effect of such treatment reduces with the passage of time.

An alternative assessment technique is to measure the contact angle of a liquid, usually water, on the paper or paperboard surface. In this method, a small liquid droplet is applied to the substrate surface, and the shape is monitored either manually on a screen or by digital image capture. From the droplet shape, its angle of contact can be calculated. Values in excess of 90° indicate repulsion; those below 90° show attraction. Oxidative treatment of a polymer surface will generally reduce water contact angles from around 100° to 40° or below.

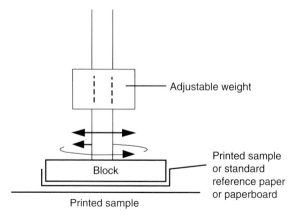

Figure 1.22 Print rub testing. (Courtesy of Iggesund Paperboard.)

1.5.2.11 Rub resistance

It is unacceptable for packages to be scuffed, smudged or marked in any way as a result of post-printing handling, transportation or use. Wet rub resistance is also necessary where packaging materials become wet as the result of contact with water or condensation, as is common with food packaging for products which are frozen or chilled. Good rub resistance is achieved by a combination of the paper or paperboard surface properties, the printing, varnishing or sealing process, and the formulation of the print and varnish or sealer. Rub resistance can be measured against pre-set standards using standard test methods (Fig. 1.22).

1.5.2.12 Surface cleanliness

A major consideration in printing is that the surface of paper and paperboard which is to be printed should be free from particles and surface dust.

Problems can be caused by loose fibres, fragments of fibres, clumps of fibres, shives (non-fibrous particles in pulp) and coating particles (Fig. 10.17). They can originate in the finishing processes of slitting and sheeting, and together with additional particles which can originate in paper and paperboard manufacture, they can cause print impression problems.

These problems are typically spots (hickies) in solid print, loss of screen definition in half-tone illustrations, ink spots in non-printing areas, etc. These problems also lead to poor efficiency in the printing operation and ultimately to waste.

There are no officially recognised methods for assessing sheet cleanliness though techniques have been developed to measure loose edge debris. Surface debris can be investigated by rolling a soft polyurethane roll over the surface and inspecting and counting particles, using magnification, from a fixed area. Wiping a surface or edge with a black cloth will also give a qualitative indication of loose material.

Whenever a problem of this nature occurs, the printer should find the particles causing the problem and identify them under magnification. Having made a correct identification, action can then be taken to eliminate or minimise the effect of the problem. It should be noted that problems of this nature may not have originated in the material. They can also arise on, or in the vicinity of, the press or as a result of problems with the inks. Hence, a correct diagnosis is essential.

1.5.2.13 Light fastness

Packaging is often placed in environments where it is subject to high levels of light exposure, mainly from artificial light although occasionally from direct sunlight. The paper or paperboard, printing ink and other components of the packaging should be chosen to withstand the lighting conditions expected in the storage or end-use environment. Incorrect choice can lead to fading, where the basic colours are maintained but lose their brilliance, or discolouration, which most often shows as yellowing or darkening of white areas. In either case the affected packaging will be viewed unfavourably, especially if it is shown alongside fresh unexposed material.

Paper and paperboard made from fibres which retain a high lignin content, such as mechanical wood fibre or unbleached chemical wood fibre, have poorer light fastness than sheets made from fully bleached chemical wood pulp. Other parts of the packaging, such as the inks, varnishes and sealers, and adhesives, must also be chosen to reflect the expected lighting environment into which the packaging is to be placed.

1.5.3 Performance properties

1.5.3.1 Introduction

Adequate performance to enable a paper or paperboard material to meet the needs of packaging manufacture and use is essential. The material must provide strength for whatever structural shape is necessary for the packaging, be it a tea bag, a label, folding carton or shipping container. Strength is necessary in printing and constructing the packaging, both in packaging manufacture, also known as conversion, and in the packaging operation, whether this is carried out manually or by machinery. Strength is also necessary for the physical protection of the goods in distribution and storage, at the point of sale and in consumer use.

Research has identified the specific features of strength and other performance needs, and tests which simulate these features have been developed so that specifications can be established. Specifications fulfil two important functions. Firstly, they provide the basic parameters for manufacture whereby paper and paperboard products are defined. Secondly, by regular testing during manufacture against the specification, the manufacturer has an accurate view of the degree of uniformity within a making and of consistency between makings. Many tests can now be performed online during manufacture, and by linking testing with computer technology a high frequency of such testing is possible. It is also possible to provide feedback within the system to automatically maintain parameters such as moisture content, thickness and grammage within the target range. This approach is also being applied to other parameters, for example colour, gloss, ash content, fibre orientation, formation and stiffness.

In testing strength and other related performance criteria, account is taken of the hygroscopic nature of the cellulose fibre, since moisture content influences most physical and optical properties of paper. The fibre absorbs moisture when exposed to high humidity and loses moisture when exposed to low humidity, so paper and paperboard will vary in moisture content depending on the RH of the atmosphere to which it is exposed.

As strength properties in particular vary with moisture content, it is necessary for specifications and test procedures to be based on samples conditioned at, and therefore in equilibrium with, a fixed temperature and RH. This is set in laboratories in Europe and

North America at 50% RH and 23°C; in tropical regions the alternative conditions are 65% RH and 27°C. It is therefore necessary to correlate specification values with the actual conditions prevailing during manufacture on the machine such that when subsequently tested after conditioning, the paper and paperboard conform with the specification.

The specific type and value of the various performance properties required will depend on the needs of the packaging concerned. Both the thinnest tissue and the thickest paperboard will have specific requirements, and the actual properties may be the same properties such as tensile strength, elongation (% stretch), tear, creasing and folding, wet strength, etc. The underlying principles and how they are achieved for each type of paper and paperboard have much in common. This is because paper and paperboard are sheet materials formed from an overlapping network of cellulose fibres. Differences in the type and value of the strength and other performance properties depend on the amount and type of fibre and its processing, whether the paper or paperboard is multilayered, together with any other ingredients, coatings or laminations which provide additional properties.

The difference between MD and CD has already been noted. Strength properties and other features show variations which are characterised by these two directions. The value of many of the test-method measurements of properties will vary depending on the direction of measurement.

1.5.3.2 Basis weight (substance or grammage)

The amount of material in paper and paperboard is measured as weight per unit area. In the laboratory, this is done by weighing an area of material which has been cut accurately. Basis weight is expressed in a number of ways – typically the units are grammes per square metre (g m^{-2}) or pounds per defined number of square feet. For a given paper or paperboard product, most of the strength-related properties increase with increasing basis weight.

This also has commercial implications as for a specific paper or paperboard, the higher the basis weight the lower the number of packs from a given weight of packaging material. Higher basis weight generally means more fibre per unit area, and more fibre requires the removal of more water and lower output on the paper or paperboard machine.

1.5.3.3 Thickness (caliper)

Thickness is measured in either microns (0.001 mm or 1×10^{-6} m) or points (1 point is 0.001 in. or one thousandth of an inch). Paper, and paperboard, is a fibrous structure; it is compressible and therefore thickness is measured with a dead weight micrometer which applies a fixed weight over a fixed area at a known rate of loading. For specific papers and paperboards, thickness increases with basis weight, and hence for a given grade, several strength properties increase with increasing thickness. However, as will be seen when stiffness is discussed, thickness can be more relevant than basis weight.

1.5.3.4 Moisture content

Moisture content is measured as a percentage of the dry weight. Many strength properties alter with changes in moisture content.

The cellulose fibres in paper and paperboard will expand by absorbing moisture in high RH and shrink by losing moisture in low RH conditions. The dimensions of fibres change

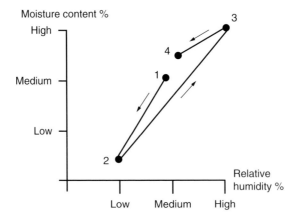

Figure 1.23 Equilibrium moisture content with RH rising (position 2/3) and RH decreasing (position 3/4 and 1/2). (Courtesy of Iggesund Paperboard.)

more in their width by swelling or contracting than they do in their length. As more fibres tend to line up in the direction of motion through the paper machine, any change in dimension across the fibres results in a greater cumulative change in the CD of the sheet. Hence, dimensional stability is more critical in the CD compared with the MD. (This can be used to determine the MD and CD of a sheet by moistening one face of a square sample where one side is parallel to an edge of the sheet. The fibres quickly swell and the moistened face expands in the CD, tending to form a cylinder, the axis of which is in the MD.)

Every paper and paperboard product will seek to achieve moisture content equilibrium with the RH of the ambient conditions in which it finds itself. This is known as hygrosensitivity. It is possible to construct curves showing how this changes over a range of relative humidities. Paper and paperboard have one set of equilibrium moisture content when the RH is rising and a different set when the RH is decreasing. This is known as the hysteresis effect (Fig. 1.23) where results are affected by previous storage conditions.

The implication of this is that the moisture content achieved in manufacture is critically important for what subsequently happens to the material in printing, conversion and use. There are therefore two aims in manufacture with respect to moisture. Firstly, a moisture content specification range which matches the equilibrium moisture content of that material over the average range of RH likely to be encountered by the product in the course of its use needs to be set. In the Northern Hemisphere, the recommended percentage of RH in which paper and paperboard are printed, converted and used on packaging lines is 45–60%. Secondly – papermakers have many techniques at their disposal to achieve this – uniform moisture content within this range during manufacture should be maintained.

The hygroscopic nature of cellulose fibre, however, also implies that the material must be adequately protected in distribution and storage. If optimum efficiency in printing, conversion and use is to be achieved, the following elements of good manufacturing practice must be observed:

- Use moisture-resistant wrappings in transit and storage.
- Follow mill recommendations with respect to storage.
- Establish temperature equilibrium in the material before unwrapping.
- Provide protection after each process.

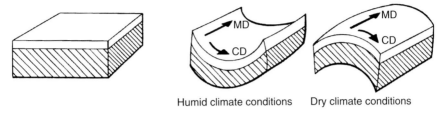

Figure 1.24 Humidity changes affect paper and paperboard flatness. (Courtesy of Iggesund Paperboard.)

Critical situations can exist when paper or paperboard is brought from a cold to a warm environment. Users should never remove moisture-resistant wrappings from paper and paperboard until the material has achieved temperature equilibrium with the room where it is to be used, for example a printing press or a packaging machine. A paperboard with a cold surface, for example after being unloaded from a lorry in winter, can cool a tacky ink, causing the tack to increase to such an extent that a severe print blister occurs during printing.

Additionally, the cold edges of a stack cool the adjacent air when moved from a cold store to a warm production area, and this can lead to the condensation of moisture on the edges. This moisture cannot be seen, but it can be absorbed, causing distortion and hence difficulty in feeding the material on a printing press or a packaging machine (Fig. 1.24). Equally, if unwrapped material is left exposed to high temperature or low humidity, it can dry out, also causing distortion.

In practice, papers and paperboards are manufactured in ways which are intended to minimise such dimensional changes (hygrosensitivity). The following of mill-recommended practices in the wrapping, storage and use of paper and paperboard by printers, converters and users are also important in order to achieve the best efficiencies in printing, conversion and use.

1.5.3.5 Tensile strength

The strength, or force, required to rupture a strip of the material is known as the tensile strength. The material shows elastic behaviour up to a certain point. This means that the force, or stress, applied to the strip is proportional to the deformation or elongation caused by the applied force. This is known as Hooke's Law and is expressed as:

$$\text{Stress (applied force)} = \text{Constant } (E) \times \text{Strain (dimensional change)}$$

This constant is known as the modulus of elasticity (E) or Young's modulus.

Up to a certain point, paper and paperboard show elastic properties (Fig. 1.25). This means that if the force is removed, the sample will regain its original shape – however, above the elastic limit this no longer applies as the material increasingly undergoes plastic deformation until it ruptures.

Specifications are based on test methods with fixed strip widths and rates of loading – the tensile strength being recorded as force per unit width. Tensile strength is higher in the MD compared with the CD.

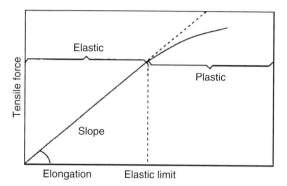

Figure 1.25 Stress–strain relationship showing elastic and plastic properties. (Courtesy of Iggesund Paperboard.)

For substrates that are to be used in the presence of liquids, wet strength resins (Section 1.2.6) may be added, in which case wet tensile strength should also be measured. In this test the strip is soaked in liquid, usually water, until it is saturated, then a tensile test is performed. The strength retention when saturated, expressed as a percentage of the original dry tensile strength, is then recorded. Typical wet-strengthened papers retain a minimum of 15% of their dry strength when fully saturated.

The tensile value at the point of rupture will vary with the rate of applying the load. When the load is steadily increased the measurement is referred to as a static tensile, and when the load is applied suddenly over a very short time interval, the measurement is referred to as a dynamic tensile.

The latter, defined as tensile energy absorption (TEA), is important in understanding the paper properties which relate to the drop-test performance of a multiwall paper sack. This test is a measure of the work done, i.e. force × distance, to rupture the sample, and it combines the features of tensile strength and percentage stretch.

1.5.3.6 Stretch or elongation

This is the maximum elongation of a strip in a tensile test at rupture and is a measure of elasticity expressed as a percentage increase compared with the original length between the clamping jaws. CD elongation is higher than MD elongation, unless the material has been creped.

1.5.3.7 Tearing resistance

Tearing resistance (Fig. 1.26) is the measured force required to promulgate a tear in the sheet from an initiated cut. In most situations, the need is to prevent damage by tearing. In some cases as, for instance, with a tear strip to facilitate opening a pack and gaining access to the contents, the requirement is for the material to tear cleanly.

1.5.3.8 Burst resistance

To test for burst resistance, the sheet is clamped over a circular orifice and subjected to increasing hydraulic pressure until expansion of a rubber diaphragm causes rupture (Fig. 1.27). It is a simple test to perform but its relevance to strength in practice is complicated. High

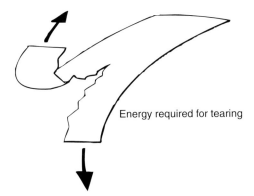

Figure 1.26 Principle of tearing resistance. (Courtesy of Iggesund Paperboard.)

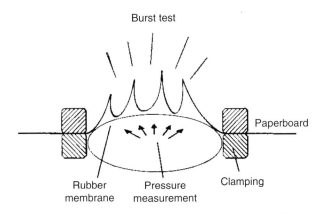

Figure 1.27 Principle of burst resistance. (Courtesy of Iggesund Paperboard.)

values, however, indicate toughness. As noted in Section 1.2.6, wet strength resins can be added at the stock-preparation stage to enable the paper to retain a significant proportion of its dry strength if it becomes wet during subsequent usage. The extent of wet strength is calculated by comparing the dry burst strength with the burst strength after the sample has been wetted in a specified way. The percentage of wet burst to dry burst expresses the extent of strength retention when wet.

1.5.3.9 Puncture resistance

Puncture resistance testing follows a similar process as that for burst resistance, although the penetrating probe is very different; also, the test is only suitable for heavyweight material, such as paperboard. The test sheet is clamped over an orifice and a pointed projectile attached to a heavy sector is swung so that the point contacts and penetrates the paperboard. The loss in potential energy, determined by the extent to which the sector continues to swing after contacting the test material, is a measure of the energy expanded in penetrating the sheet.

1.5.3.10 Stiffness

This property has major significance in printing, conversion and use. Stiffness is defined as the resistance to bending caused by an externally applied force. Stiffness of lightweight

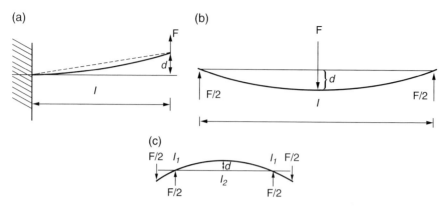

Figure 1.28 Loading principles for the measurement of paper and paperboard stiffness by the (a) two-, (b) three- and (c) four-point methods.

materials is measured by applying a force (*F*) to the free end of a fixed size piece of the material, length (*l*), which is clamped at the other end, and deflecting the free end through a fixed distance or angle (δ). This is known as the two-point method (Fig. 1.28). It is used to measure bending stiffness, bending resistance and bending moment.

For heavyweight or thick materials, the two-point method causes crushing, even at low-bending angles, which results in anomalously low stiffness values. For these materials, three-point (Fig. 1.28) or even four-point (Fig. 1.28) methods are preferred (Markström, 1988; Müller, 2011).

The MD stiffness value is higher than the CD value, and sometimes this is expressed as the stiffness ratio, i.e. MD stiffness/CD stiffness. This difference is the result of the differing fibre alignment arising as a result of the method of manufacture. Stiffness is also related to other important features, such as box compression, creasability, foldability and overall toughness. An important consideration regarding stiffness is that it is related to the modulus of elasticity (*E*) and thickness (*t*) (caliper) as follows:

$$\text{Stiffness} = \text{Constant (material specific)} \times E \times t^3$$

The cubic relationship is valid for homogeneous materials providing that the elastic limit is not exceeded. For paper and paperboard, the index is lower than 3.0 but is still significant. For some types of paperboard the index is around 2.5–2.6. Thus it is valid to claim that stiffness is highly dependent on thickness as is shown by doubling the thickness and noting that the stiffness increases by a factor of just over five.

1.5.3.11 Compression strength

When we discuss compression in the context of packaging needs, we usually mean the effect of externally applied loads in the storage, distribution and use of packed products in packaging such as cartons, cases and drums.

We can study the effect on compression of different aspects of pack design, different types and thicknesses of paper and paperboard, and different climatic conditions. We recognise the difference between static loads applied over long periods, as with palletised loads in storage, and the dynamic loads associated with high forces applied for very short periods, as in dropping and with transport-induced shocks. So we carry out compression tests on the packs at different rates of loading.

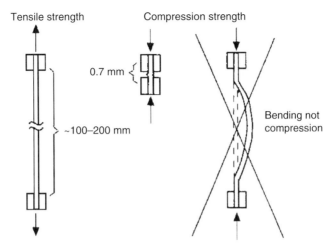

Figure 1.29 Compression strength testing – note difference in sample length compared with the tensile test. (Courtesy of Iggesund Paperboard.)

Research has shown that the inherent paper and paperboard properties involved in box compression are stiffness, as already discussed, and what is known as the short-span compression strength.

When an unsupported sample of paper or paperboard is compressed by applying a force to opposite edges in the same plane as the sample, the material will, not unexpectedly, bend. This does not give a measure of compression strength (Fig. 1.29). If, however, the sample height in the direction of the applied force is reduced below the average fibre length, say to 0.7 mm, the force is then applied to the fibre network in such a way that the network itself is compressed causing the fibres to move in relation to each other. In this situation, interfibre bonding and the type(s) and quantity of fibre become important to the result which we call the 'short-span compression strength'. It is this inherent characteristic of the sheet, in the direction of measurement, MD or CD, together with stiffness, which relates to box compression strength.

Further specialist compression tests, related to corrugated board, are discussed in Chapter 11.

1.5.3.12 Creasability and foldability

Paper and paperboard are frequently folded in the construction of pack shapes such as many designs of bags and sachets, cartons and corrugated and solid fibreboard cases. The thinner materials are folded mechanically through 180° and the resulting folds are rolled to give permanence. The thicker materials for cartons and cases require a crease to be made in the material prior to folding.

The material to be creased is supported on a thin sheet of material known as the make-ready, or counter die, which itself is adhered to a flat steel plate. Grooves are cut in the make-ready to match the position of the creasing rules in the die. When the die is closed, creases are pressed into the surface, creating a groove in the surface of the carton and a bulge on the reverse side. Crease forming in this way subjects the material to several forms of stress which are indicated in Fig. 10.29.

When the crease is folded, the top layers of fibre on the outside of the resulting fold are extended and therefore require adequate tensile strength and stretch. The internal layers are

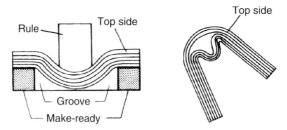

Figure 1.30 Crease forming and folding. (Courtesy of BPIF Cartons.)

compressed causing localised delamination (Figs 10.30, 10.31 and 10.32). The reverse side bulge in turn develops as a bead as folding continues to the desired angle and thus behaves like a hinge (Fig. 1.30). It is important that the bulge itself does not rupture or become distorted. Hence the layer of fibre on the reverse side also requires good strength properties.

In addition to good strength properties in the material, the geometry of the creasing operation – i.e. the width of the creasing rule, width and depth of the make-ready groove and the penetration of the creasing rule into the surface of the material – is most important. In addition to the visual inspection of creases and folds, it is also possible to measure resistance to folding and spring-back force – features which can be controlled by the creasing geometry.

In a laboratory, creasing and foldability can be assessed using either a commercial hand-creasing apparatus or a PATRA crease tester. The most basic commercial hand crease apparatus has a single fixed creasing rule and a single groove (channel); those apparatus which allow the rule and channel to be altered are preferable, as these should be chosen to suit the substrate under test. The PATRA crease tester is a very heavy and solid test instrument that encompasses the ability to use several rules and channels. Whichever apparatus is used, the test involves forming a crease, then folding the material by hand and assessing whether any cracking of the surfaces occurs.

The subsequent performance of carton creases folded during gluing is time dependent. This is important where side seam-glued cartons are stored before use on a cartonning machine. This feature can be measured as 'carton opening force'. The conditions of such intermediate storage in terms of humidity, temperature, tightness or looseness of packing and the stacking of the cases in which the cartons are stored are also important factors which can affect efficiency in packaging operations.

1.5.3.13 *Ply bond (interlayer) strength*

Ply bond strength (Fig. 1.31) is important for multilayered paper and paperboard products. It relates to the delamination of the material when subjected to forces perpendicular to the plain of the material, which cause delamination. The delaminating force is measured with the help of metal plates or platens which are attached to the paperboard surface by means of double-sided self-adhesive tape. In the TAPPI test, the platens are gripped in the jaws of a tensile tester and pulled apart; the rate of testing is relatively slow and the failure equates of a plastic yield between the plies. An alternative method is the Scott Bond test, where a pendulum strikes one metal plate, causing almost instantaneous delamination; the force transfer is rapid, and this produces an elastic failure. The two test methods tend to yield very different information.

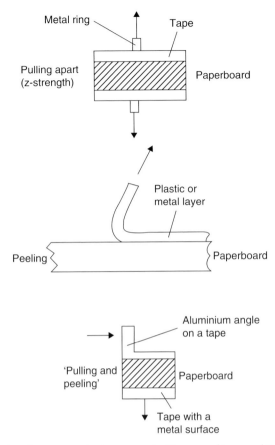

Figure 1.31 Principles of testing interlaminar strength. (Courtesy of Iggesund Paperboard.)

If delamination strength is too low, adhesive bonds may fail too easily, and if too high, the internal delamination necessary for good creasing will not occur.

1.5.3.14 Flatness and dimensional stability

Flatness is important in the sheet of paper or paperboard for its efficiency both in printing and conversion and subsequently at the packing stage. Examples of the type of problems which can occur are misfeeds, which cause stoppages, and mis-register of colour to colour and print to profile. The flatness required is built into the material during paper or paperboard manufacturing. Variations in forming, tension, drying and moisture content can cause wave, curl, twists, cockle and baggy patches (Fig. 1.32).

As discussed in Section 1.5.3.4, the hygroscopic nature of cellulose fibre requires that the material must be adequately protected in distribution, storage and use. There are requirements of good manufacturing practice which must be observed if optimum efficiency in printing, conversion and use is to be achieved. As noted, these are to do with the use of moisture-resistant wrappings, achieving temperature equilibrium before unwrapping and rewrapping where paper or paperboard is liable to be affected by storage in either high or

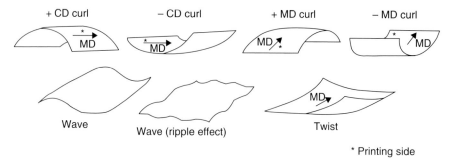

Figure 1.32 Different types of curl, twist and wave. (Courtesy of Iggesund Paperboard.)

low RH conditions. Critical situations can exist when paper or paperboard is brought from a cold to a warm environment and where the RH range is outside 45–60%.

1.5.3.15 Porosity

Uncoated papers and paperboards are permeable to air. This property is assessed using instruments that force or suck air through a defined area of sheet, using a fixed pressure. There are two main methods that measure related but distinct properties: *air resistance* measures the time taken for a fixed volume of air to pass through a sheet (e.g. Gurley) and *air permanence* measures the volume to pass per unit time (e.g. Bendtsen or Sheffield). High values of air resistance equate to low values of air permanence.

This property has implications for situations where the material is picked up by a vacuum cup so that it can be moved to another position. This occurs on printing presses, cutting and creasing machines and packaging machines. Variable porosity outside specified limits can lead to more than one sheet or piece of packaging being picked up, which, in turn, can jam the machine.

Problems can also occur when coated materials are incorrectly picked up by vacuum cups from the uncoated reverse side surface when air can be drawn in from the adjacent raw edge of the material. This, however, is a problem of either machine setting or an incompatibility between the pack design and the machine setting.

Porosity is an important factor in the speed of filling of fine powders in multiwall paper sacks where it is necessary for air to escape from the inside of the package.

1.5.3.16 Water absorbency

There are occasions when water comes in contact with paper and paperboard materials – this may happen deliberately as when water-based adhesives are used or in an unplanned way as for instance when moisture condenses on the surfaces and cut edges of a carton removed from a frozen-food cabinet at the point of sale.

Water absorbency is dealt with in one of two ways or a combination of both. Firstly, by internally sizing, whereby a water-repellent material (size) is incorporated at the stock or pulp preparation stage just prior to introducing the stock to the paper or paperboard machine. Many types of paper and paperboard are sized as part of normal production, but where a higher degree of water hold-out is required, extra hard sizing is added to the stock. In multilayer paperboards this means that each layer, including the middle layers, is hard sized.

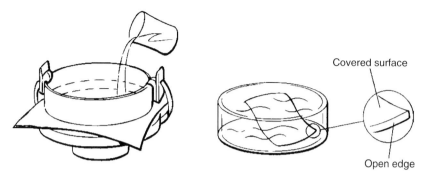

Figure 1.33 Cobb and wicking tests. (Courtesy of Iggesund Paperboard.)

The size, which may be either of natural origin (rosin) or synthetic, is deposited on the surface of the cellulose fibres making them water repellent. Secondly, a surface coating can be applied during manufacture either as a surface size or as a separate coating operation, as would be the case with an extrusion coating of PE or where a varnish or sealer is applied over print.

The simplest method of measuring the water absorbency of flat surfaces is by the Cobb test method (Fig. 1.33), which measures the weight of water absorbed in a given time over a given area. Usually the time interval is either 1 or 3 min. A note of caution must be stated that whilst this seems an obvious and suitable method of testing, it does not always correlate with what happens in practice. This is mainly due to different, usually shorter, time intervals or dwell times where water-based adhesives are held under pressure on packaging machines and where tack develops as a result of some of the water being adsorbed. However, in many cases it is an indication of functional performance.

Water can also enter through the exposed cut edges of packaging. This can also be retarded by hard sizing. The flat surfaces of samples for this test are sealed with waterproof plastic adhesive tape prior to weighing the sample and immersing it in water for the stipulated time.

1.5.3.17 Gluability/adhesion/sealing

The principles of adhesion and gluability apply in many situations where materials incorporating paper and paperboard are joined together. This occurs, for instance, when side seams are required for bags, cartons and cases and when packs are closed after filling. It is also relevant to lamination with adhesives, the manufacture and use of labels, plastic-extrusion coating and heat sealing.

The adhesives (glues) are usually either water-based with a bonding material solids content of 50–60%, where the water acts as a carrier, or wax/polymer blends which are 100% solid and applied hot in the molten state. In heat sealing, the plastic incorporated in or on the surface of the material acts as the adhesive.

Adhesion is characterised by three stages:

- open time for the adhesive to remain functional after being applied to one surface and before the joint is made
- setting time during which it is necessary to keep the joint under pressure
- drying time is the time necessary to develop a permanent bond.

Adhesives are chosen to suit the surfaces being joined, the constraints of the operation in respect of open, setting and drying times, and any special functional needs of the pack and its use, for example wet strength, direct food contact approval, etc.

A good glue bond, where at least one of the surfaces is paper or paperboard, must show fibre tear at a satisfactory strength when peeled open. Where the adhesive is applied to the surface, it must 'wet' or flow out evenly over the area of application.

Some paper and paperboards are extrusion coated with plastics such as PE on one or two sides. Packs incorporating such materials may be sealed by heat sealing either by sealing plastic to paper, as an overlap seal, or plastic to plastic. The plastic is softened and becomes tacky/molten under heat and pressure and then cools and resolidifies forming a strong seal. A strong heat seal also requires good fibre tear in the paper or paperboard material. With multilayer thicker materials, such as paperboard, the fibre tear must rupture the *internal* layers when the seal is peeled open.

Heat-sealable tea-bag tissue incorporates a heat-sealing medium such as PP in the form of a fibre dispersed within the very thin sheet.

1.5.3.18 *Taint and odour neutrality*

Some products are sensitive to changes in their taste, flavour or aroma. Examples are fat-containing food products, such as butter, vegetable fats and chocolate. Other flavour- and aroma-sensitive products are tea and coffee and a range of tobacco products. Changes can occur by loss of flavour through the packaging, the ingress of unwanted flavours from the external environment and by the transfer of flavours originating in the packaging. The first two potential causes are prevented by augmenting the paper and paperboard with additional barrier materials, such as aluminium foil, a metallised coating on plastic film, PVdC-coated OPP film, etc. Whilst all packaging associated with flavour- and aroma-sensitive products has to meet critical requirements, here we are concerned with aspects which potentially can apply to paper and paperboard packaging.

The best approach is to ensure that when odour and taint critical packaging is required, the raw materials are chosen carefully. The range of paper and paperboard materials is wide and depending on the needs of the product, many papers and paperboards can be ruled out initially from what is known, doubtful or potentially variable about their constituents.

For the most critical products virgin fibre is necessary; recycled fibre can contain a very wide range of materials, some of which are volatile and can give rise to odour and taint. The best results are achieved with chemically separated and bleached virgin pulp. For some critical products, paperboard with a mixture of this material and mechanically separated virgin pulp has also been approved and is used widely. The concern with these materials is not the pure cellulose fibre, which is inert and odourless, but other residual wood-originating compounds which may not be fully removed by bleaching and which are either not removed or only partially removed in the case of mechanical pulp. The pulp can contain residual fatty acids from the wood, and these compounds are oxidised over time producing odoriferous aldehydes.

Another potential source of odour and taint can be residual compounds originating from the chemicals which are used for the mineral coating – not so much the mineral compounds themselves but the synthetic binders (adhesives) which bind the particles together and to the base sheet of fibre.

Paper and paperboard-based packaging for use with products which have critical odour- and taint-free requirements can be checked by panels of selected people using sensory methods based on smelling and tasting. Where odour or taint is detected, the compounds responsible can be identified by gas chromatography and mass spectrometry. The concentration of compounds can be measured by gas chromatography, see Section 10.8. Many other potential sources for odour and taint arise as the result of printing, varnishing and other conversion processes.

1.5.3.19 Product safety

The basic requirement of packaging is that it should ensure the quality of the packed product however that is defined in the geographical areas of the world and the end uses involved. In some cases, this leads to the need for assurances regarding the safety of food in direct contact, or close proximity, to specific packaging materials. It also leads to a similar need with respect to the packaging used for toys.

These needs are defined in compliance regulations. In the USA, the regulations are given in the 'Code of Federal Regulations, Food and Drugs Administration (FDA)'.

In the USA, there are also the CONEG (Confederation of North Eastern Governors) Regulations which set limits for the content of lead, mercury, cadmium, chromium and nickel in packaging materials, such as paper and paperboard, in direct contact with food.

In Europe, the most widely quoted are those formulated in the 'Bundesgesundheitsamt (BGA)'. The work we are concerned with here became part of the German Federal Institute for Consumer Protection and Veterinary Medicine Regulations or BgVV in 1994 and then by the BfR (Federal Institute for Risk Assessment – Bundesinstitut für Risikobewertung) in 2002 (Heynkes, 2007). In Holland there are the 'Warenwet' regulations.

Paper and paperboard has a long and successful history of safe use in the food industry in a wide range of applications. These include applications where intimate contact is involved, such as tea bags, baking papers and filters, and direct contact packaging such as butter wrapping, sugar bags and cartons for dry and frozen foods. In addition, it has a very wide range of uses in transport and distribution packaging (CEPI, 2010).

The European Union has Directives, 76/893/EEC and 89/109/EEC, and a Regulation, 1935/2004/EC. These documents make provision for future Directives relating to specific materials, such as plastics, 2005/70/EC, but so far does not have a Directive relating to paper and paperboard, although, as already indicated, there are national provisions which paper and paperboard makers can meet.

Therefore, CEPI, CITPA (The International Confederation of Paper and Board Converters in Europe), FPE (Paper and board multilayer manufacturers) and CEPIC (Suppliers of Chemicals) worked together and developed a report which was published in March 2010 by CEPI and CITPA entitled *Industry Guidance for the Compliance of Paper & Board Materials and Articles for Food Contact* (CEPI, 2010). This is a voluntary guideline to obtaining compliance with current European food contact legislation, but it could form the basis for a specific directive in the future.

The substances permitted for use in paper and board conforming with this guideline are given in BfR Recommendation XXXVI 'Paper and Board for Food Contact'. This is an evolving subject and readers are advised to seek the latest position from manufacturers and suppliers at any given time.

For toy packaging, the regulations are embodied in EN71 Part 3 8126-3-1997 (limits to the migration of certain elements).

Where plastic coatings are involved with paper and paperboard in direct food contact, they are expected to meet the EC Plastics Food Packaging Directive 2005/70/EC and its amendments. The regulations applying to plastics concern numerous specific migration limits originating from the wide range of plastics and the materials they contain and come into contact with. Paper and paperboard is significantly different in that it mainly comprises the naturally originating polymer-based cellulose fibre and naturally occurring minerals such as calcium carbonate and natural polymers such as starch. Very small amounts of chemical additives such as sizing are used to achieve specific effects, and other chemicals, which are ultimately washed out, are used to facilitate the papermaking process.

Mills can give assurances showing that they meet the requirements of EU Regulation 1907/2008/EC (REACH or the Registration, Evaluation and Authorisation of Chemicals) which in terms of the paper and paperboard product guarantee that the content of SVCH (substances of very high concern) is lower than the stipulated limit of 0.1% in accordance with article 57 in the regulation.

Users of packaging can show diligence in ensuring that regulations are met by requiring evidence from suppliers to show which approved laboratories have checked that the materials concerned meet the appropriate regulations.

1.6 Specifications and quality standards

Having examined the appearance and performance properties of paper and paperboard packaging materials, it is clear that considerable efforts have been made to relate them to packaging needs. Test procedures which measure these properties and relate them to market needs have been developed and used in specifications.

Specifications are needed for a variety of reasons – communication of exact needs, quality assessment, resolving disputes, compatibility of competing quotations and as a basis for improvement.

The paper and paperboard market is international with pulp, paper and packaging traded on a worldwide basis. Hence there has been a need for harmonisation of test methodology. Test methods are developed locally and between suppliers and customers on a one-to-one basis. These may become national standards such as British Standards (BS), German Standards (DIN) and TAPPI in North America. In recent years, international standards have been developed through the ISO.

Tolerances which are realistic are an important requirement of specifications. Over time, the needs and expectations of customers increase – equally the abilities and achievements of suppliers have to respond to market needs. Online computer control has, in many cases, augmented laboratory testing, which by its nature is historical. This development which is progressive and ongoing results in less variability within makings and between makings, and less variability leads to higher productivity.

Equally important is the requirement that quality management systems overall must be seen to be effective. Many manufacturers now have their quality systems independently and regularly assessed under the ISO 9000 series of Quality Standards. Supplier audits are also undertaken.

1.7 Conversion factors for substance (basis weight) and thickness measurements

It will already have been noted, for example in Sections 1.5.3.2 and 1.5.3.3, that different conventions are used in different parts of the world for the measurement of basis weight and thickness.

It is, however, possible to convert measurements in one convention to another as follows:

$$\text{Substance (basis weight) measured in lb}/1000\,\text{ft}^2 = 4.882\,\text{g m}^{-2}\,(\text{grammage})$$

$$\text{Substance (basis weight) measured in lb}/3000\,\text{ft}^2 = 1.627\,\text{g m}^{-2}\,(\text{grammage})$$

$$\text{Thickness measured in 'thou' (thousandths of an inch)}$$
$$= 25.4\,\text{micron (thousandths of a millimeter)}$$

A thousandth of an inch is also known as a 'point' and, in the case of plastic film thicknesses, as a 'mil'. The SI symbol for a micron, or micrometer, is 'µm'.

References

BIR (Bureau of International Recycling), 2006, visit http://www.bir.org
BIR, 2010, *Annual Statistics*, visit http://www.bir.org
BIR, 2011, visit http://www.bir.org
CEPI (Confederation of European Paper Industries), *Annual Statistics*, visit www.cepi.org
CEPI, 2010, *Industry Guidance for the Compliance of Paper & Board Materials and Article for Food Contact*, visit www.cepi.org
Davis, A., 1967, *Packages and Print – The Development of Container and Label Design*, Faber & Faber, London.
Grant, J., Young, J.H., and Watson, B.G. (eds), 1978, *Paper and Board Manufacture*, Technical Division British Paper and Board Industry Federation, London, pp. 166–183.
Greenall, P. and Bloembergen, S., 2011, New generation of biobased latex coating binders for a sustainable future, *Paper Technology*, 52(1), 10–14.
Heynkes, 2007, for details of the BGA, BgVV and BfR and when they were changed, visit http://www.heynkes.de/isa/glossar-en.htm
Hills, R.L., 1988, *Papermaking in Britain 1888–1988*, The Athlone Press, London and Atlanta.
Khwaldia, K., Arab-Tehrany, E., and Desobry S., 2010, Biopolymer coatings on paper packaging materials, *Comprehensive Reviews in Food Science and Food Safety*, 9, 82–91.
Markström, H., 1988, *Testing Methods and Instruments for Corrugated Board*, Lorentzen & Wettre, Kista, Sweden.
Métais, A., Germer, E., and Hostachy, J.-C., 2011, Achievements in industrial ozone bleaching, *Paper Technology*, 52(3), 13–18.
Müller, G., 2011, Improving performance levels of corrugated boxes, *Paper Technology*, 52(3), 24–25.
Opie, R., 2002, *The Art of the Label*, Eagle Publications, Royston, England, p. 8.
Patel, M., 2009, *Micro and Nano Technology in Paper Manufacturing*, Industrypaper Publications, Sambalpur, Orissa, India., visit website www.industrypaper.net
TAPPI (Technical Association of the Pulp & Paper Industry) Test Methods, 2002–2003.
Zellcheming (Association of Pulp and Paper Chemists and Engineers), 2008, *Chemical Additives for the Production of Paper: Functionality Essential – Ecologically Beneficial*, Technical Committee Publication, Deutscher Fachverlag, Frankfurt, Germany.

2 Environmental and resource management issues

Daven Chamberlain[1] and Mark J. Kirwan[2]

[1]Paper Technology, Bury, Lancashire, UK
[2]Paper and Paperboard Specialist, Fellow of the Packaging Society, London, UK

2.1 Introduction

Anything *environmental* commands attention. We now know more about the effects of human activity on the environment at local, regional and global level. Society is concerned, as never before, about environmental issues. Since the 1960s, through education and the media generally, the general public has become aware of the issues. We have seen the rise of the 'green' movement in the form of non-governmental organisations (NGOs). The overall result has led to government involvement at local, national and international levels. There are regulations which require the measuring and regular monitoring of critical factors which can affect the environment.

There is a concern about the quality of life today in both the medium and the long term. These concerns have been summarised as follows (Tickell, 1996):

- population growth causing pressure on *resources*
- changes in the chemistry of the atmosphere – acidification, ozone depletion and climate change attributed to the 'greenhouse effect'
- deterioration of land quality
- pollution of rivers, lakes and seas
- limited freshwater availability
- losses of species or reduction in biodiversity.

The key components relating to world population are life expectancy, which is increasing, and birth rate, which is declining. The birth rate per 1000 is declining but because of the numbers of people overall, the population as a whole is increasing. The projected figures for 2050 have been revised several times in recent years as the overall rate of population growth reduces (Table 2.1). The fact, however, remains that more people will be using more resources.

In addition to global and regional issues, there can be local environmental issues which affect communities, such as the siting of a waste incinerator or a wind 'farm', which are

Table 2.1 World population 1950–2050 (median projection)

Year	Population (billion)
1950	2.5
1960	3.0
1970	3.7
1918	4.4
1990	5.3
2000	6.1
2010	6.9
2020	7.8
2030	8.5
2040	9.2
2050	9.7

Source: UN (2010).

examples of the NIMBY ('not in my backyard') approach. There is also concern about the effects of our activities on indigenous people and forest-based communities.

Industry is a primary user of resources and creates additional environmental impact through its manufacturing processes and the products it produces. The effect of heightened environmental concern has caused industry to balance the need to be technically efficient and profitable against the environmental needs of society which are driven by people's perceptions, resulting in a legal framework within which the industry must operate.

Paper and paperboard packaging is subject to two aspects of environmental concern. First, as an important part of the paper industry, it is subject to concerns which apply to the paper industry as a whole. Second, it is subject to the concerns which apply to all packaging.

In summary, these concerns comprise:

- pressure on resources due to the increasing demand for paper and paperboard
- wood, the main raw material, sourced from forestry
- the environmental impact caused by the use of energy and freshwater
- pollution in manufacture, particularly in fibre separation (pulping) and bleaching
- packaging which is perceived as wasteful in itself and, when discarded, after its protective function has been completed, as a waste management problem.

The increasing demand for paper-based products arises from the increase in population worldwide together with increases in per capita consumption associated with rising standards of living. The disparities in consumption shown in Table 1.3 highlight the potential for increases in per capita consumption of paper-based products.

Whilst there are environmental issues concerning forests, energy, water, impact of industrial activity and waste management that are specific to paper and paper-based packaging, such matters have also to be viewed in the context of agendas driven by world realities. The issues of climate change (global warming), dependence on fossil fuels (the 'oil economy'), competition for scarce resources such as freshwater and what to do about 'waste' all set agendas with environmental implications for society.

We are aware of the limitations of the planet Earth. We recognise 'the global village' where actions in one location or sphere of activity can have regional and worldwide implications. We used to see environmental issues in terms of a range of single issues, such as the concentration of a waste emission which when reduced was seen as a problem solved. Now we appreciate that we must consider whole systems where any change may have wider implications. Hence, there is a need for holistic solutions.

We are asked if specific products, processes or practices are 'environmentally friendly'. These are difficult questions because everything we do, cause to be done or merely observe has an environmental effect, and assessing the extent of that effect involves a value judgement. Widespread and indiscriminate use of the precautionary principle is not an option.

The guiding principle for society, when evaluating our approach to meeting the demands of today and in the future, is to ensure that we operate in an environmentally 'sustainable' way.

2.2 Sustainable development

In 1987, the World Commission on Environment and Development, also known as 'The Bruntland Commission', issued a report entitled 'Our Common Future'. This discussed, internationally, the concept and the need for sustainable development. The report defined this as 'development which meets the needs of the present without compromising the ability of future generations to meet their own needs'.

At Rio in 1992, the governments of various nations agreed that sustainability had economic, environmental and social aspects which were equally important and interdependent (United Nations Conference on Environment and Development).

This theme has been taken forward in a number of major areas including forestry with significant coordinating leadership from FAO (United Nations Food and Agricultural Organisation). Commenting on preparations for the XII World Forestry Congress held in September 2003, FAO Assistant Director-General, Mr. M. Hosny El-Lakany said:

> Nations must manage their forests in a sustainable way so that present generations can enjoy the benefits of the planet's forest resources while preserving them to meet the needs of future generations. (FAO, 2003)

Ways of measuring environmental, economic and societal performance have been developed and several leading paper companies are featured in the Dow Jones Sustainability Indexes (DJSI) (www.sustainability-index.com). These indices are based on criteria from companies which are evaluated in terms of quality of management and strategy, together with performance, in dealing with opportunities and risks deriving from economic, environmental and social developments. Such developments can be quantified and used to identify and select leading companies for investment purposes. In several annual reports, the DJSI has outperformed the market in general, indicating that sustainability priorities enhance performance overall for the companies concerned.

A similar index is the FTSE4Good Index, which also includes several leading paper companies. This index measures the performance of companies that meet globally recognised corporate responsibility standards.

There are other indices, in which paper companies are involved, that can assist in financial decisions based on a range of corporate environmental and social responsibilities. These include the Ethibel Sustainability Index, the Domini 400 Social Index and the Vanguard Calvert Social Index.

In addition, the Ethisphere Institute is a leading international think tank dedicated to the creation, advancement and sharing of best practices in business ethics, corporate social responsibility, anti-corruption and sustainability. Several leading paper companies are included in their current list.

The position today, and for the future, is that paper and paperboard:

- should not be a polluter in manufacturing, use and disposal
- should not be a drain on irreplaceable resources.

More, of course, can be said about both these points, but they are the bottom line – the key messages irrespective of what may have applied in the past when our knowledge, measurement techniques and understanding of environmental impact were much less sophisticated than is the case today.

We should also be confident of the main benefit of paper and paperboard packaging in that they protect far greater resources, in terms of the goods protected, than are consumed in the manufacture and use of the packaging. This benefit is achieved during the packing, storage, distribution, sale and consumption of food and other manufactured goods.

Effective packaging, in which paper and paperboard-based packaging is a major component, has revolutionised the safe distribution of all products, particularly food, in the advanced industrial societies and to a lesser extent so far in the developing world. It is acknowledged that packaging *reduces* waste.

Ample evidence exists of large-scale losses occurring during post-harvest and distribution operations in developing countries. The subject was first addressed at the World Food Conference in 1974. At the World Food Summit in 1996, FAO presented evidence from 51 developing countries of significant post-harvest food losses. The results have been summarised by Neil Robson (1997) and the conclusion drawn that 'the world's packaging professionals have a major task ahead if they are to convince governments and opinion leaders of the vital role which packaging should play in reducing food wastes'. A recent FAO report on global food losses concludes that roughly one-third of food produced for human consumption is lost or wasted globally and that investment is needed in low-income countries for infrastructure, transportation and packaging (FAO, 2011h).

It is therefore important that paper and paperboard-based packaging continues to be available and this has to be achieved in a sustainable manner. Sustainability for paper-based packaging concerns the availability and use of the resources needed to meet increasing demand and the minimisation of the environmental impact of manufacturing, use and disposal. In summary, it concerns forestry, manufacture of paper and paperboard, printing, conversion, packaging, distribution, consumer use and the ultimate disposal of paper and paperboard-based packaging. These subjects will now be discussed in terms of 'sustainability'.

2.3 Forestry

Forests cover around 31% of the earth's land area and are one of the essential support systems which sustain life on Earth as we know it – the other essential systems concern air, water, soil and energy.

Forests are important as they:

- reverse the 'greenhouse effect'
- stabilise climate and water levels

- promote biodiversity by providing a habitat for a wide range of animals, birds, plants and insects
- prevent soil erosion and protect watercourses
- store solar energy.

The greenhouse effect occurs when radiant energy from the Sun is unable to escape from the earth's atmosphere because of the build-up of gases, such as carbon dioxide (CO_2). These are believed to prevent the heat from escaping – acting, in effect, like glass in a greenhouse – and so cause the temperature to rise. This is known as 'global warming', which is considered to be a cause of climate change.

Carbon dioxide is the most common greenhouse gas. It is released when fossil fuels like coal, oil and natural gas are burnt to produce energy in the form of heat and electricity and when used in internal combustion engines. The world's use of fossil fuels releases about five billion tonnes of carbon into the atmosphere per annum. The proportion of CO_2 in the atmosphere is increasing by about 5% a decade and this is thought to be the main cause of global warming.

Trees grow by absorbing CO_2 and releasing oxygen. As trees grow, therefore, they remove carbon from the atmosphere and so help to reverse the 'greenhouse effect'. This is known as 'fixing' carbon and the forest acts as a 'carbon sink' or reservoir.

Growing trees process CO_2 from the atmosphere by photosynthesis (Fig. 2.1). In this process, trees, in common with all green-leafed plants, using energy supplied by the Sun, convert CO_2 and water into simple sugars and oxygen. The sugars are polymerised naturally and ultimately result in the formation of cellulose fibres and related polymers, such as hemicelluloses.

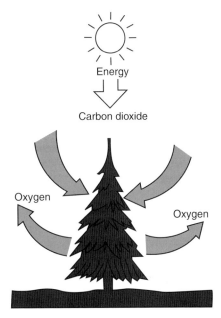

Figure 2.1 Trees grow by the combination of carbon dioxide and water, using energy from the Sun. This process, which emits oxygen, is known as photosynthesis. (Courtesy of Iggesund Paperboard.)

A mature forest, however, also contains dead trees and other vegetation. As this organic material decays, it actually gives off CO_2. Also, growing trees emit oxygen and decaying trees absorb oxygen. Hence, where the biomass (collective term for trees and other vegetation) stays approximately constant in a mature forest, the emission and absorption rates for oxygen and CO_2 remain in balance, and it is incorrect to refer to such mature forests as being 'the lungs of the world'. In expanding forests, on the other hand, where there is an annual incremental net addition of new wood volume, CO_2 is effectively removed from the atmosphere.

With respect to carbon 'fixing', area alone does not reflect the complete picture. Location, climate, age of trees, soil, rainfall, competing species, proportion of dead and decaying trees and other biomass, etc. are all important factors. Large land masses and vegetation, or lack of these, including large areas of forest, are known to be the major determinants of climate. Other factors are solar energy, temperature, humidity, precipitation, condensation, atmospheric pressure and wind, which all interact dynamically.

Forests prevent soil erosion and protect watercourses from silting – the destruction of forests has resulted in humanitarian disasters through landslides and flooding. Forests are important for the maintenance of the traditional form of life of forest-based communities in the developing world. It is self-evident that forests provide a habitat for many species, viz. animals, birds, trees, other types of plants and insects, etc. It is also a fact that species evolve and some become extinct. Biodiversity has entered the environmental debate with respect to the extent that human interference in forestry is causing species to be lost by extinction. Some exceedingly high losses have been *predicted*. However, the UN *Global Biodiversity Assessment* (UNEP, 1995) states that 'the rate at which species are likely to become extinct in the near future is very uncertain', taking note of 'the discrepancy between field knowledge and predictions'.

The importance of forests as a store of solar energy is less controversial, if not fully appreciated. Statistics show that between 1990 and 2005, about 80% of the wood felled in Africa and Asia was used as fuel; this compares with around 50% in South America, and only 15–20% in Europe and North and Central America. Globally, woodfuel is estimated to account for 50% of the wood felled annually (FAO, 2011a). In the European Union (EU), about 3% of the energy used is derived from wood though the proportion is likely to increase as the EU seeks to replace fossil fuel energy with energy from renewable sources, which include wood and biomass.

Forests, particularly in the industrialised countries, are also important in providing an environment for leisure and for the commercial viability of small communities working in forest-based activities.

Forests are therefore important for many reasons. FAO defines four types of forests by ecological zones as follows:

- Boreal (33% area) – these are forests in the northern parts of the Northern Hemisphere, i.e. northern parts of Canada, Scandinavia and the Russian Federation, including Siberia. The alternative name 'taiga' is used for the boreal area of Russia. These forests are predominantly coniferous, i.e. with pine, spruce, larch and fir, with a proportion of deciduous birch and poplar. Some analyses refer to these forests as temperate coniferous.
- Temperate (11% area) – these forests are mainly in North America and Europe; they contain mixed coniferous and deciduous species.
- Subtropical (9% area) – also referred to as warm temperate. An important area for this type of forest is the south-east of the USA.

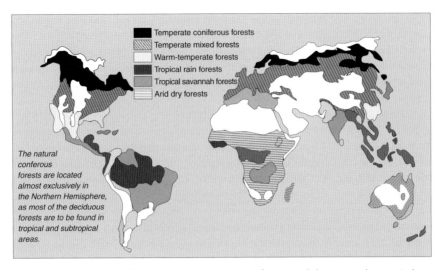

Figure 2.2 Forest locations. (Reproduced with permission from Swedish Forest Industries Federation.)

- Tropical (47% area) – there are three types of tropical forests. The rainforest is given the most publicity in areas such as the Amazon basin, Congo basin and South-east Asia. However, the area loss in the arid dry forests and savannah forests of South America, Africa and India are a greater cause for concern, where the use of wood for fuel and unsuitable agricultural practices can lead to desertification. The trees are predominantly deciduous (FAO, 2001a).

Over the decade 2000–2010, most deforestation occurred in the tropical regions whilst afforestation occurred in boreal and temperate zones (FAO, 2011b). Figure 2.2 shows six categories of forest area which give a general indication of where forests are located. FAO has defined the various types of forest area based on tree size and concentration.

In its most recent report, FAO produced a far simpler representation of forest types, distilled into three categories, as follows:

- Primary (36% area) – these are forests where native species dominate, there are no visible signs of human activity and the ecological processes remain undisturbed.
- Naturally regenerated (57% area) – this forest shows signs of human intervention, and the species present are a mix of native trees and introduced species.
- Planted (7%) – forest established solely through planting or seeding.

Over the preceding decade, the area of primary forest stayed almost constant; planted forest showed an annual increase of almost 5% whilst naturally regenerated forest is where the greatest decline in forested area occurred (FAO, 2010a).

The main source for coordinating global statistics for forestry is FAO, and reference should be made to its website for comprehensive details. They issue a 'Forest Resources Assessment' every 10 years. In the latest report, of 2010, it is stated that total forest area is just over 4 billion hectares, roughly 31% of the global land area (FAO, 2010b). (It is thought that between one half and two thirds of the land area was covered by forest, thousands of years ago.)

Deforestation in the developing world has been driven by the demand for agricultural land and fuel, though FAO states that the direct link between population growth and the conversion of forests to agriculture is not consistent across continents. In particular, Africa has experienced lower deforestation than would be anticipated given population growth, whilst in South America there is a more direct link between deforestation and agriculture (FAO, 2011c).

In the 1990s, the global forest area was diminishing in area by a net 8.3 million hectares pa (per annum). FAO (2010b) shows that the estimated net loss of forest area for the 10 year period of the 1990s as a whole was 83 mha – an area around the size of Pakistan. Data for the most recent decade, from 2000, show the rate of deforestation has fallen to 5.2 million hectares pa, or 52 mha, over the 10 year period, which equates to an area the size of Costa Rica (FAO, 2010b).

There was, however, a contrast in the change in area between forests in industrialised countries and those in developing countries. Over the decade from 2000, forest area was *increased* in Europe and Asia and remained virtually unchanged in North and Central America; however, Africa, South America and Oceania all saw a net *decrease* in forest coverage (FAO, 2010d). The fact that forest area is diminishing is a cause of concern. It is also a fact that forests are the main source of raw material for paper and paperboard. It has become a popular, but totally inaccurate, perception that these two facts are *connected* and that the paper industry is, somehow, responsible for the loss of forest area, particularly the rainforests.

Most of the wood used by the paper industry is sourced from industrial countries where forest coverage is either stable (North America) or increasing (Europe). In addition to the increase in the area of forest in the industrialised countries, another important positive factor is the difference between annual net incremental growth and the volume of wood harvested. For example, although rates of harvesting are increasing, data for Europe show that the volume of new wood growth (annual increment) continues to *exceed* the volume of wood felled (harvested) (EEA, 2010).

The fact that there are large differences between the volume of wood harvested and the incremental growth is an indication of the value of these forests as 'carbon sinks'.

There are other differences between the forests in industrialised and developing countries. Remrod (1991) stated that in the former, roughly 71% of trees are coniferous and 29% deciduous whilst in the latter approximately 94% are deciduous. This difference in tree types still holds today, although it must be noted there are now significant areas of coniferous forest in parts of Asia and North Africa (FAO, 2010e). Also, most of the deciduous trees in the natural forests of the developing countries are unsuitable for pulpwood.

As stated earlier, plantations make up only about 7% of all forest. Plantation forestry increased at an annual rate of 4.5 million hectares pa during the 1990s; by the 2000s this rate had increased to 5 million hectares pa (FAO, 2010f). Much of this new plantation was located in Asia: by 2000 it already had 42% of the world's plantation cover, which had increased to 46% by 2010 (FAO, 2011d).

In tropical areas, eucalyptus and acacia are the most common species; and in temperate and boreal zones, pine and spruce are predominant. An important advantage of plantations in the tropical zone is that they provide income and, hence, relieve pressure to remove timber from natural forests (FAO, 2001b).

Many of the forests in the industrial countries are managed. The general aim of sustainable forest management is to 'reduce the differences between virgin forest and cultivated forests' (Holmen, 2002).

Forest management is carried out in many ways depending on local factors, such as the nature of the soil, climate, height above sea level, etc. The specific aims are to:

- ensure replacement of harvested trees
- provide habitats for animals, plants and insects
- promote biodiversity
- protect watercourses
- preserve landscape
- maintain rural employment
- create leisure facilities.

Forest management today meets commercial, social and environmental needs. Forests, to an increasing extent, can be independently certified for environmental performance in relation to approved objectives. Certification of timber products can ensure a transparent chain of custody so that the eventual customer knows the source of the wood and has authentic confirmation of the certification.

The two largest international certification schemes are the Programme for the Endorsement of Forest Certification (PEFC) and that of the Forest Stewardship Council (FSC), although many countries have their own national programmes which may or may not be affiliated to either of these organisations. In 2010, PEFC certification covered 234 million hectares of forest, or roughly 6% of the total world forest by area, concentrated mainly in Europe though also covering parts of North and South America and Oceania. Meanwhile FSC accounted for around 141 million hectares (just under 4% of the world forest by area), although in addition to the countries covered by PEFC they do include parts of Asia and Africa. The ethos of certification is certainly increasing and being appreciated by both the public and forest product companies – for example, between 2008 and 2010 the area of forest certified by FSC grew by 30% – but there is still a very long way to go before a significant proportion of the world's forest is given adequate protection.

Apart from the use of wood as fuel, it is used industrially for building, e.g. housing, sheds, fencing, etc., wood products, e.g. tools, ornaments, transport, furniture, etc., and as pulpwood for the paper industry (Fig. 2.3). The pulp industry has been criticised for harvesting wood by clear felling though today this is practised in a more environmentally supportive manner. An illustration of this is the provision of 'corridors' of forest cover which allows animal movement. Another example of better management is where a statistical analysis is made of the various species being felled and regeneration by seeding is carried out so that the original proportions of the various species in the forest are maintained. Natural seeding and thinning lead to a wide range in the ages of trees, in a managed forest.

Pulp is also made from trees which have been thinned in managed forests, thereby allowing other trees to grow to maturity for other purposes. There are plantation forests which are not suitable for thinning where, owing to location, prevailing winds and soil depth, thinning would result in the loss of the remaining trees. It is hoped that after several rotations, the depth of soil will build up to the point where the trees will be more deeply rooted and thinning will become possible. Pulping can also use the tops and branches of large trees felled for building-timber as well as sawmill waste, both materials which otherwise have little commercial use other than as fuel. In this way, the paper industry contributes to the overall commercial viability of managed forestry.

The answer to the question as to whether the paper industry is responsible for forest depletion in the rainforests, and elsewhere, is a resounding negative. Only around 18% of

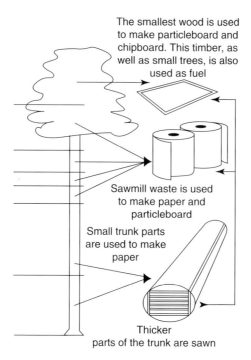

Figure 2.3 How the tree is used. (Courtesy of Iggesund Paperboard.)

the pulpwood sourced in 2008 came from the tropical zone, and this is almost all derived from sustainable plantations (FAO, 2011e). The pulp which is produced in the tropical zones, for example in Brazil and Indonesia, is derived from plantation wood.

Data show the vast majority of wood removed from most of the tropical region was used as fuel; only in South America, which accounts for roughly half of the wood pulp derived from tropical countries, was the split between industrial roundwood and woodfuel evenly distributed (FAO, 2010g). Indeed, worldwide data show that whilst woodfuel constituted only 47% of wood removal in 2005 (FAO, 2010g), by 2008 it had leapt to 55% (FAO, 2011f); during this period the total volume of wood extracted from forests worldwide stayed relatively constant. Consequently, use of wood for fuel constitutes the main threat to forests worldwide, particularly in Africa and Asia; together, these two areas alone account for 73% of extracted woodfuel.

Pulpwood is mainly produced in the industrialised countries of the temperate, boreal and warm-temperate zones, and in these areas forest cover is increasing and there is a significant annual increment in which new growth exceeds the amount harvested.

It can be said that because of the demand for paper, investments in forestry have occurred. The paper industry is in the business of using, rather than saving, trees and thereby saving forests! Yet there are still many who claim that the increased demand for paper and other wood products is turning forest destruction into a global disaster (Haggith, 2008). The authors still meet, and read about, people who think that using paper 'destroys' rainforests.

One way of quantifying the effect of the paper industry is to equate the pulp used with the total volume of wood extracted in a year. FAO figures for 2008 show the total volume of wood felled amounted to nearly 1.87 billion m^3; in the same year, the total amount of wood felled for use in the paper and paperboard industry came to 193 million tonnes (Mt) (FAO, 2011g). If we assume an average unseasoned wood density of 800 kg/m³, this means the total amount of wood felled was approximately 1.5 billion tonnes. Wood pulp therefore accounted for only around 13% of the total wood felled in 2008.

Another way of looking at the situation is to calculate the area of forest needed to produce, *in new growth per annum*, the volume of wood used for paper and other wood products. This is relatively simple because FAO has worked out the total annual volume of wood used for paper and other wood products as well as the area of the global forest. Using the average growth rate per hectare, it is possible to calculate the area of forest needed to produce that volume of wood. Growth rates vary considerably depending on factors such as climate, soil and tree species. A range was published in the Dobris Assessment (EEA, 1995), giving average growth rates in European countries.

The range included Norway at 2.7 m³/ha, Finland at 3.5 m³/ha, Sweden at 4.1 m³/ha, UK at 5.0 m³/ha, France at 5.3 m³/ha and Ireland at 8.8 m³/ha. On the basis that the Nordic areas of Sweden and Finland are significant producers, the area required would be derived from the *growth* of roughly 10% of the global forest area. In practice, many forest areas would have higher growth rates than the Nordic rate, and the true area would probably lie between 5% and 10% of the global forest area.

The conclusion to be drawn is that the area of forest necessary is a small proportion of the global forest and, as we are looking at annual incremental growth, no forest depletion is involved.

Many pulp and paper companies worldwide have forest interests, and their policy is to manage these forests in sustainable, environmentally sound ways. It is in their best interests to protect their investments and secure their main raw material with sustainable forest management.

Under the guidance of FAO and national governmental woodland supervision, sustainable forestry is progressing on a worldwide basis. There are issues which are being addressed to alleviate the problems of forest loss in the developing world, but the position from the point of view of the paper industry is of a naturally sustainable future.

2.4 Environmental impact of manufacture and use of paper and paperboard

2.4.1 Issues giving rise to environmental concern

Issues giving rise to environmental concern are:

- extraction of wood from forests
- fibre separation by chemical and mechanical pulping processes
- bleaching, i.e. whitening, of pulp
- paper and paperboard manufacture
- printing, conversion and packaging
- logistics – storage, distribution and sale of packed goods.

Environmental impact arises through the use of resources:

- energy
- water
- chemicals.

Environmental impact also arises as a result of:

- emissions to air, water and the generation of solid waste during extraction, manufacture and distribution
- packaging waste.

2.4.2 Energy

Energy is used in fibre separation, in bleaching and in paper, paperboard and packaging manufacture and use. Energy comprises roughly 25% of the cost of manufacturing paper and paperboard.

The energy source and the amount used depend on the manufacturing process, printing, conversion, packaging, distribution and the locations of energy use.

In chemical fibre separation, chemicals are used to separate fibre by dissolving the non-fibrous constituents of wood. The separated non-fibrous material is used as fuel to generate electricity and steam. The wood is therefore the energy source for both pulping and bleaching. In integrated mills, where paper or paperboard is made on the same site as the pulp, the wood by-products also provide energy in the form of electricity and steam for the manufacturing process. This energy source, biomass, is, therefore, naturally renewable and sustainable. In Europe this results in 55%, and in the USA 64%, of paper and board being made using energy derived from biomass (ICFPA, 2009).

In mechanical and recovered fibre separation, i.e. recycling of paper and paperboard, energy in the form of electricity is used to operate the processes used to separate fibres. Mills using mechanical and recycled pulp generate their electricity and steam on-site. Electricity is used to pump suspensions of fibre in water, drive the machinery, remove water by suction, and press and dry coatings using radiant heat. Steam is used to heat drying cylinders and provides other on-site heating needs. The energy source for all these processes has traditionally been fossil fuel, which in Europe today usually means natural gas. However, larger mills are increasingly installing biomass generating capacity because this gives greater price stability in what has become an increasingly unstable and volatile energy market. In CEPI (Confederation of the European Paper Industry) countries, the steady rise of both biomass and natural gas as the chosen fuels has meant the proportion of energy based on oil, coal and other fossil fuel has fallen significantly, from 29% in 1990, to 15% in 2000 (ICFPA, 2002a) and just 9% in 2008 (CEPI, 2009a).

In paper and paperboard mills processing mechanically separated fibre and recovered fibre, and in printing, conversion and packaging, the main energy source is electricity. For those manufacturing units too small to install their own energy generation supply, this is obtained from external sources and therefore the supply, and sustainability with respect to energy, depends on the way electricity is generated for society as a whole in the locations concerned. In some locations, hydroelectric power dominates (2% of global energy); another source is nuclear energy (6% of global energy). In most countries, however, the energy source for most of industry, including paper and paperboard packaging, and society is fossil fuel (81% of global energy) (IEA, 2010).

We live in an oil-dominated economy which had managed to keep the price of crude oil under $30 per barrel since the 1880 s, until relatively recently, apart from a minor blip from 1973 to 1985 when Petroleum Exporting Countries (OPEC) engineered higher prices by restricting output. However, from 2004 the price of oil has risen and shown great volatility, reaching in excess of $145 in July 2008; at no point during this period has the price fallen back to the sub-$30 per barrel levels experienced historically.

Previously, the low oil price restricted the commercial development of alternative energy sources. It is also the reason why known oil reserves were projected on a relatively short timescale, since exploration is expensive. However, more oil is constantly being discovered within this timescale and technology is extracting a higher proportion from existing oilfields.

Nevertheless, in time the earth's resources of fossil fuels on which we currently depend for the greater part of our total energy consumption, such as gas, oil and coal, will be exhausted. However, technology applied to known reserves of oil, gas and coal, together with the discovery of new sources, continues to extend this timescale. Additionally, there are significant reserves in tar sands and oil shales, and every modest rise in the price of oil makes exploitation of these resources more commercially viable. For example, it has stimulated production from Canadian tar sands. In the USA, shale gas production showed an annual growth rate of 48% between 2006 and 2010 (EIA, 2011), whilst in the UK shale gas drilling commenced at Blackpool during 2011, though it was halted after two minor earthquakes were blamed on the controversial 'fracking' method employed to extract gas (FT, 2011).

Data relating to conventional fossil fuel reserves can be confusing, because new deposits are discovered each year. The World Energy Council suggested at the consumption rates prevalent in 2007 that coal reserves would last around 150 years and gas a little under 60 years (WEC, 2007). Oil reserves were far harder to pin down; however, the reserves known at the time (of around 1.2 trillion barrels) were dwarfed by the estimated amount of shale oil available (2.8 trillion barrels).

The use of coal as fuel is less favoured today because of the high cost resulting from the removal of noxious emissions. Oil and natural gas will eventually be more difficult to find and use and therefore become more expensive at which point other fossil fuel reserves and alternative sources of energy, which today are more expensive than oil, will become commercially attractive alternatives. Renewable energy sources are by definition sustainable, and it is possible that other, environmental, priorities will be applied to the choice of energy source within the timescale for oil to become more expensive.

A more likely scenario therefore is that we will be considering non-fossil alternatives sooner rather than later in order to reduce carbon emissions. Indeed, the EU has a target of increasing the proportion of energy derived from renewable sources to 20% by 2020.

Currently, the main non-fossil alternatives which are used are themselves subject to constraints for expansion. For environmental reasons, hydroelectric schemes are unlikely to be significantly expanded, and the same conclusion, though for different reasons, applies to nuclear energy.

In industrialised countries, only a small proportion of energy is being supplied from other, renewable, sources such as wind, waves, solar, geothermal, landfill gas and biomass which includes wood, though the proportion is rising steadily. In the EU, during 1998, this proportion was 10%, but by 2008 it had already increased to almost 18% (Eurostat Pocketbook, 2010).

The subject of wood (biomass) as an energy source for society in developed industrial societies is of interest to the paper and paperboard industry as it would then be competing for supplies of wood. It is, however, realistic to encourage the use of wood for energy in a sustainable way. New coppiced plantation woodland of poplar and willow, where a crop is taken every 3–5 years by cutting trees to ground level, and stimulating regrowth, is a traditional practice which is being revived in a modern way.

Geothermal energy using heat from the interior of the earth can be harnessed where it is readily available. This is the major energy source in Iceland, but there are commercial and technical problems in exporting such energy over long distances in the form of electricity to major potential users, such as the UK. Geothermal energy is used in other countries particularly in western areas of the USA and in New Zealand – there is even a small operation in the UK which derives heat from an 1800 m deep drilling (Southampton Geothermal Heating), and more such operations are planned in the near future at Newcastle and Cornwall. However, there are commercial and technical constraints in expanding these resources. As regards use by pulp and paper mills, geothermal energy is used by companies in New Zealand.

Wind, landfill gas and solar power are being harnessed to an ever greater degree, as is a small amount of wave power; one pulp mill has even invested in a pilot plant that uses osmosis to drive a turbine (PT, 2009). Historically, the poor rate of development of renewable energy has been due to commercial reasons, principally the low cost of oil. The realisation that fossil resources are finite, that their exploitation causes pollution, and that fuel prices are both volatile and rising, has given impetus to the development of renewable resources. Currently, wind power is the fastest developing source of electricity. There are technical challenges, with cost implications, and it competes in an energy market subject to economic, fiscal and political pressures.

Non-fossil sources are also being exploited for incineration; these include municipal waste, which is discussed in Section 2.5.3.3, wood and other biomass, and specialist waste materials including scrap tyres, hospital wastes, poultry litter, meal and bone and farm waste. Of these solid fuels, municipal waste forms the greatest resource in the UK, where during 2009, 22 power stations produced around 392 MW; plant biomass came second with almost 280 MW of generating capacity (DECC, 2010).

Organic solid and liquid wastes can also be used to generate fuel. Food waste can be converted by anaerobic digestion to methane; experiments have shown that sludge from paper mills, which tends to be rich in organic matter, can also be used as a useful fuel for digestion (Wood et al., 2009; Hart and Evers, 2011), as can plant crops (Frigon and Guiot, 2010). Meanwhile, plant residues (Johnson et al., 2009) and paper mill sludge (Kang et al., 2010) can also be hydrolysed to alcohol (bioethanol); UK data show an increase from 85 million litres in 2005 to 317 million litres in 2009 – between 2008 and 2009 alone the production rate increased to 54%! Other organic liquids, such as waste cooking oils, can also be converted to biodiesel; the rapid rise in oil prices during the last decade has stimulated research in all these areas, and many are now being exploited commercially. Although they are all still at a relatively small scale compared to oil production, it is certain they will all continue to grow given the political and economic uncertainly that continues to surround oil production.

Meanwhile, in the developing world, traditional energy sources such as wood, charcoal and other animal and vegetable wastes, including bagasse from sugar cane processing, continue to be used. These forms of energy provide 25% of the energy needs of developing countries (Lomborg, 2002).

In summary:

- Over 50% of the energy currently used in paper and paperboard manufacture is renewable.
- An increasing amount of renewable energy is being used in the general electricity supply, though the impression should not be that this will significantly reduce our current dependence on fossil fuels over the short term.
- In terms of sustainability, there is no short-term concern for fossil fuel supplies, though the cost will increase as less-accessible sources are exploited.

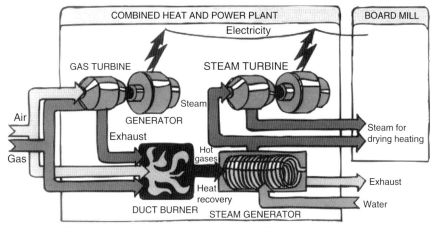

The new combined heat and power plant in Workington gives the paperboard mill much greater energy efficiency, compared with electric power from the public network. The natural gas turbine provides electricity. The temperature of the turbine's exhaust gases is raised further in a gas-fired duct burner. The hot gases from the burner produce steam in a heat exchanger. The steam goes to a steam turbine which generates electricity, and the excess heat is used both for drying the paperboard and for heating.

Figure 2.4 Principle of a combined heat and power (CHP) plant. (Courtesy of Iggesund Paperboard.)

Another way of looking at energy sustainability in the paper and paperboard industry is to use current energy sources more efficiently.

Paper mills have traditionally produced electricity and steam on-site by heating water to produce high-pressure steam. This is used in turbines to produce electricity, and the resulting low-pressure steam is used to heat the steel cylinders which dry the paper and paperboard. The critical requirement is to produce enough steam for drying.

An important way in which the paper industry has contributed to a saving in energy and a reduction of CO_2 emission is by using natural gas in a combined heat and power (CHP) plant (Fig. 2.4). In this process, natural gas is used in a gas turbine to produce electricity. The temperature of the already hot exhaust gases is raised further by heating with additional fuel and used to produce steam which is then passed through a steam turbine to produce more electricity. The exhaust steam is then used to provide heat in the paper or paperboard mill. This is a more efficient use of fuel compared with the traditional boiler and steam turbine approach and more efficient than using electricity from a national grid. CHP generation has several benefits. CHP consumes less fuel for a given energy output and can balance the output of electricity and steam.

The energy saving is of the order of 30–35% compared with conventional boilers (ICFPA, 2002b). Moreover, as a consequence, there is a lower gaseous emission (CO_2, SO_2 and NO_x) per unit of electricity and any excess electricity can be supplied to other (local) users, thereby saving the use of less efficiently, remotely produced electricity. The capital cost of CHP plant is high and commercial effectiveness is dependent on the price obtained for the electricity exported from the mill.

Energy saving is a major area for development as this can have a significant impact on cost and emissions. For an industry that is so energy intensive, there are numerous ways that savings can be made, from preventing wastage of heat by, for example, installing adequate thermal insulation and ducting to swapping over-specified pumps for smaller more energy-efficient models. Vacuum systems, mechanical pulpers, refiners, presses and steam systems are all major sources of energy usage, and significant advances have been made in recent years to improve both their efficiency and to understand the fundamental processes involved in their operation (EU-China, 2009; Hutter and Jung, 2010; Austin et al., 2011).

In the year 2000, 90% of electricity used in European mills and 88% used in US mills were produced by CHP (ICFPA, 2002b); by 2008 this had risen to 94% in Europe (CEPI, 2009b). Meanwhile, energy consumption per tonne of product in the manufacture of paper and paperboard continues to fall. In the period 1990–2008, the reduction per tonne of product in Europe was 17% (CEPI, 2009c).

2.4.3 Water

In 2001, European mills consumed, on average, around $48\,m^3$ of water per tonne of paper and paperboard, having already achieved a reduction of one third in the previous decade (CEPI, 2002a). However, according to CEPI, production of paper and paperboard over the same period increased by 36.5%. Since then major changes have occurred, so that currently the average European mill only uses around $12\,m^3$ of water per tonne of paper or paperboard and some use as little as $2\,m^3$ per tonne of product (PPT, 2010). High volume producers, such as packaging mills, tend to have water consumption figures at the low end of this range.

Water is used in mechanical and chemical pulping to separate, bleach, wash, refine and transport primary (virgin) fibres, i.e. fibres made directly from wood or primary vegetation. Water is used in recycling to separate, wash, refine and transport recycled (secondary) fibres recovered from waste paper and paperboard. It is also used in the de-inking process.

In manufacturing paper and paperboard, water is used to refine or prepare fibres for use on the paper or paperboard machine. The sheet is then formed from a dilute water suspension of fibres from which water is then removed by vacuum, pressing and drying during which interfibre and, in the case of multi-ply paperboard, interlayer bonding takes place. The role of water in the consolidation and the development of bonding is of fundamental importance in the manufacturing process. Water is used to transport the active ingredients in surface sizing and mineral pigment coating and is converted into steam to generate electricity and heat drying cylinders.

Increasingly, the source of water is becoming an issue, and the concept of 'Water Footprint', which is analogous to 'Carbon Footprint' for energy, is being put forward (WFP, 2011). Water can be categorised as 'blue', 'green' and 'grey' – the first is that extracted from surface and groundwater sources; the second is water stored within soil; and the third is polluted water. Historically, most paper mills have used 'blue' water; typically they will extract from rivers or sub-surface aquifers. However, with current pressures on water extraction, mills are increasingly being forced to review their use of water. Some have closed their water systems and increased the level of process water they reuse; this has required them to install increasingly sophisticated water purification equipment on-site. A few other mills have been more adventurous, and are sourcing 'grey' water as their primary resource (Mubin, 2004; Holmen, 2010). In the future, increasing use of 'grey' water is likely to be forced upon the paper and paperboard industry.

After use, the process water is treated to meet the local regulations and returned to nature, i.e. rivers, lakes or tidal estuaries. As already noted, there is a trend to use less water. Water is recycled and condensate is recovered. Increased closure of water systems reduces the environmental impact of water use. Improved effluent water treatment and the move away from bleaching with elemental chlorine has resulted in a reduction in water usage.

In the printing and conversion of paper and paperboard-based packaging, water is used in many ways:

- as the vehicle in liquid inks
- in emulsion-based functional coatings
- in lithographic printing (treatment of non-printing areas of the plate)
- as vehicle in adhesives
- in cooling systems
- in drying systems
- in processing printing plates
- in cleaning systems.

Many reports in recent years have predicted severe shortages of freshwater. *Global Environmental Outlook 2000* from the United Nations Environmental Programme states, 'the declining state of the world's freshwater resources, in terms of quality and quantity, may prove to be the dominant issue on the environment and development agenda of the coming century' (UNEP, 2000). This theme is continued in the follow-up report *Global Environmental Outlook-4*, where it is stated that two thirds of the world could be subject to water stress by 2025 (UNEP, 2007).

The paper industry is therefore managing water usage with care in respect of the quantity used, proportion recycled and the quality of water returned to nature. The paper industry has reported significant reductions in water consumption and significant improvements in the quality of returned water are continuing.

The use of water in printing, conversion and packaging is much smaller than the amounts used in paper and paperboard manufacture. Users are, however, subject to metering in respect of consumption and controls in respect of effluent (discharges). Overall, it is therefore considered that water management in the manufacture and use of paper and paperboard-based packaging is managed in a sustainable way.

2.4.4 Chemicals

Process chemicals used in the manufacture of paper and paperboard comprise 3% of the raw materials. A further 8% of the overall raw material usage is accounted for by mineral pigments used in surface coatings and fillers (Zellcheming, 2008). (Note: fillers, which comprise small particle size white mineral pigment, are used to improve opacity, printability and smoothness.)

A few chemicals, like starch, which is used for surface sizing and promotion of fibre bonding and strength, and inorganic filler and coating pigments, are of natural origin; however, most are synthetic. Chemicals are used to separate (release) fibres in wood. In the case of the sulphate process, most of the chemicals used are recovered and reused in the process. Bleaching chemicals were discussed in Section 1.2.4.

This leaves an important group of chemicals which are used to enhance the product, improve the efficiency of the manufacturing process and meet ecotoxicological demands. This group would include coating ingredients, such as binders; sizing agents, including speciality chemicals that impart specific properties, such as grease resistance; optical bleaching agents (OBA), also known as fluorescent whitening agents (FWA); wet-strength additives; retention aids; defoamers; dyes; and anti-fungal and anti-microbial aids which keep the process system in the mill clean and efficient.

Waste chemical materials from the applications described are waterborne and are subject to effluent regulations applying in the areas where mills are located. Reference to environmental reports from paper mills indicates that process chemicals are carefully chosen and that environmental impact, health and safety are key criteria. From the standpoint of sustainability, it should be noted that some of these chemicals are already derived from sustainable resources, for example starch. Whilst many of the synthetic polymers currently used are based on monomers derived from oil or coal, other renewable feedstocks are technically possible and are just starting to be made available commercially (Chamberlain, 2010).

A wide range of materials is used in printing, conversion and package use:

- resins, vehicles and pigments used in inks
- adhesives for both structural design and lamination – a wide variety can be used including starch-based, polyvinyl acetate (PVA), cross-linking resins and hot melts based on resins and waxes
- plastic extrusion coatings (polyethylene (PE), polyethylene (PP), polyethylene terephthalate (PET or PETE))
- functional coatings for grease resistance, water repellency, non-slip, release, etc.

All these materials must be suitable for food packaging where they are required to be in contact with, or in close proximity to, food. Printers and converters should check the status of all materials with their suppliers with respect to the prevailing regulations. In addition, the processes of making printing plates and cylinders also involve the use of chemicals.

2.4.5 Transport

Packaging and packaged goods are, predominantly, transported by road. The movement of logs, pulp, paper and paperboard is by road, rail and sea. Reference to the environmental reports issued by major forest product companies indicates that they are working on reducing the energy usage and the emission of volatile organic compounds (VOCs) in transportation.

2.4.6 Manufacturing emissions to air, water and solid waste

So far we have discussed the resources necessary for the manufacture of paper and paperboard packaging. Of equal concern from the point of view of environmental impact and sustainability are the emissions and solid wastes which occur during manufacture and the disposal of packaging when its packaging function has been completed.

2.4.6.1 Emissions to air

We have already noted that the combustion of fossil fuel releases CO_2 into the atmosphere, and this is considered to be one of the main causes of the 'greenhouse effect' which leads to global warming. Around 35–40% of the fibre raw material for the paper and paperboard industry is made from chemically separated fibre where the energy needs are provided by biomass, and hence the main emission to air is in the form of CO_2. The balance, made up of recycled fibre and fibre separated mechanically from wood, uses energy which is often derived from fossil fuel.

Printers, converters and users of packaging all use electricity and, dependent on the energy sources, can also be assessed for the amount of CO_2 which results from their usage of electricity. Reference has been made to the work companies undertake to reduce the environmental impact of transportation. Reductions in the use of energy in this area are accompanied by reductions in the gaseous emissions associated with the consumption of fossil fuels.

Producing energy from fossil fuels results in the emission of carbon dioxide (CO_2), sulphur dioxide (SO_2) and nitrogen oxides (NO_x). These gases can result in 'acid rain' and cause environmental damage to trees and lakes. The amounts which are released of these three gases can be calculated in kilogram per tonne of product.

The use of natural gas as the preferred fossil fuel and the installation of CHP plants for the production of electricity and steam result in lower emissions. In some locations, mills have access to hydroelectric power which does not produce gaseous emissions, and installation of large-scale biomass power plant in mills is becoming more common.

The separation of pulp by chemical means results in sulphur-based malodorous gases, which although present in very small amounts are noticeable in the areas close to the mills because of the sensitivity of the human nose to the odours concerned. Reference to pulp-mill environmental reports indicates that measures to reduce these emissions are ongoing. These acidic and malodorous gases may be burnt in the energy recovery boilers and the flue gases passed through an alkali scrubber. Other possible emissions to air comprise soot (carbon particles) and dust, which can be removed by electrostatic separators.

The printing industry has had to reduce organic discharges to the atmosphere from the drying of solvent-based ink, varnishes and coatings. There is increased usage of water-based inks for flexographic printing; water-based sealers have also replaced oil-based varnishes for most lithographic applications. Also in lithography, the use of isopropyl alcohol, which reduces the surface tension of the fount solution, has been declining over recent years, as health and safety legislation has reduced the concentration of VOCs allowable in the workplace. Indeed, dampening systems on most modern lithographic presses have been improved, such that alcohol-free founts, which contain other surface active agents, are now becoming common.

2.4.6.2 Emissions to water

Water is collected after use in both pulp mills and paper mills. Some processed water is immediately recycled, for example white water containing fibre, fillers and dissolved chemicals, from the wet end of the paper machine.

Wastewater is treated to render it suitable for reuse or discharge which is usually directly back to nature or via an urban sewerage system. Mill discharges are subject to external regulation. In Europe, the regulations are operated locally within the guidelines of the EU Integrated Pollution Prevention and Control (IPPC) Directive.

The contents of wastewater depend on the type of mill and the processes operated. The processing could involve chemical or mechanical pulping, bleaching, recycling of recovered fibre, de-inking, paper and paperboard manufacture, mineral pigment coating, etc. Wastewater can contain fibre, fines (very small fibrous particles) and other solid matter such as particles of bark, fillers and mineral pigments. It could contain colloidal materials such as starch and soluble inorganic salts. Pulp-mill wastewater can contain lignosulphonates, hemicellulose, organic high molecular weight soluble fatty acids and degradation products from bleaching.

Wastewater is processed to achieve acceptable standards in respect of its:

- total suspended solids (TSS) – this material would otherwise form deposits in natural waterways
- biological oxygen demand (BOD) – it is necessary to remove material which would compete with other forms of aquatic life, such as fish, for oxygen
- chemical oxygen demand (COD) – slowly decomposing organic material would also compete with other forms of life
- total phosphorus (P) – this is a nutrient and if the concentration (eutrophication) builds up, algae and microscopic forms of life develop which, in turn, prevent oxygen absorption and light penetration into water
- total nitrogen (N) – eutrophication, as with phosphorus
- temperature – the temperature of water emissions to nature must be regulated
- adsorbable organic halogens (AOX) – this measures any organic chlorine present as a result of the bleaching process, though some is naturally present in wood itself.

AOX is not a measure of the toxicity of chlorine (halogen) compounds present. This is an important point as the most common chlorine chemical used today is chlorine dioxide where the by-products are simple compounds which are not persistent in the environment and are similar to those which occur in nature.

Particles in suspension are removed by settlement. This can be assisted by coagulation and dissolved air floatation. BOD reduction can occur naturally in shallow lagoons though this may take some time. If space is limited, aeration can be employed to accelerate the process. The process of BOD reduction is also quicker if 'activated' sludge is added (sludge containing bacteria which breaks down the organic material). The treated effluent is subjected to settlement after which some of the sludge is removed for reuse and the excess is removed as solid waste.

Overall, the requirement to reduce not just waterborne emissions but also volume of water used per tonne of paper has led in recent years to large-scale investigation into cleaning techniques for water. A number of viable technologies have been proven to provide in-line purification of process water from paper machines (Fig. 2.5): ultrafiltration; anaerobic, aerobic or combined biological treatment; chemical coagulation; reverse osmosis; and distillation (Hubbe, 2007).

Significant improvements have been made in reducing the levels of the key emissions. The reductions in Europe are shown in Table 2.2.

Regulations should not be based on arbitrary standards for BOD, TSS, etc. Local conditions and environmental impact should be taken into account when setting standards for non-toxic measurements of paper mill emissions. The levels set must take into account the environmental impact and long-term environmental sustainability, especially in coastal waters and estuaries, due to the dynamic nature of such environments. For example, a marine biologist who studied a specific estuary environment for 30 years concluded that changes which occur in the disposition and dominance of different species can be wrongly attributed to discharges from a paperboard mill over that period. He found that, for

Environmental and resource management issues **71**

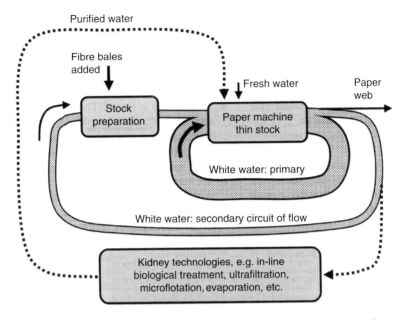

Figure 2.5 Generalised diagram showing how an in-line water purification system can be incorporated into a paper or paperboard mill process. (Reproduced with permission from the Paper Industry Technical Association (PITA))

Table 2.2 Reduction in mill water emissions 1990–2008

Feature	Reduction 1990–2008 (%)
BOD	84
COD	76
AOX	95

Source: CEPI (2009c).

example, a colony of one species could move as the result of a storm disturbing its environment. Without such knowledge, its demise could be associated with the discharge from the mill. In another example, a successful colony of one species can attract the attention of a natural predator and disappear as a result. Again, the mill could be blamed for the disappearance. A conclusion of this study was that observations had to be made regularly over a long period – conclusions based on random observations could be misleading (Perkins, 1993).

It is accepted, for example, that the environmental impact of cellulose residues is less critical when discharged into open sea than when it is discharged into shallow lakes and rivers. In the former, it would not be an environmental benefit to use energy, which in most cases would be derived from fossil fuel with its associated emissions to aerate the residues using pumps, when the same effects of dilution, dispersion and aeration would be performed in nature through the action of the wind and tides.

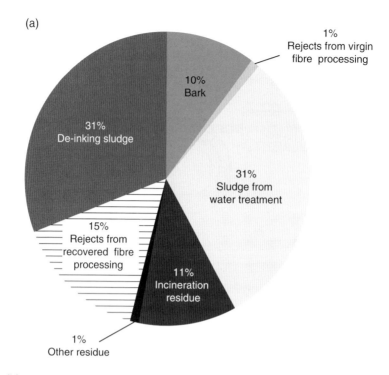
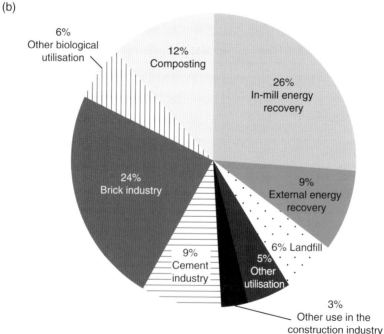

Figure 2.6 (a) Waste from the German pulp, paper and paperboard industry in 2001. (Redrawn from Hamm, 2006.) (b) Utilisation and ultimate disposal of pulp, paper and paperboard mill residues originating from Germany in 2001. (Redrawn from Hamm, 2006.)

2.4.6.3 Solid waste residues in the paper industry

Solid waste from pulp mills and paper mills comprises:

- sludges from wastewater treatment
- soda dregs from pulp-mill chemical recovery
- bark (biofuel or soil conditioner)
- residues from washing and screening pulp in suspension
- waste from mineral pigment coating
- de-inking sludge
- residues from the recycling of recovered paper which can be contaminated with plastics, staples, etc.

CEPI (2009d) data show that a reduction in mill waste disposal by landfill of almost 80% has been achieved over the period 1990–2008. This has mainly been realised through an increase in the amounts composted and used in land spreading. The main method for disposal of mill residues is by incineration with energy recovery; the resulting ash can be used as a soil improver, especially if it has a high lime content – it also finds use as a sewage sludge stabiliser, a desiccant and steriliser in cattle sheds, and as a raw material in the manufacture of cement or bricks. Bark is used as either fuel or as a soil conditioner.

Data from Germany for 2001 (Fig. 2.6a,b) show both the composition of an estimated 3.6 Mt of pulp, paper and paperboard mill residues, and how the industry dealt with the material (Hamm, 2006).

Mills will continue to seek other ways of diverting waste from landfill, especially if they provide either new income streams or supply a new raw material for reuse within the mill: for example, recovery of pigment-coating waste and its reuse as filler in papermaking (Stora Enso, 2002) or recovery of usable fibre from sludge (Krigstin and Sain, 2005).

2.5 Used packaging in the environment

2.5.1 Introduction

Natural resources provided by the environment are used in many ways but ultimately everything is returned as waste to the environment. Some value has always been extracted from waste by recovery, reuse and recycling, but ultimately waste has been disposed of to water, for example rivers and tidal estuaries, buried in landfill or burnt (incineration). We are now more aware of the implication of using resources and of the environmental impact of the various forms of waste disposal.

The main areas from which wastes in society arise are:

- agriculture and horticulture
- mining and quarrying
- construction and demolition
- sewage treatment and dredging
- commerce and manufacturing, i.e. offices and factories
- the service sector, for example distribution, retail, hospitals, transport and education
- municipal solid waste, mainly from households.

Cumulatively, total waste, measured by weight in the UK in 2008, was 289 million tonnes (Mt); although high, the figure has been decreasing steadily over recent years. The used packaging component was around 10.7 Mt (4% approx.), of which about 3.1 Mt was the paper-based component that was ultimately recycled (DEFRA, 2011a). Similar proportions would be expected in other developed countries, yet the idea exists that waste is synonymous with used packaging!

Whilst used packaging may not be a large proportion of total waste, it is highly visible in places where people live and work. Many aspects of the 'throw-away society' are accepted but society is concerned at the volume of used packaging which accumulates.

In addition to comprising part of domestic waste, used packaging arises in distribution, commerce, manufacturing and catering as well as the packaging disposed of during travel, leisure, education, in hospitals and all types of work. This highlights one of the waste management problems associated with packaging – packaging waste arises in relatively small amounts in a large number of locations.

2.5.2 Waste minimisation

The first principal of waste management is that waste should be minimised. This can be applied to paper-based packaging in a number of ways:

- Pack-design changes can optimise the amount of material used whilst maintaining strength. For example, combined plastics/paperboard tubs can optimise the protective properties of both materials whilst at the same time minimising the amounts used.
- Reducing weight per unit area ($g\ m^{-2}$) of material for the same pack design and material ensuring that protection is not compromised. It may be possible to make a net saving by reducing the material in the primary pack by increasing strength in the distribution pack, or vice versa.
- Changing the material specification. Some paper and paperboard materials offer the same strength with weight savings of at least 20% when compared with alternatives, for example extensible kraft, corrugated board and by the use of virgin fibre.

A note of caution here is to make the point that 'higher recovered paper utilisation in a paper substrate is generally equal to greater weight', since recycling tends to reduce both strength and stiffness compared with virgin fibre (CEPI, 2002b).

2.5.3 Waste management options

2.5.3.1 Recovery

There are several disposal options once the weight of packaging has been minimised, but they all require recovery once the primary function of protecting the product throughout its distribution, point of sale and consumer use has been completed.

Recovery can be achieved in several ways:

- Segregation at the point of disposal, for example in the home, followed by collection or by being taken to a central collecting point, after which the material content of particular waste streams can be recycled.

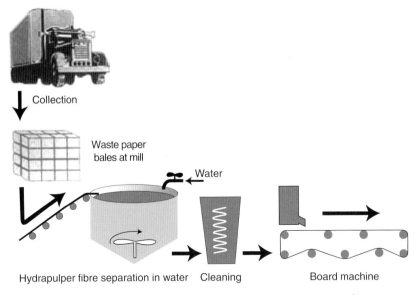

Figure 2.7 Recovery and recycling of waste paper and paperboard. (Courtesy of Pro Carton.)

- Collection in the form of mixed waste which is sent for segregation at a sorting facility after which the material content can be recycled.
- Collection in the form of mixed waste which is sent to a plant which can use the waste in that form, for example an energy-from-waste (EFW) incinerator, anaerobic digester or composting plant.

Paper and paperboard packaging presents several disposal options since the basic material is recyclable, combustible and biodegradable.

The reuse of paper and paperboard packaging as a bag, box or other container in its original form is not widely practised. An example would be a rigid solid board case which can be folded flat after the contents, such as folding cartons, have been removed by the packer, so that the case can be returned to the converter for reuse.

2.5.3.2 Recycling

Recycling is defined as reprocessing material in a production process for the original purpose or for another purpose. Another purpose, here, could include organic recycling (composting). Energy recovery by incineration is usually excluded in official definitions of recycling.

In general there is a 'feel-good' factor about recycling. There is a consensus that it is a good activity to promote. 'Recycling is one of the most tangible symbols of the commitment to do the right thing' (Akerman, 1997). Unfortunately, society often uses the word 'recycling' when it should use 'recovery' or 'collection'; as already noted, these actions must *precede* recycling. The recovered sorted material is then sent to a reprocessing facility, which for paper and paperboard waste would be a paper or paperboard mill.

Material recycling of paper and paperboard is a major business (Fig. 2.7). Recovered fibre contributes 54% of the fibre used in the paper industry worldwide, currently over 210 Mtpa.

Indeed, the major expansion of the paper industry during the last decade has been as a direct result of the increased use of recovered paper. It is a significant and essential resource and the proportion is likely to increase, though its use, measured as a percentage of consumption, will ultimately be limited by factors which will be discussed later.

The proportions *recovered from consumption* and the proportions of *recovered fibre used in paper and paperboard manufacture* in the USA and Europe for 2011 are shown in the following text.

Market	% Recovered	% Recycled in production
USA	66.8	37 (AF & PA, 2012)
Europe (EU)	72.2	54 in 2009 (CEPI, 2011)

A significant proportion of the recovered fibre in these markets is exported. In 2011, exports were 42% and 18% of the totals recovered for the USA and Europe, respectively. The main export destination for recovered paper is currently Asia, especially China.

Waste management infrastructures have been operating during the last 100 years. Merchants collect recovered waste paper, which is then made available to paper and paperboard mills.

There are three important facts about paper recycling which should be noted:

- Some fibres by virtue of their use cannot be recovered – food-contaminated packaging and tissues used for hygienic purposes, cigarette paper, paper and paperboard materials used in building, and materials used in books, maps, documents and archives. Some estimates put the proportion which cannot be recycled as 15–20%, including those products, such as books, which cannot be recycled within a relatively short timescale, and some as high as 30%.
- Cellulose fibres cannot be recycled indefinitely. The fibre loses length, and this is a major cause for a relatively high fibre loss, as sludge, during reprocessing. Interfibre bonding potential is also reduced each time it is recycled, due mainly to loss of fibre flexibility. For this reason, the processing of recovered fibre is an important subject for technical development.

    Estimates of the number of cycles a fibre can survive and still be useful suggest that five to ten cycles are possible depending on the type of paper or paperboard product and the life and usage of that product. The number of cycles is somewhat theoretical because the logistics of the situation show that even at a high recovery rate (50% approx.) from the market, only 6% of original fibre survives to the third reuse due to losses in production (20% approx.) and unrecoverable waste (50% per cycle).
- Not all fibres are equal. Fibres in waste comprise those which have been chemically and mechanically separated and fibres which have already been recycled through one or more cycles. There are also differences which relate to the tree species from which any particular fibre has been derived, i.e. long and short fibres, and other species-specific features.

For technical reasons, all fibres cannot be satisfactorily used for all paper products. The differences between fibres are taken into account in the way waste papers and paperboards are classified and sold to the mills. In the European classification, EN643 (2001), there are now five groups of waste which are subdivided into 67 specific types (CEPI, 2009e). Other parts of the world use similar classifications. Each subdivision has a price which reflects the quality of the fibre and hence its value to the papermaker. Many people think of waste paper solely as 'waste paper' and not as a whole range of products from industry each with their own characteristics and price/value.

The highest value waste paper is white, unprinted and woodfree – meaning that it only comprises chemically separated bleached pulp, i.e. no mechanical pulp or recycled pulp. Such waste arises as trimmings in the production of books. Paperboard waste derived from industrial and commercial premises is usually clean, i.e. contains no contraries or foreign matter, and easy to collect regularly in reasonable quantities. Post-consumer waste is more difficult to manage. It may be sorted at source, such as newspapers and white-printed waste in the home or brown corrugated cases at the supermarket. Waste which is mixed has lower value and is more likely to contain contraries: inks, adhesives, other materials laminated to the paper and, in the worst cases of unsorted mixed post-consumer waste, broken glass and other solid contamination. It has been found that as more waste is collected, i.e. recovered, it tends to contain higher proportions of unsorted post-consumer waste. This can reduce the cost of recovery/collection but lower the value of the waste for material-specific recycling.

De-inking is a process whereby ink is removed from recovered printed papers. Firstly, the fibre is dispersed in water and then treated with surfactants which aid removal of ink particles. The fibre is separated from the ink particles by a cascading, flotation process based on the difference in hydrophobicity between the two materials. Finally, a mild bleaching treatment may be performed to increase the brightness of the pulp. This process is not widely used for packaging products though some may contain a proportion of de-inked fibre.

Organic recycling may be described as 'the aerobic (composting) or anaerobic (biomethylisation) treatment under controlled conditions and using micro-organisms, of the biodegradable parts of packaging waste, which produces stabilised organic residues or methane' (UK Packaging (Essential Requirements) Regulations 1998).

Waste paper and paperboard can be organically recycled because cellulose fibre is biodegradable. It can be broken down into natural substances by organisms in the environment and, in particular, by bacteria using microbial enzymes to convert organic material into CO_2, water and humus or compost. Compost is used in agriculture and horticulture as a soil conditioner.

2.5.3.3 Energy recovery

In the past, burning or incineration was carried out as a hygienic way of reducing the volume of waste. When emission regulations were brought into force, municipal incinerators were closed down and more waste was sent to landfill.

Although the technology to burn waste safely has become available in recent years, the plants require large capital investments. There is also a deep public mistrust of incineration due to lack of knowledge and information.

Paper and paperboard waste retains solar energy because the fibre originates from plant sources such as wood. This energy can be recovered in an EFW incinerator. The minimum calorific value of paper and paperboard is 10 mJ/kg and 2 tonnes yield the energy equivalent of approximately 1 tonne oil or 1.5 tonne coal. We cannot, however, discuss the incineration of waste paper and paperboard in isolation from the disposal of mixed municipal waste. It makes no sense to segregate paper and paperboard waste, which is unsuitable for recycling, from mixed solid waste (MSW), and incinerate it in isolation. We must therefore examine the issue of the disposal of MSW by incineration with energy recovery as part of a holistic approach to MSW management.

The benefits of recycling with energy recovery are as follows:

- The main benefit is the recovery of solar energy. For example, the South East London Combined Heat and Power (SELCHP) collects around 420 000 tonnes of MSW from around one million people and provides electricity, around 35 MW, or enough for 48 000 homes.
- The energy recovered is non-fossil-based – fossil fuel, by contrast, is non-renewable, contributes to the greenhouse effect and produces noxious emissions.
- In cities incineration greatly reduces lorry traffic to out-of-town landfill sites.
- Incineration reduces landfill need by reducing the volume of waste by 90%.
- Diversion of waste from landfill reduces methane production – methane being a much more harmful greenhouse gas than CO_2.
- Recovery of other materials is possible, for example ferrous residues.
- Hygienic benefit – organic waste is reduced to less biologically active materials.
- Incineration is to be preferred for highly flammable, volatile, toxic and clinical wastes which should not be landfilled.

When carton board and other types of paper-based packaging are incinerated, the cellulose component reverts to CO_2 and water vapour. The emission to air should not contain carbon particles, or carbon monoxide, as this would indicate incomplete combustion. The main community concern is, however, more emissions which may arise from other components of mixed waste.

Incineration is now carried out under carefully controlled conditions to minimise potentially harmful effects and to meet the very tight emission regulations. This is achieved using high temperatures of, approximately, at least 850°C and extensive flue-gas cleaning. Processing comprises the use of:

- fine particles of activated carbon to absorb dioxins, furans and other organic compounds, along with heavy metals
- sprays of lime water mixtures to neutralise acidic gases such as sulphur dioxide (SO_2) and hydrogen chloride (HCl)
- bag filters to remove dust particles down to 10 μm size
- a high chimney stack of 100 m.

Another local community concern is the possibility of noxious odours in the vicinity of the EFW plants, especially in hot weather. This is counteracted by running the waste handling plants under negative pressure, i.e. drawing air into the plants which is then used in the combustion process.

Emissions from combustion are monitored by appropriate authorities. In Europe, a European Directive, 1989, lays down minimum standards, though individual national standards may be tighter still. Incineration is widely practised in countries such as Germany, France, Sweden, Denmark and Switzerland, where environmental issues have a very high profile. Many of the incinerators today produce heat for local communities, in addition to electricity.

The UK Government view is that incineration can provide environmental benefits, but that a sustainable approach to the overall use of recovered material is essential (DEFRA,

2011b). It accepts the implementation of the EU waste directive (EU, 2006), which states that a 'waste hierarchy' should be followed in which priority is given first to waste prevention, then reuse, recycling, energy recovery and finally disposal, for example landfill. The UK Government's stated commitment is to 'support energy from waste where appropriate, and for waste which cannot be recycled', and this includes both the use of incineration and installation of anaerobic digestion facilities as a means of dealing with food and similar biodegradable waste.

2.5.3.4 Landfill

Disposal to landfill of waste, including paper and paperboard waste, is perceived as the least environmentally sound option. There are several objections:

- waste of agricultural land – this is not the case where waste is disposed of in disused quarries but such sites are not usually located near the source of the waste.
- emissions and waste of energy in the use of road or rail transport.
- dangers such as contamination of groundwater by leaching, scavenging and health hazards.
- release of methane which is a much more harmful 'greenhouse' gas than carbon dioxide.
- shortage of space in convenient locations.

In many locations, however, it is a low-priced option and the government has to use other means in order to divert waste from landfill. In Europe, a Landfill Directive 99/31/EC has set targets for the reduction of biodegradable waste.

2.6 Life cycle assessment

Life cycle assessment (LCA) is an audit (list or inventory) of the resources used by a product in its manufacture and an assessment of the environmental consequences (impact). LCA can also be applied to a process. The first stage is to define the scope of the study. This sets the boundaries across which material resources and energy pass into the cycle and from which the products and waste emissions to air, water and solid waste leave the cycle. The study can include extraction of raw material, manufacture, transportation, use up to the point of ultimate disposal or recovery – hence the description 'cradle to grave' and even 'cradle to cradle'.

The 'cradle to grave' approach to an environmentally sound life cycle starts with the source and supply of raw material and works through the processing, manufacturing, conversion into packaging, distribution, disposal, recovery and waste management, etc., of the product. The 'cradle to cradle' or C2C model seeks final disposal which is beneficial as discussed later.

The life cycle audit is factual and specific but must be carried out in conformance with ISO (International Organisation for Standardisation) standards. The second phase is the assessment of environmental impact. Whilst the criteria used in the audit is objective, assessing effect is more difficult as the threshold level for an effect may be different in one location compared with another for a variety of reasons. An example already noted was the case of receiving waters, which can be as different as a shallow river compared with a tidal estuary.

The ISO standards for LCA have been drawn up after international liaison coordinated by the Society of Environmental Toxicology and Chemistry (SETAC) and the European Commission.

The following standards have been issued:

ISO 14040:2006 LCA Principles and framework
ISO 14044:2006 LCA Requirements and guidelines
ISO 14047:2003 LCA Examples of application of ISO 14042
ISO/TS 14048:2002 LCA Data documentation format
ISO/TR 14049:2000 LCA Examples of application of ISO 14041 to goal and scope definition and inventory analysis
ISO 14050:2009 Environmental management – Vocabulary

LCA is useful in quantitatively highlighting points in a life cycle where environmental impact is high, and for evaluating the effects of life cycle changes by, for example, substituting one process with an alternative process (Gaudreault et al., 2010). One such case study indicating several life cycle changes has been published, 'Towards sustainable development in industry – an example of successful environmental measures within the packaging chain'. It is the result of cooperation between Tetra Pak, Stora Enso and the Swedish Forest Industry Federation. It includes paperboard minimisation and changes in printing, plastic extrusion coating, distribution, recovery and recycling, all of which have reduced environmental impact in the life cycle of a 1 L milk carton (SFI, 2002).

A major review of European LCA work in the paper industry was published by the European Environment Agency. It provides a summary of all the relevant reports on the subject. The main conclusion is that the recycling of waste paper has a lower environmental impact than the alternative waste disposal methods of landfilling and incineration. It also concludes that where a mix of wood, fossil fuels and other energy sources are used in both virgin and recycling systems, paper recycling systems tend to be favoured environmentally because of lower energy use (EEA, 2006).

All those involved in packaging of any kind are encouraged to appreciate the environmental implications of their proposed choice of materials, processes and pack designs and to take into account the environmental impacts throughout the life cycle of the products concerned. Sustainable design not only takes into account environmental impact but also social and economic impact. The sustainability of paper and paperboard packaging performs very well when all these environmental issues are taken into account. The principles of sustainable design include consideration of materials, energy and toxicity – use of renewable energy – that all emissions to air, water and land are safe, and issues of the efficiency in the use of materials, energy and water along with social issues are taken into account. The principles of this holistic approach are discussed in Datschefski's 5 Principles (Datschefski, 2001).

Another approach to promoting sustainable design is the cradle to cradle or C2C concept (Braungart and McDonough, 2002). This is a more sustainable approach to design than cradle to grave because the end of life products are considered to be useful as either 'biological' or 'technical' environmental nutrients. Biological nutrients are materials which re-enter the environment whilst technical nutrients have an industrial recycling use. Paper and paperboard are C2C products. At the end of life, paper and paperboard can be recovered and the fibre recycled. Alternatively, the energy in paper and paperboard may be recovered in an EFW process or the fibre may be composted for use in agriculture and horticulture. Even fibre that is difficult to recover on account of the way it is used, for example hygienic tissue in sewage sludge and food-contaminated paper and paperboard waste in municipal solid waste which would otherwise be sent to landfill, may be a source of energy by means of anaerobic digestion (Environment Agency, 2010).

2.7 Carbon footprint

The paper and paperboard industry is well placed to help in combating climate change resulting from an increase in greenhouse gases in the atmosphere. The total carbon dioxide emission, together with the carbon dioxide equivalents of other greenhouse gases associated with a product, process or service, is known as its carbon footprint. This is calculated by studying the emissions at each stage in the life cycle of the product, process or service.

There are Life Cycle ISO Standards for life cycle studies. However, the setting of the boundaries for the life cycle studies is discretionary. Whilst a particular process may be examined to investigate possible improvements in emissions reduction, there is also a current trend for the carbon footprints of different products to be compared, and this requires conformity of approach for comparisons of the data to be meaningful.

In Europe, CEPI has developed a common framework that enables its members to calculate carbon footprints. The framework defines the various sources of carbon dioxide and carbon dioxide equivalents to be investigated and how they should be calculated. Companies and industry sectors are able to include other elements which address their needs whilst ensuring that the information is transparent and well presented to paper and paperboard users (CEPI, 2010).

The ten considerations, referred to as the 'toes' of the carbon footprint discussed in the CEPI report, are discussed later.

2.7.1 Carbon sequestration in forests

Trees grow by removing carbon dioxide from the atmosphere. This is known as *carbon sequestration*. It transfers carbon from the atmosphere to a carbon pool, sink or biomass where the carbon is stored in trees. Trees grow in forests and fulfil a number of commercial, environmental and community needs.

The demand for wood in Europe, North America and the Russian Federation has ensured that forest area is high and is increasing. The volume of new wood grown per year in these areas has exceeded the volume harvested by a significant amount, as discussed in Section 2.3. According to the European Union Greenhouse Gas Inventory (GHG) document, forests of the EU-15 are a net carbon sink, with net CO_2 removals by forests having increased by 27% between 1990 and 2004 (CEPI, 2010a). The industry therefore stimulates carbon dioxide retention in the forests.

The biomass carbon in the wood harvested for paper and paperboard can be calculated. A lot of this wood stored energy is, depending on the pulping process, used during manufacture and the carbon returned to the atmosphere. If this biomass carbon is returned to the atmosphere as fast as additional carbon is removed, there is no effect on the atmosphere. Some remains bound in products so that carbon is removed from the atmosphere faster than it is returned so that there is a net positive effect. The rest is bound in the product but ultimately at the end of life in either landfill or EFW incineration, it is also returned to the atmosphere.

When wood is harvested on a sustainable basis, i.e. the amount harvested is less than or equal to new growth, there is no depletion in the forest carbon. Also during recycling and source reduction, for example when the same number of packages are based on a lower weight of wood, a 'credit' can be applied for increasing the forest carbon stock (US Environment Protection Agency, 2006).

Quantitatively therefore bioenergy carbon has a low, or even zero, emissions effect for some products in the footprint calculation. It is, however, of considerable importance in that it:

- stimulates environmentally sound forest management
- ensures sustainability in wood availability
- encourages societal support for forest communities
- supports the use of wood in other areas where otherwise products of a higher carbon footprint such as fossil fuels, alternative building materials and alternative packaging materials would have to be used.

Biomass carbon sequestration is an environmental benefit for paper and paperboard which is not shared by many other materials. As there are several ways in which wood is processed together with the way in which paper and paperboard products are recycled, there is no simple single method for calculating the effect of different paper and paperboard products on forest carbon. The CEPI approach indicates various types of information which can be sought (CEPI, 2010b).

2.7.2 Carbon stored in forest products

The extent of time during which carbon storage in paper and paperboard is retained is dependent on the product. Forest product wood used in buildings lasts for a very long time – also products used in legal documents, libraries, galleries and museums.

With packaging the lifespan is much less. Some 40% of paper and paperboard produced is used in packaging. In the main, particularly with food and beverage packaging, the pack is disposed of after use when it is either recovered for recycling or disposed of in landfill or in an EFW facility. When recycled, the bioenergy in the fibre is retained, and when used to produce energy it replaces the use of energy from another source which may well be fossil based. In landfill, the bioenergy emissions of methane occur, although there are installations which collect and use landfill methane for energy.

2.7.3 Greenhouse gas emissions from forest product manufacturing facilities

Manufacturing occurs in pulp mills, paper and paperboard mills and the facilities used to print and convert these products into packaging. The greenhouse gas emissions we are concerned with are those arising from the use of *fossil* fuels and not those from *biofuel* or, as it is also called, *biomass*. Pulp mills and pulp mills operating alongside paper and paperboard mills where the cellulose fibres are separated chemically use little or no fossil fuel as most of the energy required is bioenergy derived from the non-cellulose components of the wood. Mills separating fibres by mechanical means use electricity derived from fossil fuel though some have moved or are moving to replace fossil-fuelled plants with those using biomass material (Iggesund, 2011).

2.7.4 Greenhouse gas emissions associated with producing fibre

These emissions include those from harvesting wood and forest management and those from the collection, sorting and processing of recovered waste paper and paperboard.

2.7.5 Greenhouse gas emissions associated with producing other raw materials/fuels

These emissions are those concerned with the manufacture of chemicals and other additives in paper, paperboard and packaging manufacture and in the manufacture of the fuel which is used. They also include emissions involved with the electricity purchased for the manufacture of the chemicals and additives.

2.7.6 Greenhouse gas emissions associated with purchased electricity, steam and heat, and hot and cold water

These emissions are from purchased electricity, steam and heat used to manufacture paper, paperboard and packaging.

2.7.7 Transport-related greenhouse gas emissions

These emissions are from all the forms of transport used from taking wood to the pulp mill or chipping plant to the transport of waste at the end of life.

2.7.8 Emissions associated with product use

These are usually zero for forest products.

2.7.9 Emissions associated with product end of life

These are mainly those from anaerobic decomposition of forest products in landfill. They do not include biogenic energy emissions released in an EFW plant.

2.7.10 Avoided emissions and offsets

This category is optional and its use in emission balance sheets is controversial since it involves the calculation of emissions which are avoided by the use of forest products. An example is when a mill exports electricity from biomass, where if this did not occur the electricity would have had to be made by using fossil fuel. Another would be to calculate the emission avoided if mill bark waste is used on farmland, where if it were not used a chemical fertiliser based on fossil fuel would have had to be used. It is therefore important that the assumptions and methods used to calculate avoided emissions are transparent and explainable to those interested (CEPI, 2010b).

2.8 Conclusion

Basically, we are where we are today with respect to all paper and paperboard issues for reasons other than those based on environmental considerations. Paper recovery and recycling was established at least 100 years ago for sound technical and commercial reasons. In 2002, recovered paper and paperboard provided around 45% of the world's demand for fibre; in 2010 it accounted for 55% of a greatly increased world production tonnage. The amount of fibre recovery and recycling is rising for several reasons:

- Fibre demand is rising as the production of paper and paperboard increases.
- Higher proportions are being recovered for societal and waste management reasons.

Benefits can be stated for each of the three main fibre sources:

1. *Chemically separated fibre* – flexible fibre giving products with high strength; when bleached it is chemically pure cellulose, making the fibre odour and taint neutral and the best choice for the packaging of flavour- and aroma-sensitive products; process chemicals are recovered and reused; and the energy used in manufacture is renewable as it is derived from the non-fibrous components of wood.
2. *Mechanically separated fibre* – stiff fibre giving high bulk, i.e. high thickness for a given weight of fibre (g m^{-2}), which results in products with high stiffness compared with products incorporating other fibres, high yield from wood, may be given chemical treatment to lighten colour, and is sufficiently odour and taint neutral for the packaging of many flavour- and aroma-sensitive products.
3. *Recovered fibre* – recovered fibre is functionally adequate and cost-effective in many applications. The quality depends on the paper or paperboard from which the fibres were recovered. The use of recovered fibre in making recycled paper and paperboard would rate highly for societal and economic reasons.

Recovered fibre retains the original solar energy and the energy expended in the manufacture and use of the virgin fibre. There are however fibre losses in paper and paperboard manufacture when using fibres derived from recovered waste paper and paperboard, and as equivalent recycled products incorporate a higher weight of fibre because they are generally less bulky, weaker and less stiff than products made from virgin fibre, more water, proportionally, must be evaporated during manufacture.

Logistically, we need recovered fibre for recycling. It would be impossible to replace recovered fibre with virgin fibre over a short period, and the economic constraints of the market together with society's waste management needs will ensure further expansion in the recovery and use of waste paper. This is important as sustainability depends on economic and social needs as well as environmental implications.

The respective benefits of the different types, and combinations, of fibre become more clear-cut when particular papers and paperboards are considered for specific applications. All fibres are not universally interchangeable. It is, therefore, not practical to insist on a mandatory minimum level, or declaration, of recovered fibre content.

Industry requires virgin fibres to meet the functional needs of many applications. It also needs virgin fibre to maintain the quality of recovered fibre and the total quantity of fibre in the system required by the industry overall. Virgin fibre is required to replace ('top up') recycled fibre which is lost in reprocessing. Fibres cannot be recycled indefinitely, and reprocessing reduces fibre length to the point when it eventually accumulates as sludge.

Hence for all practical purposes, it is correct to say that virgin and recycled fibres are both necessary.

Sustainability, as we have seen, depends on social and economic aspects as well as environmental criteria. Many commentators have said that the environmental debate has moved on from single issues, such as the virgin/recovered fibre debate, to a more holistic environmental consideration of whole systems, which comprises:

- the procurement of raw materials
- using energy to make paper/paperboard
- converting the material into packaging
- meeting regulations at all stages applying to air and water emissions and solid waste
- meeting needs of the food or other product to be packed during packing, distribution, point of sale or delivery to customer and use by the ultimate consumer
- the end-of-life management of the packaging where the options are to reuse, recycle, incinerate with energy recovery or dispose to landfill.

The whole system must be sustainable in environmental, economic and societal terms and should include procedures to ensure continuous improvement. The issues discussed provide evidence that this is the approach currently used in the manufacture and use of paper and paperboard-based packaging.

The wood supply for the paper industry is sustainable. Independent forest certification is established in many regions including North America and Europe. Over 50% of the energy used in paper and paperboard manufacture is produced from renewable sources. This includes many larger mills not having biomass directly available as part of their process, but which are starting to install biomass or EFW boilers to protect themselves against the otherwise volatile commercial energy market. Other smaller mills and factories where electricity is imported are in the same situation as society at large as regards the resources used. At the present time, the greater proportion of that energy depends on fossil fuels but the proportion of renewable energy is increasing; some packaging factories and warehouses are installing solar panels to provide energy for basic heating and lighting, and some mills are installing wind turbines. However, pulp and paper manufacture is highly energy intensive, and solar, wind and most other renewable resources provide insufficient energy from a given land footprint to be considered as viable for providing all the power needed to operate a pulp or paper mill. The two exceptions are biomass and EFW incineration, both of which are becoming increasingly common power sources for modern pulp, paper and paperboard mills.

In addition, the mills have increased their efficiency in energy usage (CHP), and where fossil fuel is still used they have reduced emissions by switching from coal and oil to natural gas. Reductions have been made in water consumption, and the quality of water returned to nature has been improved. The amounts of paper and paperboard recovered and the proportion of recovered fibre used in the manufacture of paper and paperboard has increased such that it now exceeds virgin fibre as the principal fibre source worldwide.

By their activities in all these areas and by having independent assessments to internationally accepted environmental and quality management standards, the companies involved in the manufacture and use of paper and paperboard packaging continue to demonstrate their commitment to sustainable development and continuing improvement.

Finally, an important feature of the paper and paperboard industry which underpins its claim to sustainability is the part it plays in the carbon cycle.

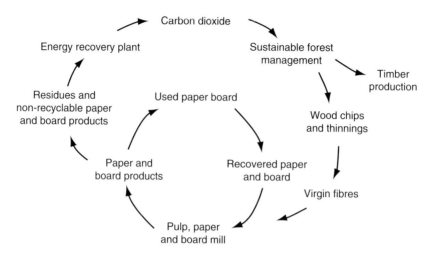

Figure 2.8 The paper and paperboards carbon cycle. (Reproduced with permission, copyright Confederation of European Paper Industries.)

The carbon cycle links the atmosphere, the seas and the earth in a fundamental way (Fig. 2.8). All life depends on carbon in one form or another. Paper and paperboard are involved because:

- CO_2 in the atmosphere is absorbed and transformed into cellulose fibre in trees
- trees cumulatively form forests
- forests are essential to climate, biodiversity, etc., because they store solar energy and CO_2
- the principal raw material for paper and paperboard is wood
- non-cellulose wood components provide just over 50% of the energy used to manufacture paper and paperboard and this releases CO_2 back into the atmosphere.
- some paper and paperboard in longlife usages such as books, together with timber, act as carbon sinks which remove CO_2 from the atmosphere
- when paper and paperboard are incinerated after use with energy recovery and even when they biodegrade in landfill, they release CO_2 back into the atmosphere.

The overall effect is that the paper industry invests in forests. This results in an accumulation of new wood as a result of incremental growth exceeding the amount of wood harvested by a good margin. Furthermore, the amount of the CO_2 used to produce the wood harvested exceeds the amount given back by the use of biofuel in the manufacture of paper and paperboard and at its end of life when it is incinerated with energy recovery or biodegraded.

Hence the paper and paperboard industry effectively promotes forest development and removes CO_2 from the atmosphere, features which support the, desirable, aim of sustainable development.

References

AF&PA (American Forest and Paper Association), 2012, *Information About Recycling*, visit http://www.afandpa.org

Akerman, F., 1997, *Why Do We Recycle?* Tufts University, Medford, Massachusetts.

Austin, P.C., Mack, J., McEwan, M., Afshar, P., Brown, M., and Maciejowski, J., 2011, Improved energy efficiency in paper making through reducing dryer steam consumption using advanced process control, *Tappi PaperCon 2011*, Covington, Kentucky, pp. 1122–1132.
Braungart, M. and McDonough, W., 2002, *Remaking the Way We Make Things*, North Point Press, New York.
CEPI (Confederation of European Paper Industries), 2002a, *Environmental Report 2002 – Working Towards More Sustainability*, p. 30, visit http://www.cepi.org
CEPI, 2002b, *Environmental Report 2002 – Working Towards More Sustainability*, p. 29, visit http://www.cepi.org
CEPI, 2009a, *Sustainability Report*, p. 22, visit http://www.cepi.org
CEPI, 2009b, *Sustainability Report*, p. 23, visit http://www.cepi.org
CEPI, 2009c, *Sustainability Report*, p. 20, visit http://www.cepi.org
CEPI, 2009d, *Sustainability Report*, p. 42, visit http://www.cepi.org
CEPI, 2009e, *The European List of Standard Grades of Recovered Paper and Board Based on EN643, 2001*, visit http://www.cepi.org
CEPI, 2010, *Framework for the Development of Carbon Footprints for Paper and Board Products*, p. 22, visit http://www.cepi.org
CEPI, 2011, *Annual Statistics*, visit http://www.cepi.org
Chamberlain, D., 2010, Satino black – setting the benchmark for sustainable toilet tissue. *Paper Technology* 51(4), 24–25.
Datschefski, E., 2001, *The Total Beauty of Sustainable Products*, Rotovision, Hove, UK, p. 21.
DECC (UK Government Department of Energy and Climate Change), 2010, *Digest of United Kingdom Energy Statistics*, pp. 195, 206, visit http://www.decc.gov.uk
DEFRA, 2011a, *Waste Data Overview*, visit http://www.defra.gov.uk
DEFRA, 2011b, *Government Review of Waste Policy in England 2011*, visit http://www.defra.gov.uk
EEA (European Environment Agency), 1995, *Europe's Environment, the Dobris Assessment*, p. 494, visit http://www.eea.europa.eu
EEA, 2006, Paper and cardboard – recovery or disposal? Review of life cycle assessment and cost-benefit analysis on the recovery and disposal of paper and cardboard, *European Environment Agency Report No. 5*, pp. 10, 34, visit http://www.eea.europa.eu
EEA, 2010, *The European Environment – State and Outlook 2010 – Land Use*, p. 7, visit http://www.eea.europa.eu
EIA (US Energy Information Administration), 2011, *Annual Energy Outlook 2011 with Projections to 2035*, p. 2, visit http://www.eia.gov
Environment Agency, 2010, *Anaerobic Digestion and Environmental Permitting*, R. Wheaton, Environment Agency, visit http://www.environment-agency.gov.uk
EU, 2006, *Directive 2006/12/EC of the European Parliament and of the Council of 5 April 2006 on Waste*, visit http://eur-lex.europa.eu
EU-China, 2009, EU-China Energy and Environment Programme (EEP), *Pulp and Paper – A Reference Book for the Industry*, visit http:www.chinaeci.com/admin/upload/20090817022427.pdf
Eurostat Pocketbook, 2010, *Energy, Transport and Environment Indicators*, European Commission, p. 37, Table 1.3.1a, visit http://epp.eurostat.ec.europa.eu
FAO, (Food & Agricultural Organization of the United Nations), 2001a, *Global Forest Resources Assessment 2000*, p. xxii, visit http://www.fao.org
FAO, 2001b, Food & Agricultural Organization of the United Nations, *Global Forest Resources Assessment 2000*, p. 28, visit http://www.fao.org
FAO, 2003, M. Hosny El-Lakany, Assistant Director-General, Forestry Department, *News Release*, Rome, 24 July , visit http://www.fao.org/english/newsroom/news/2003/20888-en.html
FAO, 2010a, *Global Forest Resources Assessment 2010*, pp. 9–44, visit http://www.fao.org
FAO, 2010b, *Global Forest Resources Assessment*, p. 10, visit http://www.fao.org
FAO, 2010d, *Global Forest Resources Assessment*, p. 18, visit http://www.fao.org
FAO, 2010e, *Global Forest Resources Assessment*, pp. 35–36, visit http://www.fao.org
FAO, 2010f, *Global Forest Resources Assessment*, p. 27, visit http://www.fao.org
FAO, 2010g, *Global Forest Resources Assessment*, p. 101, visit http://www.fao.org
FAO, 2011a, *State of the World's Forests 2011*, pp. 2–27, 128–136, visit http://www.fao.org
FAO, 2011b, *State of the World's Forests 2011*, p. 3, visit http://www.fao.org
FAO, 2011c, *State of the World's Forests 2011*, pp. ix–xi, visit http://www.fao.org

FAO, 2011d, *State of the World's Forests 2011*, p. 9, visit http://www.fao.org
FAO, 2011e, *State of the World's Forests 2011*, pp. 137–146, visit http://www.fao.org
FAO, 2011f, *State of the World's Forests 2011*, pp. 126–136, visit http://www.fao.org
FAO, 2011g, *State of the World's Forests 2011*, pp. 128–146, visit http://www.fao.org
FAO, 2011h, Jenny Gustavsson et al., *Global Food Losses and Food Waste*, July, Swedish Institute for Food and Biotechnology, Rome, visit http://www.fao.org
Frigon, J.-C. and Guiot, S.R., 2010, Biomethane production from starch and lignocellulosic crops: a comparative review, *Biofuels, Bioproducts and Biorefining*, 4(4), 447–458.
FT, 2011, Earthquake fears halt shale gas fracking, *Financial Times*, 1 June, visit http:www//ft.com
Gaudreault, C., Samson, R., and Stuart, P.R., 2010, Energy decision making in a pulp and paper mill: selection of LCA system boundary. *The International Journal of Life Cycle Assessment*, 15(2), 198–211.
Haggith, M., 2008, *Paper Trails: From Trees to Trash – The True Cost of Paper*, Virgin Books, London.
Hamm, U., 2006, Environmental aspects, in: Holik, H. (ed.), *Handbook of Paper and Board*, Wiley VCH, Weinheim, Germany, pp. 422–445.
Hart, P.W. and Evers, D., 2011, "Green" natural gas from paper mill sludge, *Paper 360°*, 6(2), 41–42.
Holmen, A.B., 2002, *Environmental Report 2002*, p. 10, visit www.holmen.com
Holmen, A.B., 2010, *Holmen and Sustainability 2010*, p. 23, visit www.holmen.com
Hutter, A. and Jung, H., 2010, Practical ways to save energy in paper production, *Paper Technology*, 51(5), 10–13.
Hubbe, M.A., 2007, Water and papermaking 3. Measures to clean up process water, *Paper Technology*, 48(3), 23–30.
ICFPA (International Council of Forest and Paper Associations), 2002a, *Sustainability Report*, p. 15, visit http://www.icfpa.org
ICFPA, 2002b, *Sustainability Report*, p. 14, visit http://www.icfpa.org
ICFPA, 2009, *Sustainability Progress Update*, pp. 12, 14, visit http://www.icfpa.org
IEA (International Energy Agency), 2010, *Key World Energy Statistics*, visit http://www.iea.org
Iggesund, 2011, *Iggesund Investing in Lowering Fossil Carbon Emissions*, Press Release, 29 November 2012, visit http://www.iggesund.com
Johnson, B., Johnson, T., Scott-Kerr, C., and Reed, J., 2009, The future is bright, *Pulp Paper International*, 51(10), 19–22.
Kang, L., Want, W., and Lee, Y.Y., 2010, Bioconversion of kraft paper mill sludges to ethanol by SSF and SSCF, *Applied Biochemistry and Biotechnology*, 161(1–8), 53–66.
Krigstin, S. and Sain, M., 2005, Recovery and utilization of fibre from recycled papermill sludge, *Paper Technology*, 46(7), 37–42.
Lomborg, 2002, *The Skeptical Environmentalist, Measuring the Real State of the World*, Cambridge Press, Cambridge, UK, p. 130 (based on WRI (World Resources Institute)), *World Resources 1996–97* and Botkin, D.B. and Keller, E.A., 1998, *Environmental Science: Earth as a Living Planet*.
Mubin, F., 2004, Outlook of the paper industry in GCC, *Pulp & Paper International* 46(11), 25–27.
Perkins, A., 1993, '*Do Board Mills Produce Harmful Emissions?' Iggesund/Packaging News Environmental Seminar*, London, October.
PPT, 2010, A new standard for water use in paper mills, *Pulp and Paper Technology*, Winter 2010, 24–29, visit http://www.cavendishgroup.co.uk
PT, 2009, Osmotic power from sea and fresh water, *Paper Technology*, 50(5), 10.
Remrod, J., 1991, *Forest of Opportunity*, Shogsindustrierna, Stockholm, Sweden (based on FAO data).
Robson, N.C., 1997, *More and Better Packaging – A Solution to World Food Waste*, International Trade Centre, Geneva, April, Packaging in Action Factsheet No. 5.
SFI, 2002, *Towards Sustainable Development in Industry – An Example of Successful Environmental Measures within the Packaging Chain*, Swedish Forest Industries Federation, 20 August, visit http://www.forestindustries.se/pdf/sustainable_development.pdf
Stora Enso, 2002, *Environment and Resources 2002*, p. 22, visit Publications at http://www.storaenso.com
Tickell, C., 1996, *Greenery & Governance*, Burntwood Memorial Lecture for the Institution of Environmental Sciences, November, published in *Environmental Studies*, 53(3), 1997.
UNEP, 1995, *Global Biodiversity Assessment*, in: Heywood, V.H. (ed.), Cambridge University Press, Cambridge, UK, p. 12.
UNEP (United Nations Environmental Programme), 2000, *Global Environmental Outlook 2000*, visit http://www.unep.org/geo2000/english/index.htm

UNEP, 2007, *Global Environmental Outlook-4*, visit http://www.unep.org/geo/GEO4/report/GEO-4_Report_Full_en.pdf

UN, 2010, *World Population Prospects: The 2010 Revision*, Population Division of the Department of Economic and Social Affairs of the United Nations Secretariat, visit http://esa.un.org/unpd/wup/index.htm

US Environment Protection Agency, 2006, *Solid Waste Management and Greenhouse Gases, A Life-cycle Assessment of Emissions and Sinks*, visit http://www.epa.gov/climatechange/wycd/waste/downloads/fullreport.pdf

WEC (World Energy Council), 2007, *Survey of Energy Resources 2007*, pp. 57, 94, and 147, visit http://www.worldenergy.org

WFP, 2011, *Water Footprint Assessment Manual*, Earthscan, London, September 2007, visit http://www.waterfootprint.org

Wood, N., Tran, H., and Master, E., 2009, Pretreatment of pulp mill secondary sludge for high-rate anaerobic conversion to biogas, *Bioresource Technology* 100(23), 5729–5735.

Zellcheming (Association of Pulp and Paper Chemists and Engineers), 2008, *Chemical Additives for the Production of Paper: Functionality Essential – Ecologically Beneficial*, Technical Committee Publication, Deutscher Fachverlag, Frankfurt, Germany.

3 Paper-based flexible packaging

Jonathan Fowle[1] and Mark J. Kirwan[2]

[1]Innovati Partners, Shepton Mallet, Somerset, UK
[2]Paper and Paperboard Specialist, Fellow of the Packaging Society, London, UK

3.1 Introduction

Flexible packaging is one of the fastest growing type of packaging, driven by good functional and aesthetic properties, competitive cost and by being relatively lightweight compared with rigid packaging.

Paper has been used for many decades as part of a laminate construction, often combined with high barrier substrates and a heat-seal layer. Dehydrated soup pouches are a classical example.

Paper has been used to make sugar and flour bags for over a century and, when coated with a heat-sealing polyethylene (PE) layer, for around 60 years.

The main properties of paper are good rigidity/stiffness, a good printing surface (depending upon the papermaking technology) and good deadfold. That is the ability to remain flat when folded.

Furthermore, paper is relatively easy to tear. When combined with aluminium foil, the contents of pouches/sachets can be accessed quite readily.

Paper is very accommodating to all normal print processes used in flexible packaging, such as photogravure, conventional flexo, ultraviolet (UV) flexo, litho, letterpress and digital.

Despite being around for over a century, paper is experiencing something of a renaissance in flexible packaging applications, due to its consumer perception as being a sustainable material.

Many examples are hitting the retail shelves where plastic films have been engineered to look and feel like paper, adding traditional authenticity to the brand's profile.

It is estimated that the European market for Flexpack Papers is worth approximately 1.3 million tonnes. Machine-glazed papers represent about two thirds of this, one side coated papers about 20%, greaseproof papers 10% and calendered papers 5%.

Paper-based flexible packaging comprises sachets, pouches, bags made on form, fill, seal equipment and overwrapping materials. It is also used as lidding and to provide product protection and as tamper evidence when used as a membrane (diaphragm) in rigid plastic, metal and glass containers. Paper-based flexible materials are also used for bag-in-box lining material, multiwall and single-wall bags.

Paper-based flexible packaging is usually lighter in weight than the other forms of packaging. Paper-based flexible packaging is a part of the overall flexible packaging market. This market in the USA comprises the following:

- flexible packaging at $26.4 billion is the second largest type of packaging in the USA (18% in value of the total US $143 billion packaging market)
- the largest market for flexible packaging is food accounting for 56% of shipments to the retail and institutional markets
- medical and pharmaceutical packaging usage is 8%
- usage for industrial applications is 9%, retail non-food 12%, consumer products 10% and institutional non-food 5% (FPA, 2010).

The overall position in Europe is similar with flexible packaging of all specifications accounting for nearly one third of total annual expenditure on consumer packaging at around €10.4bn in 2009 as reported by PCI Films Consulting (European Plastics News, 2010).

Flexible packaging based on paper uses composite structures in which the properties of the paper are combined with those of other materials, such as plastics, aluminium foil, wax and other materials by means of coating, lamination and impregnation. Depending on the combination of materials used, paper-based flexible packaging can provide the following properties:

- barrier to moisture and moisture vapour
- barrier to gases, such as oxygen, carbon dioxide and nitrogen, making flexible packaging suitable for vacuum and gas-flushed packaging and for the protection of products sensitive to flavour and aroma loss or contamination
- resistant to products, including those containing fat
- barrier to light
- sealable by heat sealing and use of cold-seal coatings
- medical packaging applications place special requirements on paper-based packaging, such as to be sterilisable by gas, irradiation and steam, to provide satisfactory pack seals, using both heat and cold sealing, which peel without generating loose fibres, and to be permeable to a sterilising gas but impermeable to the entry into the pack of microbiological contamination
- strength for packing, distribution and use
- printable by several printing processes.

Flexible packaging is usually supplied on rolls for forming, filling and sealing by the end-user/packer although some pre-made formats such as gusseted bags, stand-up pouches, three-side seal pouches, shrink sleeves, die-cut labels and lids are also supplied, particularly for small lot sizes where the volumes do not justify high speed, automated filling and sealing lines.

Flexible packaging is used for the packing of:

- solids in the form of free-flowing products such as powders, granules and agglomerates
- solids in the form of blocks, slices, bars and tablets
- individual items and groups of items

- liquids and pastes
- multipacks
- bottle and jar decoration, for example beer labels and mineral water labels
- medical devices, kits and consumables such as dressings and surgical gloves.

Sachets and pouches may be single-portion, or single-dose, packs and they can also be multiportion packs, for example for coffee, with a reclosable feature. These packs may be sealed under vacuum or may be gas flushed.

The main materials, or substrates, used in flexible packaging until the middle of the twentieth century comprised paper, aluminium foil and regenerated cellulose film (RCF). Paraffin wax-coated paper was widely used as a barrier to moisture, moisture vapour and volatiles, and it has product-release properties. Subsequently, the uses and properties of waxes were enhanced with microcrystalline waxes and plastic additives which provided better heat sealing, barrier properties and made them more versatile for use as a laminating adhesive, for example paper to paperboard, paper to aluminium foil and RCF to plastic films.

From the 1950s onwards, plastics began to play a larger role in packaging. The change was particularly significant in flexible packaging. Plastics became available as films, emulsion coatings and in extrusion coatings and laminations. Plastics were also used as adhesives and heat-seal coatings. New conversion equipment, ancillary materials and higher performance adhesives became available, and printing processes and inks were modified to suit their use with plastics. The development of plastics themselves and new techniques in blown film, cast film, biaxial orientation, co-extrusion, lamination and coating significantly increased the use of plastic-flexible packaging such that by 2010 approximately 75% of flexible packaging was *solely* plastic in composition.

The availability of plastics has, however, widened the opportunities for paper-based flexible packaging. The use of wax has diminished though its use in several types of application is still an option, and other non-food-packaging materials, such as bituminised kraft papers (kraft union), whilst still available in some markets, have been replaced by PE film.

However, paper and aluminium foil are still widely used in flexible packaging. Papers are used in the form of laminates and as a substrate for coating and impregnating.

Paper is used in flexible packaging because:

- paper is a strong material available in reel form in a range of widths and substances (grammages or basis weights), which meets the needs of printing, conversion and packaging machinery and also, thereafter, in distribution and consumer use
- paper is a suitable substrate for providing functional packaging properties. This may be achieved in paper manufacture, for example greaseproof or glassine paper, or by addition of other materials by subsequent conversion in the form of a coating, lamination or impregnation
- paper imparts stiffness in the packed sachet or pouch
- paper can be printed by all the main, commercially available, inks and printing processes
- there is a wide choice of the print quality which can be achieved depending on the colour, surface finish, smoothness and gloss of the paper surface and whether the surface has a mineral pigmented coating
- the surface promotes adhesion in a range of conversion processes including those using water and solvent-based adhesives, wax and curing (cross-linking) 100% solid adhesives as well as plastic extrusion coatings and laminations

- fibre choice, stock preparation and papermaking can be varied to achieve specific product characteristics, for example porosity and heat sealing in an infusible tissue for a tea bag, grease resistance in a greaseproof, glassine or size press treated for grease resistance, mould resistance, etc.
- paper-based medical packaging meets specific requirements for sterilisation, sealability, peelability, porosity and microbiological barrier
- paper is cost-effective when used in flexible packaging; it is environmentally benign and naturally sustainable.

Many brand owners, in an attempt to support their sustainability credentials, are seeking accreditation from their supply chain to Forest Stewardship Council (FSC).There are two parts, the first ensures the forest is maintained to a high standard embracing social, environmental and economic factors. The other does a full trace of the timber from the forest through all stages of processing and distribution. This is known as chain of custody.

Aluminium foil is used in flexible packaging as a barrier to moisture, moisture vapour and common gases, and it is also a light barrier. Some applications for aluminium foil have been replaced by metallised paper, metallised film and ethylene vinyl alcohol (EVOH).

RCF is usually discussed within plastic packaging, but as it has its origin in bleached chemical woodpulp some mention is justified here. It is a clear transparent film. In its various specifications with respect to coating and additives, it has a wide range of applications because of its barrier properties and composition together with its suitability for lamination to paper, aluminium foil and plastic films. All the properties listed earlier can, however, be matched in most respects with lower cost plastic films, particularly oriented polypropylene (PP) film, and this has led to a significant fall in usage. Today, its marketing is based on the fact that it is derived from a renewable resource, i.e. forestry, and that it has excellent stiffness in packaging machineability and point-of-purchase (POP) display. RCF can also be produced specifically for applications where high twist retention is required, for example wrapping of sugar confectionery. Coloured RCF is available.

3.2 Packaging needs which are met by paper-based flexible packaging

It should be noted that this discussion is limited to paper-based flexible packaging as used for food products in various forms together with, for example, confectionery, pharmaceuticals, tobacco, horticulture, agrochemicals, DIY and electrical goods. The needs of medical packaging will be discussed separately in Section 3.4.

3.2.1 Printing

Flexible packaging must carry information, including safety information where appropriate, about the product and its use. In many instances, depending on the product and how it is marketed, flexible packaging must provide visual impact at the point of purchase.

The main processes used in printing paper for flexible packaging applications are flexography and gravure. Several systems are available including water based, solvent based, UV cured and digital printing. The print quality can be varied depending on the surface and colour of the paper. In terms of colour, the choice is usually white, i.e. based on chemical-bleached pulp, or brown, i.e. based on either unbleached chemical pulp or dyed recovered fibre or mixtures of these two types of pulp. The surface may be either machine finished (MF), machine glazed (MG), supercalendered (SC), on-machine coated or cast coated. The whiter and smoother coated papers will give the best print reproduction – sharper dots in colour illustrations, fine text, good contrast and solid uniform colour.

3.2.2 Provision of a sealing system

A sealing system may be used to construct the pack, for example a sachet which is sealed on two or three sides prior to filling, and to seal the package after filling. Where the package has to provide product protection through the provision of barrier properties, it is necessary for this sealing to be pinhole free.

The main method for providing sealing to paper so that it can be formed and sealed is through the application of a sealable material. Examples are by applying:

- Polyvinylidene dichloride, PVdC, from an aqueous dispersion, also acrylics. Another vinyl-based heat-seal coating is VMCH (Dow Chemical Co.) which is applied from a solvent-based solution. Where a higher solids or lower viscosity coating is required, Dow recommends VMCC. (*Note*: VMCH and VMCC are carboxyl modified vinyl copolymers. They are functional terpolymers of vinyl chloride, vinyl acetate and maleic acid [Dow, 2005].)
- Hot melt coating by roll application, or slot die, based on a wax blend containing plastic polymers, such as ethylene vinyl acetate (EVA) and other additives which improve initial tack and seal strength.
- Extrusion coating with PE and EVA-modified PE. Other plastics can also be heat sealed but they would not be chosen purely for their heat sealability on grounds of cost. They may be chosen where they perform an additional function such as that of an oil or grease barrier, for example an ionomer such as Surlyn® (DuPont), and heat resistance, where PP, polyethylene terephthalate (PET or PETE) or polyamides (PA) would be used.
- Cold seal, based on either natural rubber or synthetic latex on a smooth paper surface, such as glassine or bleached kraft. Sealing is activated by pressure. Cold sealing can be carried out at high speed and is particularly favoured for the packaging of chocolate and other heat-sensitive products.

3.2.3 Provision of barrier properties

3.2.3.1 Introduction

Barrier properties are necessary in paper-based flexible packaging to protect the product. It is always necessary to define the nature of the product protection, for example texture, flavour, aroma, etc., required to achieve the desired shelf life in the expected environment(s) of storage, distribution and use. Examples of the barrier properties required for specific food products are shown in Table 3.1.

It is also necessary to determine the type and amount, in terms of thickness and coating weight, of the barrier material(s) required to provide the product the protection it needs.

There are several types of protection relating to moisture, texture, flavour and aroma, and the choice of barrier materials which are available to provide that protection will be

Table 3.1 Barrier requirements for various types of food product

	Moisture	Grease	Oxygen	Odours	Mould	Flavour	Aroma	Light
Biscuits	x	x	x	x				
Yellow fats	x	x	x					x
Cereals	x		x					
Chocolate		x				x	x	x
Coffee	x	x	x	x		x	x	x
Baked goods	x	x				x		
Dried foods	x		x					
Dry pet food	x		x			x	x	x
Soap					x		x	
Tobacco	x			x		x	x	x
Detergent powders	x	x					x	

Table 3.2 Typical paper-based flexible packaging uses

Specification	Application
Paper/PE	Flour bags
	Sugar bags
	Detergent powder bags
Paper/PVdC/PE	Oxygen and grease sensitive bag-in-box products, e.g. stuffing mix, gravy powder
Paper/cold seal	Ice cream bars
	Chocolate bars
	Sticking plaster wraps
Paper/grid heat-seal lacquer	Medical applications requiring sterilisation, e.g. needles, gowns, bandages
Paper/PE/aluminium foil/PE	Dehydrated soup, sauce and dessert powder sachets
	Instant coffee sachets
	Biscuits
Paper/metallised polyester/PE	Salted snacks, confectionery bars
Metallised paper	Cigarette liners, beer labels
Aluminium foil/paper	Yellow fat wraps, e.g. butter, margarine
Greaseproof paper	Yellow fat wraps

discussed in this section. See examples in Table 3.2. It must be appreciated that the materials are not necessarily interchangeable because the degree to which each provides a particular barrier property varies, for example aluminium foil compared with PVdC or PE.

Furthermore, the amount or degree of any particular form of protection will vary with the amount of material applied in terms of thickness or coating weight. Both these causes of difference have commercial implications in that the weight or thickness necessary to achieve the required performance may rule out the use of a particular material where the cost is uncompetitive.

Examples shown in Table 3.1 should be used only as a guide as some product groups will vary in their requirements. For example, instant coffee granules are not usually as sensitive as ground coffee.

The packing conditions, distribution requirements and shelf-life necessities must always be taken into account.

Whilst paper as a stand-alone substrate has no inherent barrier properties to speak of, there are various technologies which the paper industry and converters can employ, either 'in line' during papermaking or 'off line' as a subsequent operation. Examples are shown in Table 3.2.

3.2.3.2 Barrier to moisture and moisture vapour

The effect of moisture on packaged products depends on the product. For some it is necessary to maintain the moisture content at a high level to prevent the product from drying out. For others the reverse may be the case as by absorbing moisture the product may loose texture (e.g. crispness in a snack food) and metal products containing ferrous iron can rust when exposed to high humidity and oxygen.

Every food product has an optimum moisture content with respect to its stability as well as its texture and flavour. Non-food products, such as tobacco, also have an optimum moisture content range within which the quality is satisfactory. There will, however, be a moisture content level at which the quality of the product concerned would be deemed unsatisfactory. Assuming that the product is at the correct moisture content when packed, it is the aim of the packaging to ensure that an unsatisfactory moisture level is *not reached* within the intended shelf life in the recommended storage conditions.

Food products affected by moisture will, however, gain or lose moisture until equilibrium is established with the relative humidity (RH) of the atmosphere to which they are exposed. In the case of a packed product, this environment will be that existing within the sealed package.

It is therefore necessary to know the moisture content specified at the point of manufacture/packaging and the equilibrium relative humidity (ERH) at that moisture content. The next important consideration is the RH of the environment in which the packaged product will be stored, distributed and merchandised. In order to simulate temperate conditions, testing is carried out on sample packages stored at 25°C and 75% RH where the product will gain moisture and 40% where it would lose moisture. For tropical conditions, 38°C and either 90% or 40% RH is used.

At 25°C/75% RH, there will be a tendency for packaged products with a low ERH to gain moisture from the external environment and for those with a higher ERH to lose moisture to the external environment. It is the function of the packaging to ensure that any movement of moisture either into or from the product still ensures that the product does not achieve a moisture content, within the specified shelf life, at which the product is judged unacceptable.

Many flexible packaged ambient storage food products are required to have shelf lives of 6–18 months depending on the product and where they are marketed and used. This would be an inconvenient length of time to carry out a shelf-life test and therefore laboratory studies have been developed to give guidance on shelf life and barrier based on accelerated storage tests (Paine, 2002).

Dried foods, such as instant coffee and potato chips (crisps), have a typical moisture content of around 3% and the ERH is 10–20%. These products require a high, or good, barrier to water vapour – note this means a *low* water vapour transmission rate (WVTR). Dried foods such as breakfast cereals with ERH 20–30% are less stringent with respect to the water vapour barrier. For dried fruits and nuts, the ERH is 30–60% and for salt and sugar, 75% and 85%, respectively. A cake has an ERH of around 90% and it maybe thought that the barrier should *prevent* the loss of moisture from the cake. However, an RH of 90% *inside* the package is an ideal condition for mould growth. So rather than trying to prevent moisture loss, it is necessary to allow some loss but to slow down the rate to prevent the creation of too high an RH within the package (FOPT, 1996).

Low water vapour permeability, i.e. a good or high barrier, may be provided in several ways:

- laminate/extrusion coating, paper/aluminium foil//PE
- laminate/extrusion coating, metallised PET/paper/PE
- dispersion coating, paper/PVdC
- hot melt coating, paper/hot melt
- wax coating, paper/wax blend (microcrystalline wax plus additives)
- extrusion coating, PE/paper/PE or paper/PE
- laminate/extrusion, SiO_x coated PET/paper/PE.

A moderate or medium barrier to water vapour would be provided by PE extrusion-coated paper.

3.2.3.3 Barrier to gases such as oxygen, carbon dioxide and nitrogen

This barrier is necessary for the protection of products sensitive to flavour and aroma loss or contamination. Products where atmospheric oxygen causes product deterioration through oxidative rancidity, for example potato chips (crisps), may be gas flushed with an inert gas such as nitrogen and carbon dioxide. Ground coffee is sensitive to oxygen and may be vacuum packed.

The barrier is provided by either aluminium foil, EVOH, PVdC, aluminium-metallised PET, SiO_x coated PET or wax. If aluminium foil is used, the heat sealing is provided by hot melt or PE. Some end-users have discouraged the use of aluminium foil on the grounds that the structure is not easily recyclable in post-consumer waste-management systems. EVOH is an excellent gas barrier, but the barrier deteriorates significantly in a high RH environment. It is therefore sandwiched, when used in a paper-based flexible material, between layers of PE or PP. An EVOH structure of this type is offered for liquid packaging.

3.2.3.4 Barrier to oil, grease and fat

This is required where the product has an oil, grease or fat content which must be retained within the pack because loss of these ingredients would be reflected in a lower quality product and because any permeation to the surface of the package would be unsightly.

This barrier is provided by aluminium foil, PVdC or ionomer extrusion coating. If aluminium foil is used, the heat seal needs to be a plastic coating with oil, grease and fat resistance. Medium-density PE, high-density PE, PP and ionomer resin such as Surlyn® all provide oil, grease and fat resistance, the degree of the barrier property being given here in ascending order, with Surlyn® providing the best barrier.

It may be sufficient to use an ionomer resin as a thin coating between PE and aluminium foil. This will provide a good product barrier and also ensure that the adhesion between the aluminium foil and the PE does not break down in the presence of the product during storage. The use of ionomer resin in this way also ensures that good PE adhesion is achieved without resorting to higher temperatures at which there would be a danger of odour and product taint.

Greaseproof paper and glassine also have oil, fat and grease resistance but will also require a heat- or cold-seal coating for flexible packaging applications.

3.2.3.5 Barrier to light

Some, particularly fat containing, products can deteriorate in light, especially sunlight, containing a UV component, which can promote oxidative rancidity. Therefore an opaque barrier is required which is only really guaranteed by including aluminium foil in the construction.

3.3 Manufacture of paper-based flexible packaging

3.3.1 Printing and varnishing

Printing is normally the first process in the manufacture of flexible packaging. Flexible packaging is printed reel to reel. The main processes used are flexography and gravure. In addition, some printers use web offset litho. Digital printing can be used, and normally it would be used on short runs. All the usual ink systems are used, including metallic and fluorescent inks. UV-cured systems are used to give a high gloss, a rub and product-resistant

Figure 3.1 Sugar/flour bags. (Reproduced with permission from Robert Bosch GmbH.)

surface together with high heat resistance in the subsequent heat-sealing areas. Whereas printing is used for text, illustrations and overall decoration, varnishing is used for surface finish, i.e. gloss, matt or satin, protection of the print for rub and product resistance and for the control of surface friction.

The print quality depends on the paper surface. To obtain the best print results, a mineral pigment-coated paper should be used. Some applications of printed paper do not require further conversion. Examples of paper products supplied, printed on reels, to end-users for forming, filling and sealing include flour/sugar bags (Fig. 3.1) and labels for use in injection moulding, for example ice cream tub lids.

Patterned heat-seal and cold-seal coatings may be applied on the reverse side of the paper in register with print on the other (outside) surface using printing techniques.

3.3.2 Coating

Coating is the simplest method of adding other functions to paper. The active functional material, for example VMCH, PVdC, PE and wax, is either applied from a solvent solution, water-based dispersion or as a solid, in the molten state.

3.3.2.1 *Solvent-based coatings*

Solvent-based coatings applied to paper by gravure mainly comprise varnishes which impart heat resistance so that the surface does not pick under heat-sealing bars. The varnish could be 100% solids UV cured and have high gloss, product and heat resistance.

3.3.2.2 *Water-based coatings*

Water-based coatings for paper include PVdC with the coating applied by roll. The coating weight and smoothness is controlled by air knife – a non-contact technique (Fig. 3.2). Several applications may be applied and dried in one pass through the coating machine, giving a total coating weight up to around 30g/m^2.

3.3.2.3 *Coatings applied as 100% solids, including wax and PE*

Wax is the oldest paper-based functional coating. Originally paraffin wax was used. From the 1950s, the main wax component has been microcrystalline wax to which polymers such

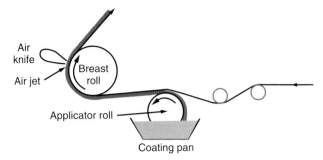

Figure 3.2 Principles of air-knife coating. (Reproduced with permission from Paper Industry Technical Association, PITA.)

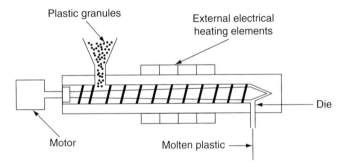

Figure 3.3 Plastic extruder.

as PE and EVA have been added by blending to improve the barrier in folded areas and also the hot tack in heat sealing. Wax is applied as either 'dry waxing' or 'wet waxing'. In the former, the wax is applied and the weight controlled by roller application. It is then passed over heated rolls so that the wax is driven into and saturates the paper. Alternatively, after application the wax can be rapidly chilled, usually in a water bath, thereby creating a high gloss and keeping most of the wax on the surface.

PE is supplied as pellets which are melted by a combination of high pressure, friction and externally applied heat. This is done by forcing the pellets along the barrel of an extruder using a polymer-specific screw under controlled conditions that ensure a homogeneous melt prior to extrusion (Fig. 3.3).

The molten plastic is then forced through a narrow slot or die onto the paper, or the aluminium foil surface of a previously laminated paper/aluminium foil laminate. As it comes into contact with the paper or foil, it is brought in contact with a large diameter chill roll with either a gloss or sand-blasted matt finish which determines the surface finish of the PE (Fig. 3.4). To render the surface suitable for print and/or adhesion with an adhesive, it is treated with a corona discharge. A coating weight of $20 g/m^2$ would be typical though higher and lower coating weights are possible. Usually the coating is confined to one side but if two-side coating is required then in order to avoid blocking in the reel only one surface may have a gloss finish. Extruders having two extruding dies can coat both sides of a sheet in one pass through the machine (Fig. 3.9).

The simplest PE application is kraft/PE as an overwrapping material – for example as a transit overwrap for 10×20 cigarette cartons. In this case, labels would be applied over the envelope end seals.

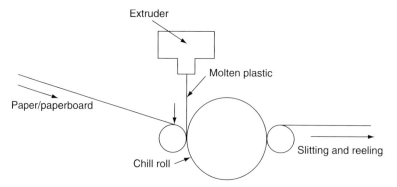

Figure 3.4 Extrusion coating of paper or paperboard.

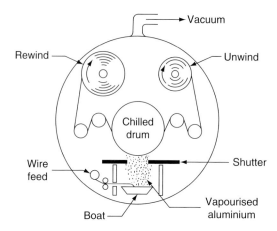

Figure 3.5 Metallising process. (Courtesy of the Packaging Society.)

There is a wide range in the choice of PE. Hot tack in heat sealing is improved by blending EVA with low-density polyethylene (LDPE). If a medium-density PE is used, it improves the puncture resistance. This is important for providing better puncture resistance when packing granules with sharp edges. Surlyn® ionomer improves adhesion to aluminium foil and provides resistance to essential oils and other oil- and fat-type products. EVOH can be sandwiched between outside and inside coating of paper-coated PE to significantly improve the barrier to oxygen, which causes rancidity in fat-containing products, and other gases.

3.3.2.4 Metallisation

Metallisation is a process whereby aluminium is vaporised in vacuum and deposited to form a thin film on the surface of a substrate (Fig. 3.5). This process has been applied to paper but it is more usually applied to film. There are three problems with paper. First, the time required to create the vacuum in the material extends the time required at a reel change and reduces the moisture content so that the paper requires rehumidification after metallising. Second, being thicker than film the length of material processed per reel is lower, and hence the area coated in a given time is lower than with plastic film. Third, the surface of paper is not as smooth as film and the barrier improvement is not sufficiently significant.

These features can be overcome at a cost. The paper surface can be clay coated and prelacquered to make the surface smoother and improve the metallised appearance. The moisture changes can be avoided by using a transfer metallising process whereby the metallising layer is first transferred to a reel of PP. The metallised layer on the PP is transferred to the paper with the help of an adhesive. In this process, the plastic film (PP) can be reused.

The advantage of metallised paper is that it can replace aluminium foil in some applications. The barrier is, however, poor compared to foil unlike the situation when plastic film is metallised. Due to the on-costs required to metallise paper, the price differential compared with aluminium foil is not significant. Metallised polyester film (PET or PETE) has been laminated to paper and paperboard as an aluminium foil replacement. A large usage of metallised paper is for the bundling of cigarettes prior to packing in a carton or soft pack.

3.3.2.5 Hot melt coatings

Hot melt coatings are also wax-type blends, but they have higher viscosities than the simpler waxes discussed under Section 3.2.3. They are either applied by roll coating or by extrusion, though the extruder would be a simpler design than the type used to extrusion coat PE.

Hot melt-coated paper is used as an in-mould wrap-around label on thermoformed plastic tubs for yoghurt.

A typical specification is:

- overlacquer for gloss and print protection
- gravure or flexo print
- bleached kraft paper (95–100 g/m^2)
- hot melt coating.

The labels are fed from the reel into the pot moulds just before the sheet of plastic, for example polystyrene, also fed from reels, is thermoformed. The hot melt coating seals to the plastic. It also seals where it overlaps the print. The coefficient of friction is an important property of the overprint lacquer to ensure efficient runnability on the machine.

One of the main applications for hot-melt-coated paper is in the wrapping of soft cheese. The packaging in this application is designed to participate in the maturing of the cheese.

A typical specification is:

- 20 μ oriented polypropylene (OPP) film reverse side printed by flexo or gravure
- adhesive laminated – sometimes in stripes – to 25–40 g/m^2 bleached kraft paper
- hot melt coating.

The OPP is microperforated after printing to allow cheese respiration and a controlled WVTR. The hot melt coating which is used in the direct wrappage of food provides:

- controlled oxygen permeability (which influences the development of the cheese bacteria and, therefore, the taste)
- good folding around the product

- heat sealing for a better pack closure
- glossy finish to the product.

This specification can be modified by metallising the OPP film over the print.

3.3.2.6 Cold-seal coating for pack closure/sealing

For specific applications in flexible packaging, an alternative to heat sealing is *cold seal*. This technology requires only mechanical pressure to bond two cold-seal-coated surfaces together, without any heat being required. Sealing at room temperature, excellent seal integrity and a wider variety of substrates are the main advantages of cold sealing compared with heat sealing. When a packing machine using heat seal stops, heat from the sealing jaws can result in significant product damage and high rejection rates, especially with a product involving chocolate. Cold seal is not particularly sensitive to the sealing dwell time and allows a higher tolerance to variations in-line speed. The main applications for cold seal are on horizontal form, fill, seal (HFFS) bar lines and overwrapping, although some vertical form, fill, seal (VFFS) operations exist as well.

Cold-seal technology is based on a coating technique. A water-based emulsion is printed onto a web substrate by means of the gravure process. The key component of the emulsion is natural rubber latex which provides cohesive features. The coatings stick preferentially to each other when two coated surfaces are brought together.

Other ingredients are water, ammonia, surfactants, anti-oxidants, anti-foam agents, biocides and an acrylic component. The latter acts as an adhesive, bonding the cold seal to the substrate.

In the wet state, cold seal has a shelf life limited to 6 months. It must be stored away from frost and heat. Below 0°C, an irreversible loss of sealing ability occurs. After printing, a shelf life, before use, of at least 6 months is guaranteed. Excessive ageing allows physical, chemical and biological degradation to occur and leads to loss of seal strength and the development of repulsive odours. After sealing, cold-seal packaging keeps its seal integrity in a freezer, for example ice cream sticks.

Typical constructions are:

- release lacquer
- printing inks
- substrate, for example glassine
- cold seal applied in a pattern

and

- release film
- printing ink
- laminating adhesive
- substrate
- cold seal applied in a pattern.

Release lacquer or film is required to allow for easy unwinding (low cling) and prevention of blocking of the material in the reel. Release lacquer consists of PA resin. Release film is a low surface tension plain OPP. Glassine, which may be coloured, for example chocolate brown, would be a suitable paper substrate. (Typical plastic film substrates would be white pearlised PP, transparent and metallised PP as well as metallised polyester.)

The cold-seal pattern provides a sealing medium in the seal areas only (fin seal and cross seal). The central area, where the contents are in contact with the wrapper, must be free from cold seal in order to minimise the direct contact between cold seal and food. Coating weights range from 2 to 6 g/m², depending on the application. Typical seal strengths are 5 N/30 mm. Apart from the seal force, as measured after flat sealing and pulling in a tensile tester, a certain coating mass must be present to ensure seal integrity in the pack folds and edges.

Frequently encountered cold-seal issues are:

- foaming in the gravure press cold-seal distribution system
- formation of lumps as a result of high shear conditions between gravure (application) cylinder and doctor blade
- quality variations due to seasonal fluctuations in rubber latex (natural agricultural product)
- lack of sealing strength
- scumming (appearance of a cold-seal ghost in the food contact area)
- misregister of the cold-seal pattern with respect to the printing
- smell, for example residual ammonia
- upon unwinding reels due to transfer of cold seal to the release side or damage to the printed side.

In recent years, a synthetic version of cold seal has been developed at the request of the market (food end-users). Main drivers for this are the elimination of natural rubber latex, suspected of causing allergy in sensitive individuals, and the reduction of the unpleasant organic smell. In the synthetic version, a reduced and different odour level was observed as well as excellent converting and sealing behaviour.

Despite the advantages of the synthetic cold seal, which is a higher cost material, it has not been widely adopted. There have been successful developments such as minor modifications in the formulation to reduce smell (flash stripping of residual acrylic monomers, improved centrifuging of latex), more consistent natural latex (rubber from large plantations with cloned trees) to allow higher converting speeds (through the use of surfactants) and changes necessary to comply with evolving legislation.

Due to its presence at the food contact side of a wrapper, cold seal is subject to very strict limitations as to the ingredients used in order to comply with food-packaging regulations.

3.3.3 Lamination

In the laminating process, the functional usage of paper is enhanced with the addition of one or more additional layers, or webs, of material using an adhesive to achieve the bonding of the materials. Different adhesive systems which differentiate the various laminating processes are discussed later.

3.3.3.1 Lamination with water-based adhesives

Water-based starch and polyvinyl acetate (PVA) adhesives are used to laminate paper with aluminium foil. Casein and sodium silicate can also be used as adhesives. The adhesive is applied to one of the surfaces and the other surface combined with it by nip pressure between two rolls. Water from the adhesive is absorbed by the paper leaving the active part of the adhesive in a tacky state on the surface. The combined material then passes through a heated

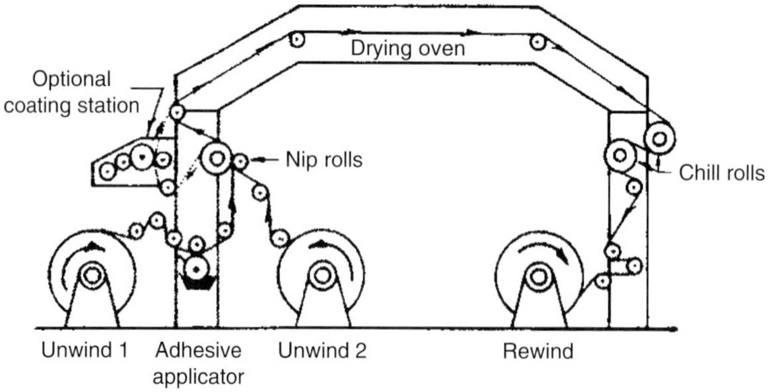

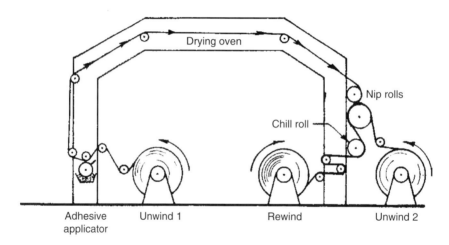

Figure 3.6 Wet-bond lamination. (Courtesy of the Packaging Society.)

Figure 3.7 Dry-bond lamination. (Courtesy of the Packaging Society.)

tunnel which dries the adhesive (Fig. 3.6). Examples of laminates which are widely used are greaseproof paper/aluminium foil for butter wrapping and bleached kraft/aluminium foil for subsequent PE extrusion coating for many types of sachet and pouch, for example for dry food products such as dehydrated soup and instant coffee.

3.3.3.2 Dry bonding

There are two types of dry bonding (Fig. 3.7) adhesive:

- There is a type where a solvent-based adhesive is applied to one substrate and passed through an oven to remove the solvent after which it is combined with the other substrate between two rolls with nip pressure. The laminating nip may be heated if the adhesive is activated by heat, otherwise the adhesive arrives at the nip in a tacky state. Dry bonding is usually associated with two plastic films or one plastic film and aluminium foil and is not a usual process for laminating when paper is one of the substrates.

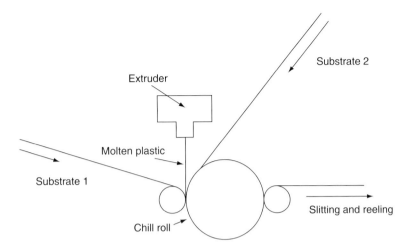

Figure 3.8 Extrusion lamination.

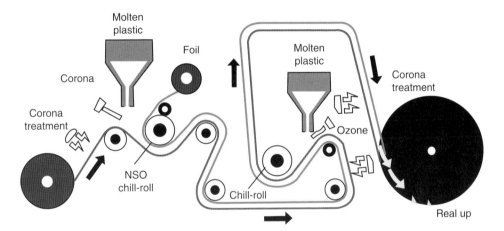

Figure 3.9 Extrusion lamination and coating. (Courtesy of Iggesund Paperboard.)

- Alternatively, the adhesive may be a two-component 100% solids system which is applied to one surface which is then combined with the second surface. Adhesion is activated by heat under nip pressure. This is suitable for bonding a printed paper with a plastic film, such as OPP or biaxially oriented polypropylene (BOPP). A typical application of this technique enables strips of paper to be laminated with film whilst still allowing product visibility in the finished packaging.

3.3.3.3 Extrusion lamination

Extrusion lamination is often used to laminate paper or paperboard to aluminium foil (Fig. 3.8).

On an extruder with two dies, it is possible to extrusion laminate the paper with aluminium foil and then extrusion coat the aluminium foil in one pass, (Fig 3.9).

Examples of paper-based extrusion-coated/laminated materials and their properties and advantages are as follows.

Aluminium foil/PE/MG bleached kraft/PE

- Material suitable for HFFS machinery
- Suitable for flow packs, sachet packs and stick packs
- Bleached MG paper suitable for simple decorations and flexo printing
- Excellent oxygen and water vapour barrier
- Suitable for powdered products
- Good tearing of, for example, stick packs

Aluminium foil/PE/coated bleached kraft/ionomer

- Material suitable for HFFS machinery
- Suitable for flow packs, sachet packs and stick packs
- Clay-coated paper suitable for advanced decoration, flexo or gravure printing
- Excellent oxygen and water vapour barrier
- Suitable for powdered products and outer wraps
- Good tearing of, for example, stick packs
- Good sealing properties through product-contaminated seal areas
- Good seal strength
- Very good hot tack
- Suitable for high-speed applications

Aluminium foil/PE/MG bleached kraft/ionomer

- Material suitable for HFFS machinery
- Suitable for flow packs, sachet packs and stick packs
- MG paper suitable for simple decors and flexo printing
- Excellent oxygen and water vapour barrier
- Suitable for powdered products
- Good tearing of, for example, stick packs
- Good sealing properties through product-contaminated seal areas
- Good seal strength
- Very good hot tack
- Suitable for high-speed applications

Aluminium foil/PE/coated bleached kraft/PE

- Material suitable for horizontal machinery
- Suitable for flow packs, sachet packs and stick packs
- Clay-coated paper suitable for advanced decoration, flexo or gravure printing
- Excellent oxygen and water vapour barrier
- Suitable for powdered products and outer wraps
- Good tearing of, for example, stick packs

3.3.3.4 Lamination with wax

Wax is used as an adhesive with barrier properties to water, water vapour and gases and odours (Fig. 3.10). For example:

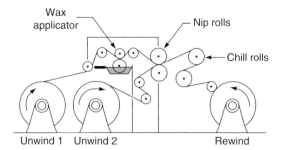

Figure 3.10 Wax lamination. (Courtesy of the Packaging Society.)

- aluminium foil/wax/greaseproof paper – the aluminium foil can be printed and embossed for use as a butter wrap
- paper/wax/unlined chipboard – the paper can be printed and this material is used for detergent powder cartons where the product requires moisture and moisture vapour protection
- aluminium foil/wax/tissue.

3.4 Medical packaging

3.4.1 Introduction to paper-based medical flexible packaging

Paper-based packaging is used to pack medical devices such as catheters, surgical instruments, operation kits and consumables such as dressings and surgical gloves. Unlike the flexible packaging discussed so far, paper-based medical packaging has the following special characteristics:

- Products are all sterilised after packaging by one of several processes, viz. steam in an autoclave or other form of steam sterilisation, ethylene oxide (EtO) gas and gamma radiation.
- Packs must thereafter maintain a microbiological barrier – in the case of paper, this is achieved by limiting the maximum pore size.
- All sealing must remain secure until the product is required for use at which point the seals must be capable of being opened in a peelable manner which does not generate loose fibre.
- There must be no possibility for resealing medical packing once it has been opened.

A wide range of flexible packaging incorporates paper, such as:

- pouches (Fig. 3.11), sachets, strip packs which are all formed, filled and sealed
- lidding for thermoformed plastic packaging on HFFS machines
- bags, pre-made pouches and die-cut lids.

Figure 3.11 A medical pouch. (Reproduced with permission from Amcor Flexibles.)

The following account of the background to the use of paper in medical packaging has been prepared by Bill Inman, formerly Technical Manager at Henry Cooke – a Billerud company specialising in the manufacture of medical packaging papers.

> Paper is used in the construction of packaging for terminally sterilised medical goods in a number of ways. The earliest large scale application was in the hospital environment, where items to be sterilised were first wrapped in a sheet of special sterilisation paper then sealed inside paper bags which were subjected to sterilisation in a steam autoclave. The bags were sealed either by a heat seal or by folding and taping the top. The contents could vary from a few swabs or dressings up to a full surgical procedure pack, so the bags were made in a range of different sizes to accommodate this. When the contents of the bag were used, the bag would be cut open and the sheet of wrapping paper opened out to form a sterile working surface. UK hospitals were pioneers in this application, and initially both the sizes and construction of the bags and the specification for the paper to be used were covered by UK Department of Health Standards, first published in 1967. In the 1980's these were converted into British Standards, which themselves were a major consideration when European Standards for terminally sterilizable medical packaging were formulated during the period 1990–2000. The special feature of the bag paper was that it had to have sufficient air permeability (sometimes referred to as porosity) to allow the rapid transfer of air and steam during the autoclave cycle but on the other hand it must provide an adequate barrier to prevent the passage of contamination. Traditionally, high strength bleached kraft papers, with wet strength and high water repellency, have been used for this application. This continues to be an important use for paper.
>
> Another important use of paper is in combination with plastic films and laminates in the construction of peel open systems, the paper forming one side of the pack, the plastic the other, with the two webs heat sealed around the edges of the pack. Whilst these are sometimes used to package goods in hospitals, their overwhelming use is by industry where large volumes of small items such as syringes, needles, catheters etc are packed at the end of production using web-fed techniques. The plastic web may be flat or heat formed into a blister shape. Sterilisation is likely to be by gamma radiation or ethylene oxide. The key in this application is to achieve a sufficiently strong bond of the two webs around the edges of the package to maintain its integrity but yet allow a clean, controlled and easy peel on opening. Much of the development in this area has been concerned with achieving this through

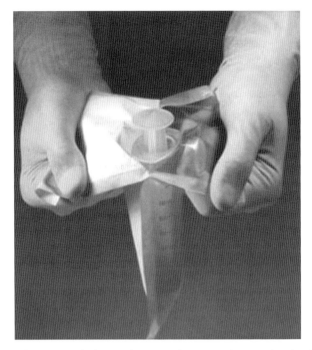

Figure 3.12 Typical syringe pack showing how a pack is opened. (Reproduced with permission from Amcor Flexibles.)

the use of coatings and lacquers. When steam and ethylene oxide sterilisation is used, the paper must meet an air permeability specification, but for radiation sterilisation this is not necessary.

Currently, materials for medical packaging are covered by the European Norm EN868, "Packaging materials and systems for medical devices which are to be sterilized". This is a multi-part standard in which Part 1, "General requirements and test methods" is a horizontal or "umbrella" standard applicable to all materials, Parts 2 to 10 defining specific requirements for individual materials.

ISO 11607, Packaging for terminally sterilized medical devices, may also need to be considered. (Inman, 2004)

ISO 11607:2006, Parts 1 and 2 deal with terminally sterilized materials, (similar to EN 868-Part 1), pack formation, validation, integrity and shelf life (James, 1999).

The importance of the requirement for seal peelability without the generation of loose fibre has already been mentioned. If a pack is torn on opening, the device may come into contact with the outer pack surface which is likely to be non-sterile. Excessive tearing can result in loose fibres being released into the environment. If these enter a wound site, there is a potential for vascular inclusion, or they can irritate sensitive tissues (Merrit, 2002).

An important aspect associated with opening some sterile packs is the practice whereby one person peels the pack open and a second person removes the sterile product (Fig. 3.12).

3.4.2 Sealing systems

There are several types of sealing systems used with paper-based medical packaging:

- *Heat-seal lacquer*: This can be applied in an all-over grid pattern to maintain the porosity of the paper for EtO sterilisation. This method of sealing is also suitable for radiation sterilisation but not for steam sterilisation as it would soften and melt at the temperatures used.
- *Cold seal*: In the 1990s, cold-seal coatings began to be used successfully in this market. Both natural and synthetic latex are used. Zone coating of cold seal can be applied on the inside of the pack in register with print on the outside. This is necessary where there is a possibility of the product being packed sticking to the inside of the pack, for example in the packaging of surgical dressings and surgical gloves. Cold-seal-coated papers run at high speed on form, fill, seal machines. The seals peel with random transfer and disruption of the seal interface. Cold-seal-coated packs can be sterilised by EtO and radiation.
- *Water-based heat seal*: A white water-based heat-seal coating can be applied to the paper using the air-knife technique for smoothing and weight control. This material seals to rigid and flexible plastic surfaces and has the added advantage that when opened the white coating in the seal area transfers cleanly, with an extremely low risk of fibre tear, to the plastic surface, thereby providing a simple check of seal integrity. This specification is suitable for sterilisation by EtO and radiation.
- *Direct-seal paper*: A type of kraft paper has been developed which can seal directly to non-corona-treated PE. This seal gives minimum fibre lift from the surface of the paper. The paper used is known as direct seal (DS) and it requires special manufacturing procedures in papermaking. These procedures relate to a good internal bond strength within the paper, surface treatment, including controlled calendaring, and a higher disposition than normal to encourage the fibres to align themselves in the MD. This latter feature is taken into account when the pack is designed so that when the pack is opened tension is applied, and hence peeling occurs, in the MD of the paper. This results in less surface disruption and the loose-fibre generation which is associated with surface disruption (Fig. 3.13). DS papers have been developed as a cost-reduction option for mature products in the medical packaging market, such as suction tubing, conventional gauze dressings and urinary catheters. The sealing mechanism is in direct contrast to other systems which rely on a coating to minimise fibre lift (Merritt, 2003).

DS papers have required changes in papermaking technology and in packaging machinery heat sealing. Seals which are too weak cannot be allowed, and if they are too strong fibre tearing and loose-fibre generation will occur. The operating window on the HFFS machines which are typically used for this type of packaging in respect of heat, pressure and dwell time is critical. Temperature control within a range of 2°C across the entire sealing surface and timers able to control dwell time to within hundredths of a second have been incorporated (Merritt, 2002). Typical basis weight is 60 g/m^2 (37 lb).

Other cost-reduction areas of interest are as follows:

- 'Peelable polymer' films, for example multilayer (co-extruded) PA films, which are designed to give minimum fibre lift when seals to uncoated papers are opened, thereby eliminating the necessity for the application of a special peelable coating to the paper during conversion.
- In-line flexographic printing on the packing line – this is more appropriate where the packing programme involves short runs of many different products.

Figure 3.13 A fully peelable package incorporating grid-lacquered paper and thermoformed nylon/PE. (Reproduced with permission from Amcor Flexibles.)

Table 3.3 Typical paper-based medical packaging structures

Structure	Sterilisation by	Application (sealing)
50 g/m² Kraft/30 g/m² LDPE 31 lb Kraft/18 lb LDPE	Compatible with EtO and radiation	Backing web to peelable materials on four-side seal machinery
108 g/m² Kraft/13 g/m² overall heat seal 70 lb Kraft/8 lb overall heat seal, white, air-knife coating	Suitable for EtO (sufficient porosity) and radiation	Lidding for thermoform, fill and seal, die-cut lids pouches
44 g/m² Kraft/29 g/m² LDPE/4 g/m² overall heat seal 27 lb Kraft/17.8 lb LDPE/2.5 lb heat seal	Radiation	Peelable for four-side sealing with uncoated paper and also lidding for thermoformed base web
60 g/m² Kraft/6 g/m² overall grid heat seal 37 lb Kraft/3.7 lb overall grid heat seal	EtO and radiation	Lidding for thermoformed base web
60 g/m² Kraft/11 g/m² overall grid heat seal 37 lb Kraft/6.8 lb overall grid heat seal	EtO and radiation	Lidding for thermoformed base web
100 g/m² Kraft/11 g/m² overall grid heat seal 61 lb Kraft/6.8 lb overall grid heat seal	EtO and radiation	Lidding for thermoformed base web
40 g/m² Kraft/12 g/m² LDPE/8 μ Al foil/14 g/m² Surlyn® 25 lb Kraft/7 lb LDPE/32 ga Al foil/8.6 lb Surlyn®	Radiation	Suitable for four-side sealing with equivalent peelable lacquer-coated material
40 g/m² Kraft/3 g/m² overall synthetic cold seal 25 lb Kraft/1.8 lb overall synthetic cod seal	EtO and radiation	Seals to itself
60 g/m² Direct-seal paper (kraft) 37 lb Direct-seal paper (kraft)	EtO and radiation	

Note: lb in this context refers to the basis weight in pounds per 3000 ft² or a ream of 500 sheets of 24 in. × 36 in.
Source: see Lockhart and Paine (1996) for pharmaceutical and healthcare packaging.

3.4.3 Typical paper-based medical packaging structures

Printing for medical packaging applications is normally by either the flexographic or gravure processes. Typical paper-based medical packaging structures are tabulated in Table 3.3.

The 'thermoformed base webs' are typically plastic laminations and co-extrusions. The cavities in which the product concerned is placed are formed from the reel on horizontal thermoform, fill, seal machines. The specification in any particular case depends on the product, its size and weight, whether it has protrusions or sharp edges, whether it is soft and compressible or hard, etc. The plastic material used may range from PE, PP, PET or PETE, ionomer, for example Surlyn®, to PA, for example nylon.

Where products in pouches and four-side sealed packs require a moisture vapour or light barrier, aluminium foil is incorporated in the lamination.

Bags and pre-made pouches are predominantly used in hospitals. They are used to a limited extent in industry for short production runs. Bags can have a tear string opening feature, side gussets and either flat- or folded-bottom seams.

Plastic-film bags can be provided with extra strength through the inclusion of a paper or Tyvek® header for use with heavy and bulky products which require EtO and radiation sterilisation. Tyvek® is a non-woven fibrous sheet material produced by DuPont. It is made from very fine, high-density polyethylene (HDPE). The process of manufacture is described as HDPE flash spinning to produce fibres which are then laid down randomly and non-directionally on a moving bed. The fibres are bonded using heat and pressure. This results in a strong sheet with high tensile, tear and puncture resistance. It has resistance to microbial penetration. It is heat sealable, and seals can be opened with lint-free, i.e. loose fibre free, peeling. Tyvek® can be sterilised by existing procedures. Tyvek® is therefore frequently used in medical packaging applications as lidding material, pouches, bags and as strengthening/breathable headers in film bags.

See Paine and Lockhart (1996) for pharmaceutical and healthcare packaging.

3.5 Packaging machinery used with paper-based flexible packaging

Flexible packaging is usually associated with form, fill, and seal machinery. VFFS machines (Fig. 3.14) are used to pack free-flowing food materials and liquids. Packs made in this way are either flat or incorporate gussets and block (flat) bottoms. Flat-bottom bags can be made in compact 'brick' designs with a wide range of top closures/reclosures (Figs 3.15 and 3.16).

Another type of machine is filled *vertically* but the pack is formed *horizontally*. It can be formed with a base gusset (Figs 3.17 and 3.18). This type of pack is suitable for a powder, granular or liquid product. Single items can be packed and cut off in multiples forming strip packs – the heat seals separating adjoining individual packs can be perforated so that the individual sachets can be used progressively.

The flow wrap-type machine both forms and fills the pack in a horizontal plane. It is used to pack single solid items, such as confectionery bars or multiple products already collated in trays, which may be made of paperboard or plastic (Figs 3.19 and 3.20).

In-line thermoforming is primarily thought of in the context of plastic packaging because the base web is made up of one or more plastics. The lidding and in-mould labelling of such packaging is, however, frequently achieved with paper-based materials. This was discussed under medical packaging though there are many food-based applications, chiefly yoghurts and cream-based desserts (Fig. 3.21).

There are machines which form bags around mandrels, sealing being made with adhesives, so that they have a rectangular cross section and a block bottom. (This type of machine can also wrap a carton around the paper on the same mandrel to form a lined carton.) Roll-wrap machines pack rows of items, for example biscuits and sugar confectionery. Individual

Paper-based flexible packaging 115

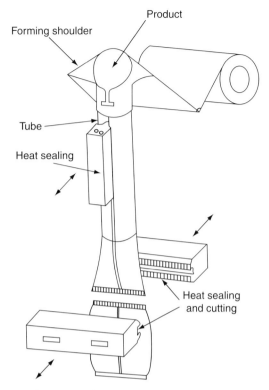

Figure 3.14 Vertical form, fill, seal machine.

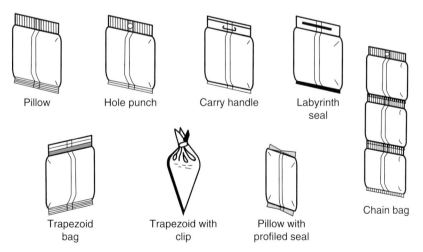

Figure 3.15 Range of typical pillow-type packs produced on vertical form, fill, seal machines, including chain or strip packs. (Reproduced with permission from Rovema Verpackungsmaschinen GmbH.)

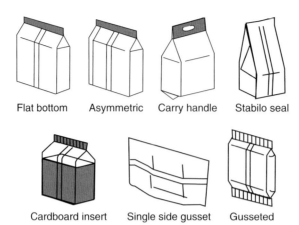

Figure 3.16 Range of typical gusseted/block bottom bags produced on vertical form, fill, seal machines. (Reproduced with permission from Rovema Verpackungsmaschinen GmbH.)

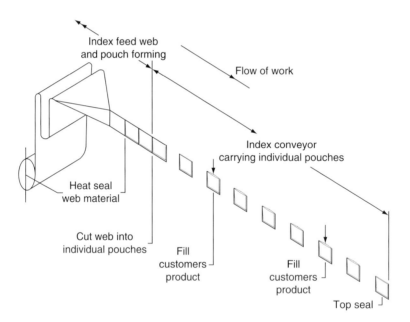

Figure 3.17 Horizontal pouch/sachet form, fill, seal machine for dry mixes (soups, sauces, etc.), pastes and liquid products.

Figure 3.18 Typical horizontally formed pouches/sachets with three-side sealing, four-side sealing and base gusset providing a 'stand-up' feature. (Reproduced with permission from Rovema Verpackungsmaschinen GmbH.)

Paper-based flexible packaging

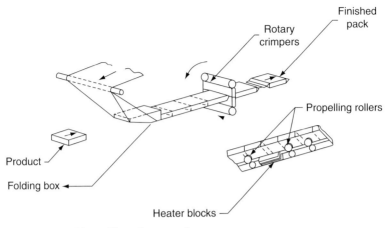

Figure 3.19 Horizontal form, fill, seal-type machine.

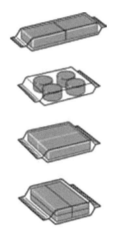

Figure 3.20 Typical horizontal form, fill, seal pack. (Reproduced with permission from Rose Forgrove (HayssenSandiacre Europe Ltd).)

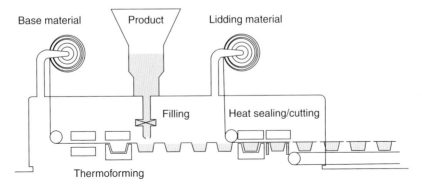

Figure 3.21 In-line thermoforming and lidding.

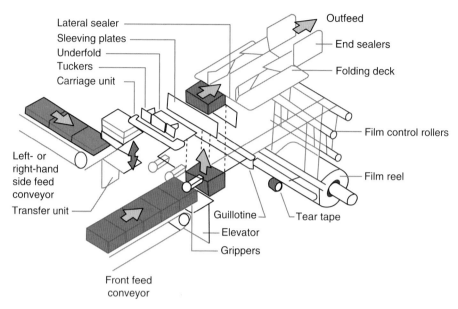

Figure 3.22 Overwrapping machine. (Reproduced with permission from Marden Edwards Ltd.)

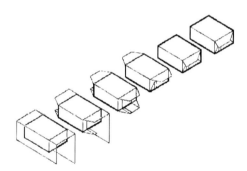

Figure 3.23 Overwrapping envelope end fold folding sequence. (Reproduced with permission from Marden Edwards Ltd.)

confectionery units may be wrapped in waxed paper for moisture protection and to prevent them sticking together. These may be twist wraps, dead folded or plain waxed liners overwrapped with a folded printed paper.

Overwrapping square or rectangular-shaped cartons with paper coated with PE or wax with neatly folded heat-sealed end flaps is also used, for example overwrapping of chocolate boxes, teabag cartons and a grouped overwrap of 10 cartons each containing 20 cigarettes (Figs 3.22 and 3.23). Another form of overwrap is the waxed paper bread wrap with ends folded progressively and heat sealed.

3.6 Paper-based cap liners (wads) and diaphragms

Paper-based flexible materials are used inside plastic caps and closures for rigid plastic and glass jars. They are referred to as discs or innerseals.

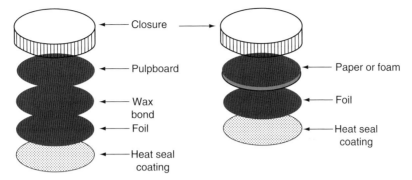

Figure 3.24 Pulpboard disks for induction-sealed cap liners. (Courtesy of the Packaging Society.)

3.6.1 Pulpboard disc

The simplest type of cap liner is a pulpboard disc made from mechanical pulp fitted inside a plastic cap. The cap liner or wad has to be compressible and inert with respect to the contents of the container, usually a food product. This liner could be faced with aluminium foil or PE where the nature of the contents requires separation from direct contact with the pulpboard.

3.6.2 Induction-sealed disc

The disc comprises pulpboard/wax/aluminium foil/heat-seal coating or lacquer (Fig. 3.24). The cap with the disc in place is applied to the container and secured.

It then passes under an induction heating coil. This heats the aluminium foil which causes the wax to melt and become absorbed in the pulpboard. It also activates the heat-seal coating and seals the aluminium foil to the perimeter of the opening of the container. When the consumer removes the cap, the adhesion between the pulpboard and the aluminium foil breaks down leaving the foil attached to the container. This seal therefore provides product protection and tamper evidence. Where subsequent contact between the contents and the pulpboard is undesirable, the pulpboard is permanently bonded to the aluminium foil. (A simpler version dispenses with the wax and replaces the pulpboard with paper.)

3.7 Tea and coffee packaging

There are many different types of paper-based tea and coffee packaging (Fig. 3.25). In the context of paper-based flexible packaging, the main interest is in tea and coffee bags and the printed envelopes, which are either simple folded designs or heat sealed with barrier protection, and the tags which are associated with them in some of the forms of marketing product presentation.

The bags themselves are made from very lightweight porous tissue. The tissue is either heat sealable or non-heat sealable. A typical heat-sealable tissue is $17\,g/m^2$ and contains a heat-sealable thermoplastic fibre, such as PP, in addition to its long fibred structure. Bags may be flat, square, round, pyramidal or four-side perimeter sealed. A 100% biodegradable and compostable tea and coffee bag tissue is now available with fibre based on Ingeo™ (polylactic acid, PLA). This product is suitable for use on tea and coffee packing machines that use ultrasonic sealing technology (Ahlstrom, 2009).

Figure 3.25 A selection of heat-sealed tea and coffee bags. (Reproduced with permission from IMA, Industria Macchine Automatiche.)

Figure 3.26 Tea and coffee bag with tag and string. (Reproduced with permission from IMA, Industria Macchine Automatiche.)

Two webs of tissue are used at high packing speeds for heat-sealed bags. Non-heat-sealable tissue is used to make a bag which is folded and stapled giving a larger surface area for infusion and using a lighter weight tissue, typically $12\,g/m^2$. All these bags are closely associated with the machinery which forms, fills and seals the bags – both types may have strings and tags (Fig. 3.26).

It is possible to link such machines with enveloping machines which can comprise paper, or paper laminated or coated with moisture and gas barrier properties. Tea and coffee bag packing machines can include, or be linked to, cartonning or bagging machines.

3.8 Sealing tapes

Sealing tapes have much in common with labels but they can be considered a flexible packaging product as paper reels are subjected to a reel-to-reel conversion process and sometimes they are also printed.

Sealing tapes are narrow-width reels comprising a substrate and a *sealing* medium which can be dispensed and used to close and seal corrugated fibreboard cases, fibre drums, rigid boxes and folding cartons. Sealing tapes are also used by the manufacturers to make the side seam join on corrugated fibreboard cases and tape the corners of rigid boxes, thereby erecting or making-up corner-stayed boxes.

A traditional and commonly used substrate is hard-sized kraft paper, both unbleached (brown) and bleached (white). Where higher strength is required, the kraft is reinforced with glass fibre, and up to four progressively stronger specifications are typically available from some suppliers. Reel widths start at 24 mm, though 50 mm width tape would be a typical width to seal the flaps of an average-sized corrugated fibreboard case.

In the case of gummed tape, adhesion is achieved by coating the kraft paper with a modified starch adhesive; animal glue has largely been superseded. The adhesive is then dried and the reels are slit to size. In subsequent use the adhesive is automatically, and evenly, reactivated by water in a tape dispenser. Tape dispensers which can cut pre-set *lengths* for specific taping specifications are available.

The advantages of gummed paper tape are that it is permanent and provides evidence of tampering, it can be applied to a dusty pack surface without loss of adhesion, is not affected by extremes of heat and cold and does not deteriorate with time. Pressure-sensitive tapes, on the other hand, are used on all types of packaging from paper and paperboard to metal, glass and plastic containers. Pressure-sensitive adhesive can be applied to several types of substrate, including moisture-resistant kraft paper which has been coated on the other side with silicone to facilitate dispensing from the reel.

Heat-fix tapes are based on kraft paper where the adhesive is applied as a thermoplastic emulsion, dried and subsequently reactivated by heat when applied to the sealing surface with pressure. Sealing tapes used are plain, preprinted or printed at the point of application.

3.9 Paper cushioning

Paper can be converted into cushioning materials which are resilient and pliable. These materials can be moulded around products in boxes to restrain their movement and by filling voids in the box. They are shock absorbing and can prevent damage to contents in transit. They are produced on relatively simple machines at the point of packing from one or more reels of paper. The paper can be based on recovered fibre and the cushioning can be recycled once its packaging protection function is completed. The cushioning is produced by crimping, folding and twisting to form a variety of shapes which retain their structure in packaging and use (Fig. 3.27).

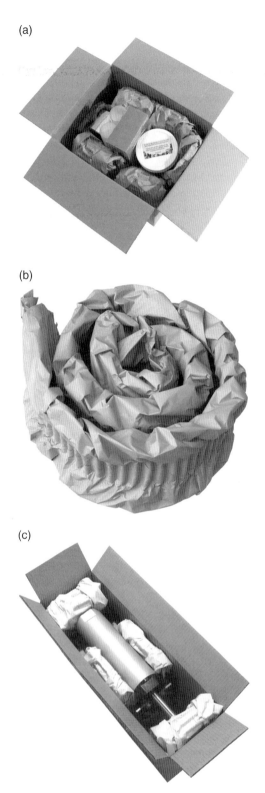

Figure 3.27 Examples of paper cushioning: (a) cushioning of items in box, (b) paper cushioning, and (c) product located in box for distribution. (Reproduced with permission from Easypack Ltd.)

References

Ahlstrom, 2009, *Biodegradable Solution for Infusion Products (Tea and coffee)*, Press Release, 18 May 2009, visit http://www.ahlstrom.com

Dow, 2005, Dow Chemical, visit http://www.dow.com/svr/prod/

European Plastics News, 2010, 1 November, visit http://www.europeanplasticsnews.com

FOPT, 1996, Walter Soroka, Packaging functions, in: Anne, E. and Henry, E. (eds), *The Fundamentals of Packaging Technology*, revised UK edition, The Institute of Packaging, Naperville, Illinois, pp. 30–31.

FPA (Flexible Packaging Association (US)), 2010, *Industry Information, Facts and Figures*, 18 June, visit www.flexpack.org

Inman, B., 2004, Private communication to M.J. Kirwan from Bill Inman, former Technical Manager, Henry Cooke Ltd, Cumbria, UK.

James, R., 1999, Packaging for sterile medical devices, Seminar, April, Association of British Healthcare Industries, London.

Merritt, J., 2002, *Medical Device Technology*, January/February.

Merritt, J., 2003, *Medical Device Technology*, March.

Paine, F.A., 2002, Prof. Frank A. Paine and Eric Corner, *Market Motivators – The Special Worlds of Packaging and Marketing*, CIM Publishing, Cookham, Berkshire, UK, pp. 161–164.

Paine, F.A. and Lockhart, H., 1996, *Packaging of Pharmaceuticals and Healthcare Products*, Blackie Academic and Professional, London, pp. 50–89, 120.

Websites

www.ahlstrom.com
www.amcor.com.
www.billerud.com
www.boschpackaging.com
www.dow.com
www.dupont.com
http://www.easypack.net
www.glatfelter.com
www.hayssensandiacre.com
www.interflexgroup.com
www.ima.it
www.mardenedwards.com
www.multivac.com
www.rovema.com
www.walki.com
www.bobstgroup.com/rotomec

4 Paper labels

Michael Fairley

Labels & Labelling Consultancy, Hertfordshire, UK

4.1 Introduction

The use of labels in a format that would be recognised today can be traced back to the early 1700s. At that time, labels were used mainly to identify products such as bales of cloth and medicines. Labelling of products for sale using printed labels had not evolved to any significant extent at that time, largely because few of the population could read or write and because there was no trade of mass produced and packaged goods.

All that began to change with the advent of education and mass literacy programmes in the 1800s, together with the evolution of the mass production of food and drink in bottles and cans – the need to put descriptive and brand information on products and to produce labels in volume quantities, now became a necessity.

By the later part of the nineteenth century, the introduction of the papermaking machine and the invention of lithography provided the means to produce economically long runs of the same quality (label) papers and to print them, in sheeted form, in colour. It was the advent of cost-effective colourful labels produced in this way that enabled manufacturers to take advantage of the opportunity to individually package and label their products themselves, rather than having them individually weighed and wrapped by the provision merchant. Product labelling, as we know it, had arrived.

Initially labels were printed onto plain paper and cut to a rectangular size by guillotine – even scissors or punches, if a special shape was required – before being applied with a wet-glue or gum. Early methods of labelling used gums and adhesives which were brushed on to the label and then the label applied by hand. These first slow labelling methods eventually evolved into full-blown applicator machines and systems which applied a wet-glue to the back of each label and then applied the label to the bottle or can. Wet-glue paper labelling in a form that would still be recognised today became a reality. Pre-gummed paper labels for application to paper and paperboard packaging were a later evolution, helping to speed up address labelling for the dispatch of transit packaging.

With these developments, the product manufacturer was able to use the label as a means of contact with the customer so as to promote a product or brand, add information regarding contents or usage and provide the name and address – even claim all kinds of (unproven) health, digestive, beauty or other benefits of the product. Indeed, it was the widespread growth of such claims that led governments to start introducing legislation to protect the consumer from misleading claims.

The dominance of the wet-glue-applied label for product labelling of glass bottles and cans, and gummed paper labels for addressing and shipping labelling of paperboard containers, continued until the early 1970s. At that time, wet-glue paper labelling (70% of the market) and gummed paper labelling (22%) were still the only significant labelling technologies.

Self-adhesive labels, invented in the mid-1930s, started to make an impact on the product decoration and labelling market in the late 1960s, initially as a means of applying price labels to products and then eventually evolving to become a high-added value method of labelling cosmetics, pharmaceuticals and shorter-run applications with both primary and secondary labels.

Since then, many factors have changed the world of labels:

- rapid growth and emergence of computer-generated and printed address labels
- bar code labels and variable information, text and graphics imprinting
- major explosion and rapid volume growth in the use of plastic containers
- relatively low growth in the use of glass bottles and cans
- new demands for anti-theft, tamper-evident and anti-counterfeit labels
- new requirements for promotional, booklet and leaflet labels.

More recently, there have been exciting innovations in smart, smart-active and smart-intelligent labels that can track and trace products, monitor temperature change, indicate food freshness, provide anti-microbial and anti-bacterial functions or provide guidance on best temperature for consumption.

The demand for shorter and shorter-run lengths, more variations and versions of labels, just-in-time (JIT) or on-demand production, personalisation and unique coding has seen rapid growth in the latest digital electrophotographic or ink-jet printing technologies which are already making up some 15% of annual new narrow-web label press installations – and having an impact on the label paper supplier.

Also making a major impact on label production and usage today are the global pressures on environment, sustainability and waste which, in turn, bring changes in label materials, recycling and recyclability, alternative non-tree papermaking fibres, non-polymer films and even new developments in linerless self-adhesive labels.

All of these innovations and developments have brought many changes in the use of labels during the 1980s and 1990s and, in particular, over the past 10 years since the millennium.

In addition, when one considers the changes which have taken place in consumer lifestyles, leisure activities – in computers, communications technology, the internet – global travel, legal requirements (consumer, health and safety, environmental legislation), the growth in label usage and the technologies used become even more dramatic. Changes in consumer lifestyle include:

- the need for more convenience in the home, leading to a decline in the purchasing of preserved food in cans and jars in favour of the purchasing of freezer packs and ovenable/microwaveable/reheatable meals and ingredients
- packaging of carry-home beers and soft drinks in cans rather than bottles

- regular eating out, resulting in a decline in home cooking and increase in packaging for the catering market
- leisure life that includes overseas travel, gardening, DIY, sports, healthy lifestyle and 'keep-fit' activities
- social networking internet sites and a growing interest in internet-active labels.

These changes have created needs for new types of label.

Pressures in the early to mid-1980s to find new ways of labelling very long runs of blow-moulded containers led to the development of in-mould labelling, whereby the label is placed in the mould prior to the forming of the container, thus becoming an integral part of the bottle. Originally only used for decorating blown plastic containers for hair care and under-the-sink products, in-mould labelling later extended to the labelling of injection-moulded biscuit containers and to tubs for soft spreads and margarines. More recently, in-mould labelling has been used for the labelling of thermoformed tubs, again for soft spreads.

Also in the 1980s came the development of plastic shrink-sleeve labelling where a continuous web of film is printed, formed into a tube and then, for application, cut to the appropriate length, placed over the container and then shrunk to fit. Having the advantage of 360° decoration, sleeve labelling has created new markets and applications for labels and also added the potential to extend the shrink capability to provide a tamper-evident cap seal. An alternative and more recent technology for 360° decoration came with the introduction of stretch-sleeve labelling in the 1990s.

Other new labelling technologies developed in the 1990s, primarily for the labelling of plastic bottles, include wrap-around film labelling, cut-and-stack film labelling, roll-on shrink-on (ROSO) labelling and spot-patch film labelling. Used for the decoration of soft drinks, carbonated beverages and some beers, these newer label technologies are achieving some of the highest growth in the label and end-user market.

With the rapid expansion of supermarkets and hypermarkets in the past 20–30 years has come the development of 'own brand' labels, successfully competing with and gaining market share from brand owners. Indeed, Walmart is now the world's largest private label 'own brand' company, with Aldi, Tesco, Royal Ahold and Meto also amongst the world's leading brands and with their own-labelled products, in more and more versions and variations.

This, coupled with faster and faster store throughput, and a significant increase in pre-packed fresh produce has brought about the need for shorter and shorter run lengths, a major requirement for price-weight labels, ever reducing lead times, JIT manufacturing practices, improved supply-chain management and demands for improved quality standards. All these trends and pressures have had an influence on the types of labels used, the printing process required, pre-press technology, label application and label usage.

Apart from labels, brand owners, packaging and marketing companies can also choose from:

- direct decoration by screen printing onto glass and plastic containers
- offset printing directly onto cans
- tamper and pad printing onto pots and tubs
- transfer decoration of glass bottles
- metallic foiling directly onto containers
- direct printing onto stand-up pouches and sachets.

These technologies compete with labels and may influence the continuing change between direct decoration and label solutions – all based on quality, performance, run length, image, cost, etc. Direct decoration of one kind or another makes up about 30% of the total product decoration and label market, often providing an attractive alternative to labels.

4.2 Types of labels

Since the 1970s, there has been a major shift in the range and variety of label technologies used in packaging by the end-user, moving from dominance by wet-glue and gummed paper label technology in the 1970s to dominance by self-adhesives from the mid-1990s onwards – at least in the sophisticated markets of Western Europe and North America. Today, self-adhesive labels make up close to 50% of all label usage in these markets, although with wet-glue still well in excess of 30%.

Gummed paper labels, so widely used in the 1960s and 1970s, are now just a few per cent of usage, largely superseded by self-adhesives. Put together, all the newer labelling technologies of the 1980s and 1990s, developed primarily for the labelling of plastic bottles and containers, are now around 18% of the total label market.

The other key shift in label usage has been a steady and continuous move to the usage of non-paper labels to meet the growing demand for new ways of labelling plastic bottles and containers. Driving forces behind this growth of filmic labels include:

- high annual growth of plastic bottles (compared with glass bottles and cans)
- the requirement for compatibility of the label material with that of the bottles due to waste and recycling issues
- demand for clear, no-label-look packaging
- the ability to provide white, coloured, silky, iridescent or pearlescent surfaces
- the need to provide moisture protective, chemical- or product-resistant labels for use in demanding applications
- the requirement for recycling of the plastic bottle and the plastic label together without first removing the label.

From almost total dominance by paper labels in the 1980s, the market has evolved rapidly to a point now where non-paper film labels make up close to 25% of all label requirements and have been growing at a rate four or five times faster than paper labels. Mention is therefore made of film labelling technologies in these pages.

When looking at label types today, they all fall into one of two key categories: those that are printed on paper or synthetic substrates (film, metallic foil, metallised paper, etc.) and to which an adhesive, glue or gum is applied at the point of application and those substrates which already have the adhesive or gum on them before they are printed. This adhesive or gum is then activated by pressure, moisture or heat at the point of application. These two groupings and the range and variety of label types available can be seen in Fig. 4.1.

A brief description of the main label types, along with some of the typical properties and applications of each type of label, is as follows.

4.2.1 Glue-applied paper labels

The application of paper labels to a glass bottle or can using a wet-glue was one of the earliest methods of labelling, and for many years it was the dominant label technology. Even

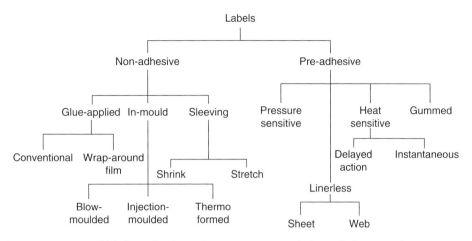

Figure 4.1 Types of labels used in the early 2000s. (Source: Labels & Labelling Consultancy.)

today, in spite of the rapid growth of self-adhesive and other label technologies, glue-applied labels are still, marginally, the main method of volume labelling bottles and cans with paper labels. Labels may be individual face, neck and back labels for application to beer, spirit or wine bottles, or wrap-around labels used extensively on canned foods and soft drinks.

Glue-applied paper labels are printed on plain paper, metallised paper or paper-laminated materials in either sheet-fed offset (most common) or web-fed gravure (some flexo) presses. They are often also varnished, coated or lacquered to provide surface protection during labelling, handling and distribution, before being guillotine-cut to a rectangular or square size in stacks of 500 or 1000 labels, or punched into special cut-out shapes, again in stacks, ready for placing in a hopper in the label-application machine.

4.2.1.1 Glue-applied paper label substrates

They include:

- one-side-coated grades and uncoated white or bleached, kraft paper 55–100 g/m^2
- metallised paper
- paper/aluminium foil laminates.

All of these have a similar base. For the one-side-coated grade, a special printable coating is applied. Foil-laminated paper is a more expensive and luxurious glue-applied label substrate and is made up of a thin aluminium foil laminated to a paper backing. Metallised paper substrates are one-side-coated papers onto which metallic aluminium has been deposited under high vacuum.

Selection of the one-side-coated, laminated or metallised paper label substrate for glue-applied labels is determined by the effects required, the performance of the finished label, the nature of the labelled product and on the label-application method. Factors such as surface smoothness, opacity, stiffness, porosity, water absorbency, wet strength, grain direction and degree of curl all need to be considered. Cost is also an important factor to be considered. Corrosion and/or mould inhibitors may be required for some applications.

The growing demand for more environmentally friendly paper label materials has also seen paper mills introduce grades containing recycled industrial waste paper and most recently, even post-consumer waste.

4.2.1.2 Label application

It is undertaken on a machine in which either a wet-glue or hot-melt adhesive is brushed or rolled onto the back of each label just before it is applied to the container, enabling the most suitable adhesive formulation for the specific application to be selected. The type of adhesive application can also be selected from options such as skip, pattern or stripe, depending on adhesion, application speed or drying speed.

The key uses for glue-applied labels are in the high-speed, high-volume, low-changeover labelling of drinks bottles and for canned foods – both human and pet food – where application speeds of up to 60 000–80 000 or more containers per hour are available.

4.2.2 Pressure-sensitive labels

Pressure-sensitive labels were originally developed in the USA in 1935 by Stan Avery and, since the 1960s, have been gaining widespread acceptance and usage to become the dominant labelling technology used today.

More commonly referred to, outside the labelling industry, as self-adhesive labels, the pressure-sensitive label offers the package-labelling market a wider range of face materials and adhesives than any other method of labelling, as well as the greatest range of in-line printing and converting options. With an adhesive which is already active and ready for immediate application, it is not surprising that they have rapidly gained popularity for a diverse range of primary and secondary labelling requirements.

4.2.2.1 Self-adhesive label substrates

These are more diverse than for any other method of labelling, using paper and board, films, synthetic papers, foils and laminates, as well as a whole range of surface treatments and top coatings to meet specific applications. The most commonly used label facestocks include:

- uncoated paper and paperboard
- on-machine coated, double-coated, high-gloss-coated and cast-coated paper and paperboard
- polypropylene (PP), orientated polypropylene (OPP) and bi-axially orientated polypropylene (BOPP)
- polyester (PET or PETE)
- polyethylene (PE) and high-density polyethylene (HDPE)
- metallic foil
- metallised paper and paperboard
- metallised film
- polyvinyl chloride (PVC)
- synthetic paper
- acetate.

Label thicknesses may vary from around 40–50 μm up to 80, 90, 100 or more microns depending on requirement and application. More recent innovations are working towards developing and using thin self-adhesive label films down to 30 or even 20 μm, which may require changes in printing press web handling and tension controls, die-cutting technology and modified label applicators that can handle such thin materials.

The release-backing paper or liner used for self-adhesive labels may be super-calendered unbleached kraft or glassine, coated with silicone or fluoropolymer. Filmic liners are used for some applications and for clear-on-clear films. Grammages can be as low as 50 g/m^2 or even lower. Recent developments in linerless self-adhesive labels using micro-perforated labels and new types of label-applicator heads are already beginning to achieve some success.

The development of a whole range of self-adhesive solutions for the variable image printing (VIP) of batch and date codes, bar codes and price-weight information using thermal overprinting equipment has further extended self-adhesive label usage into the fast-growing field of logistics, distribution, warehousing and shipping applications, as well as into retail catch-weigh labelling – solutions which no other form of labelling can readily offer. Label facestocks for VIP labels have special coatings which are thermally sensitive or surface smooth for overprinting. VIP film substrates often have special top coatings to make them print more like paper with consistent results.

With the continuing evolution and installation of digital label printing presses, predominately into the self-adhesive label printing sector, have come demands for new generations of self-adhesive label paper grades suitable for electrophotographic or ink-jet printing. Whilst dry toner and ink-jet claim to be able to print on most current labelstocks, some papers will nevertheless still require a top or primer coating. The HP Indigo liquid toner process on the other hand requires all substrates to have a special top/primer coating. Papermakers have therefore been required to develop new materials solutions for digital presses, then qualify and certify those materials for each specific digital printing process.

Although self-adhesive labels are more expensive than wet-glue, they are simple, clean and easy to apply using hand-held, semi-automatic or high-speed applicator systems that have no label change parts and the ability to apply almost any kind of label-face material at up to 300 or more containers a minute. Speeds up to 60 000 containers (bottles) per hour can be achieved with, say, a multiple (6-station) head applicator.

Almost all self-adhesive labels are made up of a sandwich construction (see Fig. 4.2) in which there is a face material (the label), a sticky pressure-sensitive adhesive and a siliconised backing paper or liner. The face of the material is printed (usually on narrow-web presses in web widths up to 300 or 400 mm wide), die-cut on the face and the matrix waste removed, slit into individual label widths and re-wound for transportation to the end-user's labelling line.

4.2.2.2 Self-adhesive label application

It is undertaken on machines in which the label is dispensed from the liner and applied to the bottle, container or pack. The waste liner is disposed of. Consequently, pressure-sensitive label materials are higher in cost than those of unsupported labels. However, the labels are only applied where needed and not as a complete wrap-around or as a whole sleeve.

The direct transfer of labels from a reel permits exact placement of labels in front, back or neck positions, top or bottom placements if required, into recesses, around corners, etc., and can be easily changed in any combination depending on the number of application

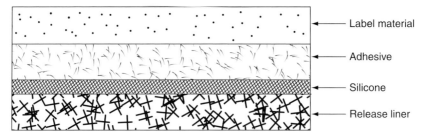

Figure 4.2 Self-adhesive label. (Source: Labels & Labelling Consultancy.)

heads on the machine. Changeover of labels from one design or one bottle type to another takes about 15 min or so. Any combination of one, two or three labels (on a 3-head applicator) can be changed at any time, thus providing extreme flexibility.

Self-adhesive labels are well regarded by industries such as cosmetics, toiletries, healthcare and beauty, pharmaceutical, foods and industrial products because of the range and variety of decoration possibilities available and by production departments due to their potential for precise application. They are also finding increasing interest in the added-value drinks markets for wines, spirits, iced beers, speciality beers, etc., particularly for iced or chill-cabinet drinks where the exclusive bottle dress of the pressure sensitives is a key attraction.

4.2.2.3 Linerless self-adhesive labels

As mentioned earlier, it should be noted that several developments have been made in the past 20–25 years – and are still continuing today – to produce linerless self-adhesive labels, the first of these being the Monoweb system developed in the mid-1980s. In this system, the web was first printed, then silicone coated on the face before finally being adhesive coated on the reverse. The web of self-adhesive labels was then wound on itself without the need for a separate release liner. A key limiting factor to this technology was that the labels had to be die-cut to shape on the label-applicator line.

Linerless self-adhesive labels for thermal printing are also produced by some label converters in-house. This can be done on a label converting or press line using coating heads to apply again a silicone face coating and an adhesive reverse coating and then winding the web on itself. Slit into narrow single label widths, the labels are then overprinted electronically, cut to length and dispensed.

More recently, developments have been taking place in which self-adhesive labelstock is coated onto both sides of a single- or double-sided release liner web. Printed and die-cut in one pass, front and back labels are carried on one backing web. A special label applicator has been developed to apply these two-sided labels. The system saves press time, reduces liner waste at the application stage and ensures that there are equal numbers of front and back labels on the reel, thereby reducing reel changes and wastage of labels.

The latest development in linerless technology has been coming from Catchpoint Ltd, who have pioneered and patented the production and application of linerless self-adhesive labels using unique micro-perforations. Working closely with founding licensee WS Packaging Inc. and applicator partner ILTI S.r.l., linerless labelling speeds and accuracy levels have now been shown to meet the demanding targets required in FMCG markets. This exciting new process is expected to facilitate and contribute to environmental targets without compromising the speed, accuracy and efficiency of FMCG production lines.

4.2.3 In-mould labels

The technology of in-mould labelling has been available for over 20 years and involves placing pre-printed rectangular- or square-shaped labels within a mould immediately prior to blowing, injection moulding or thermoforming of plastic into it to form a container, for example bottle, pot or tub. In this way, the label becomes an integral part of the finished item, with no subsequent requirement for label-application equipment or filling on the packaging line.

4.2.3.1 In-mould label substrates

Initially, the process involved placing paper labels coated with a heat-seal back layer into the mould before blow moulding. This layer must fuse to the bottle during the blow-moulding process. More recently, for recycling and performance considerations, synthetic paper materials such as Polyart or Synteape have become more common as labels for under-the-sink labelled products, as well as specially developed OPP films (which fuse directly to the container and eliminate any tendency to produce an orange-peel effect) which are used for the labelling of injection-moulded tubs for soft spreads. All in-mould label substrates must have good lay-flat properties for trouble-free feeding from the moulding-machine hoppers.

In Europe, some 80% of in-mould labelling is with injection-moulded or thermoformed tubs and lids for dairy foods, such as soft spreads, margarines, cheeses, sauces and ice cream, with 20% used with blow-moulded containers for under-the-sink products, household chemicals and industrial laundry products, personal and hair-care products and some juices. In the USA, it is in-mould labelling of blow-moulded containers which dominates the market.

In-mould labels are mainly printed by sheet-fed offset or by web-fed gravure. Narrow to mid-web flexography, letterpress and offset are also used by some printers – with off-line die-cutting to shape. Inks are critical to the process with labels protected on the surface with a ultraviolet (UV) or electron beam (EB) curable top coat. Die-cutting is also critical, particularly when labels printed in sheets are stacked and rammed through a tunnel, emerging cut to size. Edge welding may be a consequence of this process.

4.2.3.2 In-mould label application

Because of the high cost of the basic moulding equipment and moulds, plus the need to modify these to be able to insert and position labels accurately into the mould, in-mould labelling has had a somewhat limited acceptance in the market place, with little more than a 2–3% market share. Relatively long runs have traditionally been needed to make the process viable, particularly difficult when short-run, JIT label manufacturing is being increasingly demanded by the label user.

Against this, the requirement for both recyclable and/or pre-labelled containers, the potential of higher packaging line speeds and in-case filling, improved appearance and better squeeze resistance, spurred on by environmental and economic issues together with more economic methods of producing and inserting in-mould labels are all expected to aid the growth of the process in the coming years.

A recent development of in-mould labelling is with the use of single portion thermoformed pots for yoghurts and cream-based desserts where the label is applied in the mould – yet the labelstock is on a reel with a conventional pressure-sensitive adhesive. The machine used for this application forms the pot and fills and seals 12 or more pots per cycle.

4.2.4 Plastic shrink-sleeve labels

Plastic shrink labelling is not paper-based, but in order to give an overall view, it has been decided to discuss plastic shrink labelling, especially as it is a serious competitor to paper-based labelling.

Shrink sleeving was originally developed in the early 1970s as a method of combining two or more products together for promotional or marketing purposes. Adding printing to the shrink film, twin pack or multi-pack promotions soon followed, with the evolution of the technology into a high-quality 360° method of decoration for unit packaging/labelling developing in the 1980s. Today, shrink-sleeve labelling is quite widely used for the decoration of beverages, food, home and bodycare products, dairy produce and for a variety of special projects. It has achieved particular success by enabling glass bottles to be light-weighted, with the performance and strength then re-enhanced by a full-body shrink sleeve.

Sleeves are usually reverse gravure printed with photographic images, graphic design, text, colour and special finishes, and then formed into a tube – which is collapsed for re-winding, handling and shipping. Flexo printing is also used. Origination of the design and images for shrink sleeving is a special technique as the image for printing needs to be distorted, even differentially distorted, so that the final shrink image on the bottle is correct.

4.2.4.1 Shrink-sleeve label films

These were initially made from pre-stretched PVC. Now the market also uses OPP, PET and low-density polyethylene (LDPE) in a variety of surface finishes. Films are generally made using either a bubble extrusion or calendering process. Thicknesses of films range from 35 to 90 µm, combining various shrinkage and thickness variations. Additives allow UV blocking or the inclusion of UV fluorescence for security and reconciliation, as well as to provide anti-static or fire-retardant features.

Besides primary decoration, shrink sleeves offer good tamper-evidence features, using tear strips and perforations across and along the sleeve. Integrated holograms can add an additional anti-counterfeiting role.

4.2.4.2 Shrink-sleeve label applications

Compared with many other bottle-labelling technologies, shrink sleeving does tend to have somewhat higher costs. Relatively thick filmic materials are used; the bottle is completely covered by the sleeve and the sleeve has to be converted into a tube after being printed. As such, raw material costs tend to be high. Furthermore, print costs are high; even if only five colours are required on the label, it is still necessary to print an opaque white base. Gravure cylinder costs are also quite high. This gives a higher overall print cost for shrink-sleeve labels before capital costs and investment are considered.

Against this, the technology is currently achieving good success for high quality, high added-value all-round bottle decoration and, when used with glass bottles that have been light-weighted, the strength of the shrink sleeve compensates for the lower strength provided by the glass such that the performance of the pack is maintained. The cost of the sleeve can be offset against the savings resulting from the lighter-weight glass bottle.

4.2.5 Stretch-sleeve labels

Stretch-sleeve labelling is a more recent development of sleeve technology which, instead of shrinking the film tube to fit over the bottle, uses a stretchable film which is cut to size and then opened up to slide over the bottle. It then collapses elastically to fit the shape of the container.

Stretch sleeves provide advantages when applied to PET containers in that flexible sleeves compensate for the expansion of PET containers that occurs after filling. Returnable bottles can be easily de-sleeved before entering the bottle washer, since, unlike the more traditional types of label, the sleeve is not glued to the bottle.

4.2.5.1 Stretch-sleeve label films

These are LDPE materials. Like the other unsupported film labelling solutions – reel-fed wrap-around, cut-and-stack wrap-around, shrink sleeving – stretch-sleeve films are primarily printed with gravure or flexo technology. However, unlike shrink sleeving, there is no need to use image distortion techniques in the stretch-sleeve process. Again, films can be reverse-side printed to minimise ink scuffing or rubbing.

4.2.5.2 Stretch-sleeve label application

Because of the nature of the stretch-sleeve labelling process, it can only be used for body decoration of straight-sided containers. It cannot be used with tapering necks or shaped bottles – although it can be combined in an application system with, say, a pressure-sensitive neck label. No flexibility in terms of front, back or neck labelling options exists. However, stretch sleeving can be carried out at higher speeds than shrink sleeving – although it is far more limited in applications. Costs are lower than shrink sleeving.

Today, stretch sleeving is primarily used for large-size bottles for carbonated soft drinks (CSDs). The expansion of carbon dioxide (CO_2) inside the bottle is accommodated by the stretch label. Stretch sleeving is generally seen as more relevant where easy removal of body labels on returnable bottles is required. If returnability is a key criterion, then the option of stretch sleeving may have higher priority.

Roll-fed stretch sleeves also work well with straight-sided single-serve bottles and are often used with PET bottles in pints or other sizes.

Despite its limited applications, an advantage of stretch sleeving is that the elasticity of the material obviates the need for heat-shrink tunnels in the production line.

4.2.6 Wrap-around film labels

A relatively new variant of wrap-around glue-applied paper labelling, wrap-around film labelling, was developed and introduced in the past 10 years to allow the labelling of PET bottles and beverage cans with reel-fed, tear-proof plastic film labels. The labels can be reverse-side printed to provide a non-scuff, scratch-resistant print quality.

4.2.6.1 Wrap-around label films

These are plain coated BOPP films. Plain uncoated films come in thicknesses from 19 to 50 µm, whilst coated films range from 30 to 70 µm. Super-white BOPP films are made with a cavitated core and specially developed outer layers which provide a bright-white high-gloss finish. Metallised BOPP films have a vacuum-deposited aluminium layer on one side and an acrylic coating on the other.

4.2.6.2 Wrap-around film label application

Wrap-around film labelling machines incorporate a label feed system with computer-cut control, rotary cutting, gluing station and a mechanically assisted vacuum gripper for transfer of labels to the container. Large capacity reels minimise the operator workload.

In operation, glue is applied onto the leading edge of the film via mechanically driven rollers. The film is cut to length and wrapped around the bottle or can and then glued on the trailing (overlap) edge to form a complete wrap. Glass, plastic and metal containers – square, cylindrical or oval – can be handled. Label-application speeds up to 800 bottles a minute are achievable when using 2-applicator stations.

A new variant allows wrap-around labelling of returnable PET bottles with very thin plastic film labels. A small amount of a casein-based adhesive is applied to the bottle for label pickup. This cold glue crystallises after curing, which facilitates subsequent label removal from the returned PET bottle. The lap overlap is sealed with hot-melt adhesive. Label removal is carried out using a de-labelling machine, which cuts the label vertically with a knife. The label can then be flushed from the bottle using a high-pressure water jet.

The wrap-around reel-fed labelling system is primarily designed for the wrap-around body labelling of straight-sided bottles or cans. It is possible to apply a wrap-around label to a slightly contoured bottle or jar but it is not possible, using standard equipment, to apply a neck label to a bottle with a tapered neck. However, this can be achieved with a combination applicator line that can apply a wrap-around body label and a separate wet-glue or pressure-sensitive neck label on the same line.

As far as cans are concerned, wrap-around film labelling provides a number of advantages over the use of pre-printed cans, including reduced can inventory, reduced lead times and greater flexibility for sales/marketing activities (special offers, proof-of-purchase promotions, etc.).

With the addition of a second carousel in the labelling system, it is also possible to positively shrink the label film to fit contoured cans or bottles. Label technology for wrap-around or roll-on film from reels, followed by a light shrink, may also be referred to as ROSO labelling.

4.2.7 Other labelling techniques

A fairly recent development is the use of a printed paperboard wrap-around label on a tapered plastic pot for single portion desserts. The label is held in place by the plastic rim of the pot, though the label has a glued side seam. It enables easy separation of the plastic pot from the label for recycling purposes.

4.3 Label adhesives

The vast majority of labels require an adhesive to bond the label to the container or to some other surface. Being between the label material and the container or product surface, the adhesive must be compatible with both. If all the variations of label substrate (paper,

paper-backed foils, metallised materials and plastics) and container/surface type (metal cans, glass bottles, plastic bottles, wrapping films, corrugated cases, etc.) are incorporated into the adhesive requirements, the challenges for adhesive manufacturers are quite formidable.

In addition, there are many different types of plastic containers (PE, PP, PET, etc.), different metal container types – some with special coatings and varnishes – and glass bottles that also have a variety of coatings to increase surface strength and minimise breakage. There are also certain performance criteria that have to be incorporated into adhesives: cold-temperature application, wet-bottle application, food contact, removability, immersion in water, repulpability, recyclability and more. Consequently, adhesives are custom designed to meet all the required label types (wet-glue, self-adhesive, gummed), the required surfaces to be bonded and the necessary performance characteristics.

4.3.1 Adhesive types

To meet the demands of different label types, different surfaces to be bonded and different user-performance characteristics, labelling technology uses four main types of adhesive: hot-melt, water-based, solvent-based and curable adhesive systems.

4.3.1.1 Hot-melt adhesives

These are thermoplastic materials with 100% solids that are heated to temperatures above their melting point and applied to substrates in the molten state. Unlike water-based or solvent-based adhesives, hot-melt adhesives do not require drying. They have a high initial tack and set as quickly as they can cool down to their solidification temperature, which makes them ideal for picking up labels in high-speed labelling lines. They are popular because of their high setting speed and because they form a virtually invisible line on glass, metal, PET and other plastic containers.

Key criteria in the use of hot-melt adhesives are temperature – which controls viscosity – and adhesive film thickness, which affects speed of setting, tack and open time. The thicker the adhesive applied, the longer it will take to set.

4.3.1.2 Water-based adhesives

They have had a dominant place in the label industry for many years and are used in wet-glue, self-adhesive and gummed paper formulations. They are made up of materials or compounds that can be dissolved or dispersed in water to become tacky and form a bond and they dry by losing water through evaporation or by penetration into the label substrate. At least one surface must be absorbent or porous to form a strong bond.

Water-based adhesives are available in a variety of chemistries and compositions and are categorised as either natural or synthetic polymers as follows:

Synthetic polymers	Natural polymers
Polyvinyl acetate (PVA)	Casein
Acrylics	Dextrine/starch
Polychloroprene	Natural rubber latex
Polyurethane dispersions	

Key points relating to some of these water-based adhesives are as follows:

- Dextrine/starch-based adhesives were one of the first ready-to-use label adhesives. Made up from water-soluble natural polymers, they are used either cold or warm, depending on application. Resin/dextrine adhesives are used for the labelling of some types of coated glass and plastic containers.
- Casein adhesives are some of the best-known adhesive types for wet-glue labelling. Fast drying, they adhere well to cold, wet bottles, yet can be easily removed in caustic cleaning solutions when used for labelling returnable bottles.
- Water-based acrylic adhesives are usually used in self-adhesive applications where, unlike organic-solvent-based adhesives, their flame-spread resistance is an advantage. They are used in many UL applications. (Underwriters Laboratories (UL) is an independent not-for-profit product testing and certification organisation which helps manufacturers to achieve global product acceptance.)
- Polychloroprene – used for some self-adhesive applications – was developed as a substitute for natural rubber and offers a unique combination of adhesive properties:
    - outstanding toughness
    - chemical resistance
    - weathering resistance
    - heat resistance
    - oil and chemical resistance.

4.3.1.3 Solvent-based adhesives

These are noted for their fast bond strength development, good heat resistance, adhesion to a wide range of substrates and tolerance to a wide variety of production conditions – including low temperatures and high humidity. However, organic-solvent-based pressure-sensitive formulations have been widely displaced by water-based and hot-melt systems for economic as well as ecological reasons. Solvent recovery and/or incineration is now essential to meet clean air legislation requirements. This is expensive and can only be justified for large output operations.

4.3.1.4 Curable adhesives

Those that use UV or electron-beam systems to 'cure' or set the adhesive are a relatively new development and are used in some speciality-tape applications and have also found some (limited) application in the self-adhesive label sector. Curable adhesives are cross-linked during the curing process.

4.3.2 Label adhesive performance

Label adhesives are a key element in the label-application process and have to perform as required on the labelling line and throughout all handling, shipping and end-use stages.

Important features of a label adhesive are as follows:

- It must 'wet' the label surface, i.e. flow out onto the label substrate to which it is applied. It must not reticulate.
- It must have a good 'initial tack' as the label is applied to the container or product surface.

- The initial bond must be maintained, as applied, until the adhesive is fully set, and there must be no 'edge lifting' or blistering.
- It must meet any required end-usage performance criteria – food contact, chemical or product resistance, water resistance, heat or cold storage, etc.

With the wide range and variety of end-use and performance requirements of self-adhesive labels, there are, of necessity, also a wide range of special adhesive performance needs, which include:

- *Semi-permanent adhesives* that remain removable and re-positionable for some time after application before fully becoming permanent, often used for labelling high-value items.
- *Removable adhesives* that have low tack and peel values to avoid damage to labels or goods when the labels are removed. The removal must be accomplished without leaving any adhesive residue behind.
- *Filmic adhesive grades* that can be permanent, semi-permanent or removable and are used specifically with synthetic film labelstocks.
- *Freezer-grade adhesives* that are permanent adhesives and stand up to the full range of temperature extremes found in cold storage and freezer chest applications.
- *Repulpable adhesives* that are specifically designed so as not to interfere with the paper repulping process used during the recycling of paper materials. A major requirement of such adhesives is that they must not produce 'stickies' in the pulp-filtration process.
- *Permanent adhesives* that need to bond quickly to the surface of the paper and paperboard to which they are applied and also have an instant fibre-tearing bond. Permanent adhesives are also used with fragile substrates to provide tamper-evident label properties.
- *UV-curable adhesives* that have been developed to stand up to the toughest challenges and provide outstanding heat, plasticiser and chemical resistance.

As far as wet-glue label adhesives are concerned, there are three main adhesive groups:

1. *Glass-container label adhesives* are water-based casein, non-casein, synthetic or resin-based adhesives which offer high performance, superior water resistance, moisture and ice resistance throughout the label application, bottle line handling and conveying, palletising, shipping and end-use stages. If labels need to be removed after use, then a non-water-resistant adhesive is used.
2. *Rigid plastic-container label adhesives* are designed to provide good wet tack and adhesion when labelling at high speeds.
3. *Metal-container label adhesives* are hot-melt adhesives developed specially for the labelling of cans and offer hot pickup for roll-through labelling equipment and good hot-melt adhesive performance in rotary labelling equipment.

4.4 Factors in the selection of labels

Each label technology or label type has its own advantages and disadvantages in the provision of label solutions for different applications. Factors to be considered when selecting a label type include:

- cost of label material (non-adhesive or pre-adhesive, opaque or clear, metallised, etc.)
- cost of printing and converting (number of colours, in-line options, sheet or web printed)
- visual appearance and image (paper, film, design capabilities, aesthetic appeal, premium quality, tactile 'feel', no-label look)
- durability (resistance to scuffing, scratching or image deterioration)
- production flexibility (ease and speed of label line changeover)
- environmental considerations (recyclability, returnability, removability, waste disposal)
- volumes required (short, medium or long runs)
- speed of application (high speeds required?)
- capital investment of label line equipment (wet-glue high; self-adhesive lower)
- running cost/hourly rate of label line (operator and machine)
- down-time (set-up, clean-up, removal of misapplied labels)
- performance needs of labels (chemical or water resistance, labels for autoclaving or sterilising, wet strength, high or low temperature usage or application)
- information needs on label (reverse-side printed, leaflet or booklet label)
- security features (tamper-evidence, anti-theft, brand protection, hologram).

Rather than looking at any one or more criteria in isolation when selecting a label type, it is more usual today to look at the chosen solution or possible solutions in terms of total applied label cost. This is the end-of-line cost of labelling a bottle, jar, can or other type of pack. It comprises both the cost of the application technology, including investment, to achieve the desired application speed and the cost of the materials used. The materials used mainly comprise the label which must meet the needs of appearance and performance at every stage from application to end-use. The label cost will depend on the cost of the plain label material together with the costs of printing and conversion. In many cases, it will be found that there is more than one labelling solution for any one particular application.

4.5 Nature and function of labels

Many packages and containers carry more than one label. Some unit containers have two, three or more labels, each with its own specific brand or informational purpose whilst transit and distribution packaging may carry more functional types of labels for addressing, tracking and tracing or security purposes. Some of the key roles and functions of labels are set out later.

4.5.1 Primary labels

These are designed to carry the product's brand name or image identification that will attract attention on the retail shelves and appeal to prospective buyers. The labels usually promote a specific logo or brand name, incorporate special brand or house colours and may carry a picture, drawing or representation (which generally may not hide, obscure or interrupt other matter appearing on the label). For some applications, the primary face label may be reverse-side printed with secondary information. Used with clear liquids and clear bottles (glass or plastics), the information is read from the back of the container, through the container and contents, thus eliminating a separate secondary label.

4.5.2 Secondary labels

Secondary labels are usually smaller and are used to carry information such as a list of ingredients, health or safety requirements, nutritional details, instructions for use, European Article Numbering (EAN) Code, warnings, manufacturer or supplier name and address or registered office of the manufacturer or packer, prices, volume, weight or quantity and possibly promotional or special offer deals. If appropriate, the label may also contain any special storage conditions or conditions of use, and particulars of the place or origin.

Information contained on primary and secondary labels is normally required to be clear, legible and indelible, readily discernible and easy to understand, and is likely to have to conform to one or more of the regulations, standards or legislations that relate to labels and to labelling.

With one-piece wrap-around can and bottle labels, the primary and secondary elements are all incorporated into one label, rather than being two separate labels.

4.5.3 Logistics labels

Whilst primary and secondary labels are found on unit packs – the individual bottles, cans or packs – there is a need for labels on transit packages that are used to distribute goods and track and trace their movement in the supply chain to ensure they reach the correct delivery address. These logistics labels are frequently printed with computer-generated graphics, variable text and unique numerical or bar code symbols using thermal, laser or ink-jet imprintable labels.

Logistics, bar code and variable information printed labels are predominantly printed in one colour (black) only – often during the transit packaging, palletising or warehouse/distribution chain stages – and applied to cartons, trays, boxes, cases and pallet loads prior to shipment. Print-and-apply label equipment is commonly used for these applications, with the variable information entered through a keyboard or keypad.

More recently, major developments have been taking place in what are called smart, intelligent or chip labels which are finding applications in the logistics chain for identification, traceability, track-and-trace and smart logistics applications. Many of these new label solutions are based on smart label radio frequency identification (RFID) technology where the label/tag costs are now well below one US dollar. Most smart labels are in the form of very thin labels or laminates to get the price down and to make them suitable for targeted applications, which are, in the main, new ones where the label must usually be disposable.

Smart (RFID) labels in the logistics chain are used to:

- implement a step change into the supply-chain operation
- provide full visibility of the stock in the supply chain
- automate the processes associated with the management of stock.

Whilst RFID smart labels were forecast to achieve rapid label growth in the label industry, actual growth has not matched up to initial expectations with lower than anticipated take-up in key label user sectors. Alternative track-and-trace technology solutions, such as data matrix codes, have gained ground in label usage, whilst RFID has found new markets in the field of travel and transport ticketing, etc.

4.5.4 Special application or purpose labels

In recent years, an increasing need has arisen for additional product or pack labels that are used to provide unique brand protection, and the prevention of retail theft, tamper-evident or promotional solutions.

The range and variety of 'security' labels and solutions available to brand owners and end-users now includes:

- security papers and films for labels with special fibres, coatings, planchettes, threads, watermarks, etc.
- special security inks for labels from which the print cannot be erased or which may have to be fluorescent, indelible, infrared, luminescent, magnetic, photo-chromic, water-fugitive, thermo-chromic or optically variable
- security-label design software systems incorporating guilloches, fractals, special rasters, microtext, curved distortion to scrambled indicia, anti-colour photocopy features from special coated papers, void materials, reactive imaging, etc.
- DNA taggants and biocode markers on labels, used with test kits for instant authentication
- optically variable devices (OVDs) ranging through chromograms, colourgrams, micrograms, destruct foils, hologram materials, micro-embossed films, etc.

Note: Guilloches are printed security lines; the layout of the intersections and geometry is unique. Copying is inhibited by the layout arrangement of thin lines, rainbow print and the exact colour calibration. Indicia are distinct marks, signs or symbols – they are especially relevant to corporate or brand identity, and they are also used for postal-address identification.

The choice of security device(s) to be incorporated into brand protection, anti-counterfeiting or authentication labels will depend on the level of security, cost, process, application method and effectiveness. Best results can be achieved when security features are built in from the label design stage and by combining more than one security technology into the label to create a multi-layer solution that is difficult to replicate and counterfeit.

Such special labels may require purpose-built presses or converting lines incorporating silicone or adhesive pattern printing and folding units as well as special units to combine the label with the product, by either application or insertion.

Leaflet and book labels with multiple pages are used in the labelling of chemicals or agrochemicals, for some DIY products and on computer software packs. Variations may also be found in promotional applications. Tactile warning labels for packaging of products for the blind are another special type of label.

Form and label combinations – where one or more self-adhesive labels are found on a form – are used in distribution applications. One pass through a computer printer provides both the paperwork and labels that can be peeled off for pack/distribution labelling, order picking or job/production identification purposes.

Other specific label types include labels used to produce a collar or neck label. These may be pre-formed, slipped on and then locked, rather than glued in place. Tie on labels, tags or swing tickets may be used in the packaging/labelling of luxury items.

4.5.5 Smart, smart-active and smart-intelligent labels

Smart, smart-active and smart-intelligent labels today go way beyond conventional label applications so as to provide a whole range of clever label solutions that can be effectively

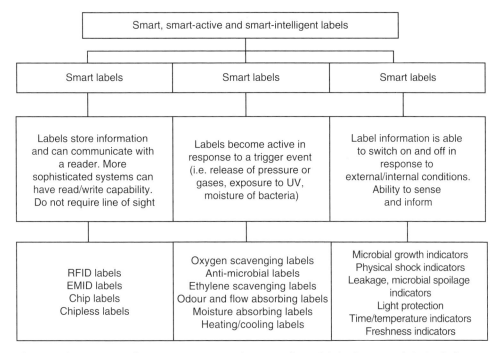

Figure 4.3 Key types of smart, smart-active and smart-intelligent labels. (Source: Labels & Labelling Consultancy.)

used in the manufacturing, distribution, sales, end-use or after-use chain and provide functions other than the standard tasks of protection, display or information carrying.

Many of these new intelligent/smart solutions may be automatically read – some remotely – some can be activated in response to a trigger event, some can be switched on and off in response to certain conditions, some can store data for subsequent downloading and some have the capability to have information added to them.

One of the key ways of assessing intelligent/smart labels is to categorise them by their action or function, as can be seen in Fig. 4.3. Even this chart does not include developments in contact or near contact smart reading technologies incorporating data matrix codes, 2D codes, quick response (QR) codes, optical or magnetic codes nor the reactive smart authentication technologies such as biocodes.

Intelligent and smart label solutions now cover a vast array of technologies that enhance pack functionality, product branding and benefit the supply chain and consumer. These include labels and packs which can indicate time or temperature, provide evidence of correct sterilisation or autoclaving, indicate freshness, detect microbial growth, absorb gases, oxygen or moisture, provide evidence of damage or shock, automatically route or track goods, deter counterfeiting, confirm authentication, reduce retail theft or provide evidence of tampering. The continuing evolution of nano-coatings, nano-taggants and nano-codes are expected to rapidly expand the range of intelligent/smart solutions available to the label converter and label user.

Adopting these smart solutions is a complex balance between the technologies that are available, the performance and usage benefits of using them and the cost factors involved. New enabling technologies and consumer demands are now driving 'smart/intelligent'

Figure 4.4 Types of tamper-evident security labels. (Source: Labels & Labelling Consultancy.)

technology even further forward. Advances are enhancing products in new and exciting ways, even becoming part of the product itself. Smart/intelligent technologies are a key to this evolution. Their potential to increase brand and product value is enormous.

For the future, labels and packaging that incorporate a disposable device or feature that interacts with an individual, a process, a personal computer, microwave or electronic shopping list are already coming to the market. These include QR codes and Snap Tags that turn brand logos into interactive marketing tools.

Such devices will have a revolutionary impact on the world of packaging and the role that it performs in a fast-changing product and consumer environment.

4.5.6 Functional labels

On-occasion labels may also be used during a packaging, processing or end-usage operation, for example to indicate successful sterilisation or autoclaving by colour change or provide evidence of significant adverse temperature change. Others may be used to provide a tamper-evident or security seal (Fig. 4.4), or can be peeled off and replaced to provide a re-closure device.

Labelling/packing lines may incorporate systems for applying a carry-home strip or label which can be applied to one or two bottles of lemonade or mineral water.

4.5.7 Recent developments

Over the past few years, 'no-label' look film labels, with matt or gloss surfaces, have become popular and part of the lexicon of labelling. Marketeers see these labels in image-enhancing terms to encourage people to buy expensive products in glass or plastic containers (hair-care products, cosmetics, near-beer drinks). This type of labelling competes with direct decoration because it achieves similar graphic effects.

Glass-clear labels and adhesives are also used in pharmaceutical labelling where glass vials have to be visually checked for contamination of the contents before use. Any particle or bubble in the adhesive could be mistaken for contamination of the contents and could lead to unnecessary rejection of the vial.

Furthermore, clear-on-clear labels eliminate pre-printing of containers for stock, offering bottle or brand owners logistics flexibility in terms of total inventories and sizes of containers. Coloured containers can also be used, with the clear label allowing the colour to show through, rather than having to match the container colour with printing ink.

Another more recent labelling development is the use of OVDs and diffractive optically variable image devices (DOVIDs), generically called holograms, which have gained considerable prominence in the fight against counterfeiting. They cannot effectively be reproduced by colour copiers or scanners or by conventional photographic and printing processes. They are versatile enough today to be bonded to labels, tags and textiles, providing complex 3D structures which provide anti-counterfeit, tamper-evident, plain, numbered or other imaging solutions.

4.6 Label printing and production

The printing and production of labels is undertaken on a wide variety of presses – from sheet-fed to web-fed – and using almost every available printing process. The type of press or process used is determined by:

- the specific label or printing requirement
- the availability and quantity of labels
- the nature and quality of the printing
- the number of colours
- whether subsequent converting operations (die-cutting, embossing, metallic foiling, laminating, waste stripping, etc.) are carried out in-line on the press or are separate stand-alone operations
- how the labels are to be shipped to the packaging or bottling line (in slit reels, cut to size or punched to shape in stacks).

Apart from the mechanical printing processes used to produce pre-printed labels – rotary and semi-rotary letterpress, flexo and UV-flexo, screen process, offset litho, gravure, hot or cold foiling – there are a range of on-demand, VIP solutions used by the label industry. These include thermal printing, laser and ink-jet printing and, more recently, complete digital colour printing presses.

The mechanical-label printing processes can be categorised by the way in which the area of the printing plate or cylinder that carries the printing ink (the image carrier) is defined when compared to the non-printing area. Printing plates and cylinders that print from a raised, inked surface are known as relief printing processes. These include letterpress and flexographic printing.

Printing processes that carry the printing ink in a recessed pattern of etched or engraved cells are known as intaglio printing. These include gravure and rotogravure. A printing process that prints from a flat, chemically defined, printing plate is known as planographic printing. The most commonly used planographic printing process used for labels is offset litho.

Each of these mechanical printing processes has undergone developments in recent years which have led to better quality, colour and speeds. As such there is no 'best process'. Each of them is capable of high-quality printing at economic speeds and each can impart a quality or characteristic to the overall design and image of the finished label. These characteristics include such things as evenness or brightness of colours, thickness of the ink film, fineness of reproduction, quality of halftones, the texture and feel of the label, etc. Some can also impart a range of special effects such as impregnated inks, which can be used with perfumes, fruit flavours or other aromas, rub-off or scratch-off inks, and colour or temperature change inks. Raised tactile images, for example Braille characters, can be printed to provide product information, including the warning of hazardous contents, for the blind and partially sighted.

An important market need which none of these mechanical printing processes can handle in a commercially viable way is that of printing variable information. Examples of such information include bar codes, batch and date codes, sequential numbers, price and/or weight information and print runs of short length, i.e. where the time to print is short compared with the time to set up the press. To meet these requirements, a number of methods of electronically printing variable images or short runs have been developed. Some can be run in line with fixed-image mechanical presses, some are stand-alone machines and some are meant for adding variable information off-line.

Most recently, digital printing technology using electrophotographic or ink-jet printing – in which the image is defined or created by computer – has rapidly begun to find a place in the label printing industry for full-colour and spot-colour short-run printing of labels. By the end of 2010, there were already more than 1400 digital presses in label plants worldwide, with new installation growing in excess of 250 new machines a year.

A brief guide to each of the key mechanical, variable information and digital label printing processes and techniques is set out later.

4.6.1 Letterpress printing

Letterpress printing was the earliest method used to print labels back in the eighteenth and nineteenth centuries and is still a key technology in the printing of quality prime and secondary self-adhesive labels today.

Modern letterpress printing (Fig. 4.5) uses photosensitive polymer plates onto which an image (ink-carrying area) is produced by photographic or direct-imaging plate-making techniques. After exposure, the plate can be treated in a special wash-out solution (chemical or water-based) to develop the image and non-image areas of the printing surface. The relief image area plate can then be affixed around a printing cylinder and placed in the printing press. During the printing cycle, the raised surface is rolled with a sticky ink and the inked image then transferred to the label substrate under pressure.

Letterpress printing roll-label presses, primarily used for self-adhesive label production, are narrow-web (200, 250, 360, 400 or 450 mm) machines and are either web-fed semi-rotary, intermittent-feed machines or web-fed full rotary printing presses – both of which can include up to six, eight or more printing units, together with all the necessary reel unwind, die-cutting, waste-stripping and re-winding capabilities. Almost all are utilising UV-curing inks which have the advantage of being fully dried on the press, thereby reducing lead times and allowing subsequent in-line finishing operations, such as lamination and embossing, to be carried out immediately. UV inks also have excellent resistance to rubbing and are chemically resistant to a wide range of products.

In intermittent-feed roll-label presses, the web of labelstock is progressively and intermittently advanced to the printing position, printed and then moved on for die-cutting and waste-stripping operations. The presses are slower than full rotary presses but do offer advantages for short (10 000 or so labels) to medium (25 000–50 000) label runs where they can provide press flexibility and quick changeover capabilities, and due to the variable feed length that can be selected on an intermittent-feed system, there are no costly changes involved in changing cylinders, gears, etc.

Rotary letterpress in-line, common impression drum (Fig. 4.6) or stack roll-label presses (Fig. 4.7), started to be installed in label printing and converting plants in the late 1970s, and

Paper labels **147**

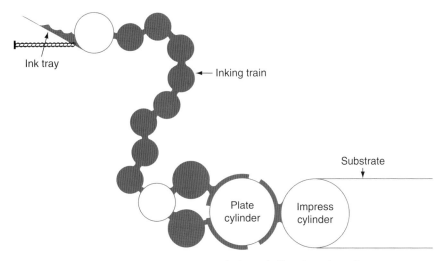

Figure 4.5 Letterpress printing process. (Source: Labels & Labelling Consultancy.)

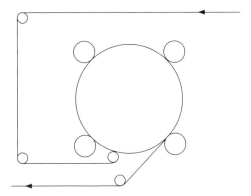

Figure 4.6 Diagram of a common impression drum label press. (Source: Labels & Labelling Consultancy.)

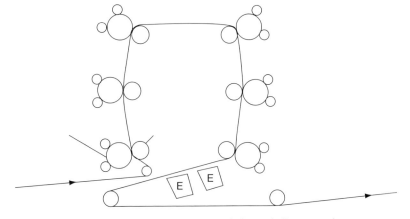

Figure 4.7 Diagram of a stack label press. (Source: Labels & Labelling Consultancy.)

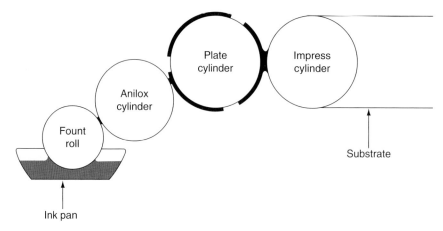

Figure 4.8 Flexographic printing process. (Source: Labels & Labelling Consultancy.)

by the mid-1980s made up almost 70% of all new roll-label presses being installed in Europe. Even today, in spite of the rapid growth of flexographic and UV-flexographic label printing, rotary letterpress still has around 30–40% of the total installed press base in Europe, much less in North America.

Compared with intermittent-feed letterpress machines, rotary letterpress offers significant market advantage in that they can produce up to three times or more output – and all to a high quality, consistent result with, probably, less dependence upon operator skills than is required by other printing processes. Rotary roll-label presses may have an in-line printing head configuration, or the print heads may be placed around a common impression drum or arranged in a satellite or stack pattern.

4.6.2 Flexography

Like letterpress, flexography is a relief printing process (Fig. 4.8) in which ink is applied only to the raised surface of the image on the printing plate and, transferred from there, to the surface of the label substrate. The key differences between letterpress and flexographic printing lie in the method of metering and applying the ink to the raised surface and to the plate itself.

Narrow-web flexographic printing is the most widely used label printing process worldwide for the production of self-adhesive labels. In the USA, it makes up more than 90% of the installed press base and around 40–50% of the installed base in Europe. It is used for production of almost all types of labels available today and has become increasingly accepted for the production of even some of the highest quality prime labels, particularly with digital computer-to-plate (CTP) imaging and the latest UV-curing flexographic press technology. Like rotary letterpress, there are three basic types of flexographic press configuration: in-line, common impression drum and stack. Any of these configurations may be used for printing a wide range of materials.

Printing plates for flexographic printing may be made from a softer, more resilient rubber or flexible photopolymer plate material than letterpress, though similar exposure and wash-out stages to that of letterpress plates have traditionally been required. However, rather than the thicker, sticky ink used for letterpress printing, the flexographic process uses a very thin solvent (declining), water-based or UV-cured (growing) ink. The thin ink requires a special type of ink metering system which uses an engraved anilox roll containing thousands of tiny recessed cells – which carry the ink – and a doctor roll or blade which removes excess ink from the anilox roll surface. The degree of fineness and the shape of the cells can be changed depending on the effect required.

The last print station on a flexographic label press is frequently used for applying a varnish so as to provide additional scuff, rub and chemical resistance. For self-adhesive-label printing, flexographic presses also incorporate a die-cutting station (with flexible dies on a magnetic carrier or a full rotary cylinder), matrix waste removal, slitting units and a re-winder.

Quality of origination, pre-press, press maintenance and good press fingerprinting are important factors in the production of good quality flexographic printed labels. It is essential to get everything correct at the beginning, as it is not easy to make modifications on the run.

In recent years, UV-flexographic label printing has rapidly gained ground to successfully compete with, and overtake, UV-letterpress label production for multicolour and half-tone work and, with the latest origination (which is closer to that used for letterpress and offset litho) and direct-to-plate imaging technology, and High Density (HD) flexo plates, which are now claimed by the best flexographic printers to come close to offset litho quality. As such, the process has become widely accepted for the printing of food, supermarket, retail and other process colour labels.

Apart from narrow-web flexographic presses used for self-adhesive label production, the process is used in wide-web formats (500, 600 mm wide and upwards) for the printing of some sleeve labels and wrap-around film labels, as well as for cartons.

4.6.3 Lithography

Lithography, mostly referred to as offset lithography or simply as offset, is the most commonly used printing process for the printing of glue-applied paper labels using a wide variety of sheet-fed presses. More recently, offset is also being used in roll-fed press formats for high-quality self-adhesive label production and in both sheet-fed and roll-fed versions for in-mould label production and for some new developments in cut-and-stack and patch film labels.

Offset lithography is a high-quality planographic process (Fig. 4.9) in which the image and non-image areas of the printing plates are on the same plane (flat) surface but are differentiated chemically in a way in which the image areas are made ink receptive and water repellent, whilst the non-image areas are water receptive and ink repellent.

Some narrow-web offset label presses make use of the dry offset process which uses a special plate and does not require water to be used in the inking system.

Plate-making for offset printing is relatively simple. Thin sheets of a grained and light-sensitive coated (aluminium) plate are exposed through a negative or positive film – or by computer imaging – to create the image and non-image areas and then treated chemically to aid ink or water receptivity as required.

Once mounted on the plate cylinder in the press (whether sheet-fed or roll-fed), the plate is repeatedly contacted after each print with damping and inking rollers – damping, which is mostly water, to the non-image areas, and ink to the image areas. UV-cured inks are widely used and many labels are UV varnished in-line wet-on-wet on the printing press.

For self-adhesive label production, web-fed in-line offset presses will also incorporate multiple print heads, die-cutting, waste stripping and re-winding in one pass. More

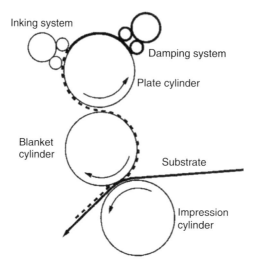

Figure 4.9 Offset lithographic printing process. (Courtesy of Iggesund Paperboard.)

commonly in self-adhesive label production, offset litho print heads are used in combination with another process, i.e. screen, hot stamping or UV-flexo varnish, for high added-value applications in, say, the cosmetics or wine-labelling sectors.

In the last few years, a new generation of mid-web (400–600 mm wide) roll-label offset presses has been developed to target the more efficient, higher added-value in-line printing and converting of traditional sheet-fed glue-applied labels for the wines and spirits markets.

4.6.4 Gravure

Gravure, or photogravure, is a true photographic process which is able to reproduce high-quality pictures, excellent colour densities and strong solid areas. It is primarily used as a long-run process for wet-glue-applied labels and for flexible packaging. Now, gravure units are also being incorporated into some narrow and mid-web presses for labels and flexible packaging where the ability of gravure to apply varnishes, lacquers, metallic inks and top coatings is a particular advantage.

The image carriers for gravure primarily consist of steel cylinders with an outer shell of copper onto which images – consisting of millions of tiny cells of varying depths and areas – are produced by chemical etching, laser etching or electromechanical engraving. Although the most expensive of all the image carriers used for printing labels, they have the advantage of being able to print very long runs in millions. For additional run life, the cylinders can be chromium plated.

Gravure printing presses are the simplest of all label printing presses (Fig. 4.10). The printing cylinder containing the cells rotates in a thin, fluid, solvent or water-based ink which fills the recesses. Surplus ink is scrapped from the cylinder surface by a flexible doctor blade, the ink in the cells (the image) then transferring to the label substrate using pressure against a rubber-covered impression cylinder.

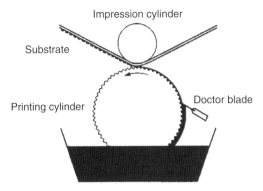

Figure 4.10 Gravure printing process. (Courtesy of Iggesund Paperboard.)

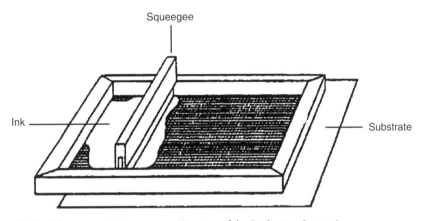

Figure 4.11 Silk-screen printing process. (Courtesy of the Packaging Society.)

4.6.5 Screen process

One of the oldest methods of printing, screen process (silk-screen, Fig. 4.11), is mainly used for the production of self-adhesive labels for cosmetics and toiletries, pharmaceutical and industrial and outdoor applications. Most frequently found in combination in a press line with letterpress, UV-flexo or offset, the screen process has the ability to print a smooth, controllable lay-down of ink and to provide durable, high-quality labels where good ink coverage is required for resistance to weather, moisture, chemicals and abrasion. It can also print a good opaque white image, something which other label printing processes find difficult to achieve.

The image carrier for screen printing is most often made from a nylon, PET or fine wire mesh onto which a photosensitive coating has been placed. Exposure to light through a film positive of the label image causes the coating to harden in the non-image areas enabling the unexposed (unhardened) areas to be washed away. This leaves a negative image of the label remaining on the screen mesh.

For printing, the screen is mounted on a frame and placed in position on the screen printing and die-cutting press. Ink is pressed through the open areas of the mesh by a blade called a 'squeegee' to produce a print on the label substrate – which in the screen process can be virtually any material. Ink drying today is mainly undertaken by UV curing.

Screen printing lays down a very thick film of ink and is the only process that allows a light colour to be printed over a dark material with satisfactory results. Very high quality work is possible and fine line or tonal work can be produced with equal success. This process is ideal for achieving visual impact with fluorescent inks and for printing raised tactile characters (Braille).

Screen presses used for self-adhesive label production either use a flat screen in a stop-and-go motion or use rotary screens in which the squeegee is mounted inside the cylinder. Flat-screen printing is a relatively slow process as it requires the web being printed to come to a stop for each print cycle. Although the rather slow, flatscreen style of printing unit gained partial acceptance with label printers, the move towards building presses in which a combination of printing processes, including screen, demanded something more – rotary screen, which is faster and has a continuous motion.

The rotary method of screen printing has been used, for many years, for the printing of fabrics and wallpaper, and this technology was developed to create narrow-web rotary screen units that could be interfaced with, say, letterpress, flexographic, offset or, today, even gravure units for high-quality and special-effect multi-process printed self-adhesive labels. In the rotary format, both the mesh carrying the photographic image and the label substrate come together as cylinders, which reduces the possible area of contact and sharpens up the image being printed. The supply of ink is retained inside the rotary screen and the squeegee is also located inside the cylinder.

Rotary screen or screen combination narrow-web presses make up around 10–15% of all new narrow-web roll-label presses installed each year.

4.6.6 Hot-foil blocking/stamping process

Foil blocking is being used more and more as either the sole printing process or as a process incorporated alongside letterpress, flexographic or offset printing in a combination process press in order to add an extra added-value capability to the printed label. It is a dry-printing process, using no inks and involving no colour mixing or matching. The process can print on a wide range of surfaces and produce bright effects from metallics, or high-opacity colours and uses relatively simple equipment. Hot-foil printing presses may be narrow-web, web-fed machines, such as those used for self-adhesive label production, or larger sheet-fed presses used for foil blocking of large sheets of glue-applied or in-mould labels.

Units designed for the hot-foil printing or decoration of labels come in a variety of configurations and widths and can be in the form of a stand-alone press or in the case of some narrow-web self-adhesive label presses in combination in-line with other printing processes, provided the press is fitted with a suitable drive and registration system.

The printing plate used for hot-foil blocking needs to be of a hard material and to have a raised image similar to that used by the letterpress process. The fact that image transfer relies upon both heat and pressure restricts plate materials to either a very hard thermoformed plastic plate for very short runs or plates produced from brass, steel or zinc for longer runs. Whilst rotary foil blocking has gained some ground, the majority of hot-foil blocking for label printing is performed in the flatbed format.

The actual hot-foil printing process is achieved by transferring either a coloured pigment or a metallic coating from a ribbon of plastic material known as the 'carrier' onto the surface of the label material to be printed. This transfer is performed through the application of heat and pressure, and the length of time the heated coating area is in contact with the substrate is known as the 'dwell time'. The balance and control of these elements is critical and must be individually calculated for the surface to be printed, and the type of ribbon or foil being used.

A more recent development of foil blocking is a cold-foil process in which a print unit is used to print a special adhesive on the label web in the area where the metallic effect is required. When foil is brought into contact with the adhesive, it adheres to it to produce the printed foil design on the label. Cold foiling is a less expensive means of achieving foiling than the hot process.

Once printed, the surface of hot or cold-foil images may be varnished, over-laminated or encapsulated in order to provide a hard-wearing, durable surface. Foiling is used to provide a luxury (metallic) look on many cosmetics, toiletries, health and beauty labels, on wines and spirits labels, and in other higher added-value label applications.

4.6.7 Variable information printing (VIP), electronically originated

The one aspect that none of the conventional mechanical-label printing presses can handle economically is the printing of variable text or images that include bar codes, sequential numbers, batch and date codes, price-weight information, lot numbers, names, mailing addresses, etc. To perform these functions using letterpress, flexography, litho, gravure or screen presses would involve stopping the press after each print, changing all or part of the printing plate, then printing the new image – an expensive and time-consuming operation.

To overcome these limitations, a number of methods of electronically producing VIP have been developed. These methods include ion deposition, laser printing, direct thermal, thermal transfer, dot matrix and ink-jet. Some of these operate as stand-alone printers or presses, some run in-line with fixed-image printing presses, or they are added on to some types of label application or print-and-apply systems. Brief descriptions of these VIP methods are described later.

4.6.7.1 Ion deposition

Ion deposition is an electronic printing system that uses an electrographic, non-impact imaging process in which a latent image is created on a sensitive drum by bombarding it with ions. Powder toner is applied to the latent image – the powder printed from the drum to the substrate using pressure and then the drum cleaned ready to receive the next image. Ion deposition printing is capable of printing bar codes and other information onto substrates such as paper, boards, vinyl, PET and other filmic materials at reasonably high speeds. It may need to be over-varnished or laminated to improve durability, rub or scuff resistance.

4.6.7.2 Laser printing

Like ion deposition, laser printing requires a latent image to be created on the surface of a sensitised drum (this time by light). Electronically charged particles of a powder toner are deposited on the image and the toner transferred to the substrate using heat and some pressure. Because of the heat involved in laser printing, there may be limitations on the label substrates that can be used. There are also the so-called 'cold' lasers or cool lasers which require less heat and can print on a wider range of materials.

4.6.7.3 Direct thermal printing

Direct thermal printing of labels is the main process used for adding price-weight information, product description and bar codes to supermarket fresh-produce labels – meat, fish, cheese, fruit, vegetables, etc. – which are weighed and priced on the pre-packaging and labelling line or, in some cases, printed and labelled in-store.

The print head for direct thermal printing consists of numerous styli (needles) in the form of a grid or matrix that are heated and cooled selectively by a microprocessor controller. A special heat-sensitive coated paper is required which, when heated by the styli, changes colour within the areas of contact to form the required letters, words, numbers or codes.

As the special thermally printable coating is heat-sensitive, direct thermal printing is primarily used for fresh and chill-cabinet products that have a short shelf life in store of several days. It is not normally used for long shelf-life-labelled products or for labelling for use in warm or hot conditions.

4.6.7.4 Thermal transfer printing

Thermal transfer printing also makes use of styli which are heated and cooled selectively. However, this time, rather than using a special thermally sensitive coated paper, the styli come into contact with a thin film one-pass ribbon which carries a heat-activatable ink or coating on the underside. The required image is therefore created by transferring the heat-activated ink coating from the film carrier to the substrate, according to the pattern or shape of the heated styli.

Thermal transfer printing is used for printing variable information, such as batch codes, date codes, sequential numbers, text, diagrams and bar codes onto pallet, carton or box-end labels, for warehousing and distribution requirements, for bakery labels and for DIY and industrial labelling. Printers may be incorporated into packaging and/or weighing lines or be stand-alone. Some are also print-and-apply systems.

4.6.7.5 Dot matrix printers

Dot matrix label printing involves the firing of pins or hammers against an inked ribbon and thus transferring ink onto a label-face material, rather like the early typewriter techniques. Each pin or hammer produces a fine dot, and these can be positioned in a grid to produce letters, words, numbers, codes or simple graphics.

Impact printing systems such as dot matrix printers have applications for printing onto card or for form/label combinations where an image may need to be created through more than one layer of material. They are far less common today than in the past.

4.6.7.6 Ink-jet printers

Ink-jet printing for VIP uses minute jets of ink which are activated and fired at a label surface by means of electrical charges to form the desired image. Ink-jet is a non-contact method of VIP.

The various ink-jet printer manufacturers use a variety of techniques to form the image. These include a single continuous jet, a multi-head jet, impulse jet and a system using solid ink which is converted to liquid at the moment of transfer.

Ink-jet printing can operate at speeds which are compatible with presses, label-applicator line or packaging line speeds and are therefore often used in in-line operations to produce batch or date codes, sequential codes, batch codes and product or delivery details.

4.6.8 Digital printing

The means to print labels digitally in four or more colours direct to paper or synthetic materials began to emerge in the early 1990s and has now become a label printing technology in its own right with around 1400 digital label presses installed worldwide by the early part of 2011. These presses are predominantly electrophotographic dry toner (Xeikon) or Electroink liquid toner (HP Indigo) technology, as well as new generations of digital colour ink-jet machines coming from companies such as EFI Jetrion, Agfa.dotrix, Durst, Domino, etc.

In the toner-based digital printing process, a latent image of one of the colour separations is created on a drum. That image is then developed for printing with the dry powder or liquid toner before being transferred to the label substrate as a printed image. Each colour is printed in turn. In the HP Indigo machine, each image is first transferred to a rubber blanket before printing from the blanket on to the substrate. It is therefore a digital offset printing process.

The most recent digital press innovations are those using colour UV or LED-cured ink-jet, such as on machines from EFI Jetrion, Agfa:dotrix, Durst, Domino and others. Several hundred of these machines have already been installed in the label industry in recent years, with the technology expected to grow rapidly in the future.

All these digital printing systems require the keyboard input of the typeset text or label – basically a file of instructions describing line work in the form of vector graphics, text, fonts and images in a page description language which specifies all the images of the layout. This file of instructions then passes to a digital front end (DFE), which is a combination of hardware and software, encompassing the entire suite of software modules necessary to process the files to be printed and to control the print engines. The modules or processes typically impacting output quality are calibration, colour management and screening.

The reproduction of the electronic original (the input) is finally printed by the digital printing engine in the digital press. The quality of the output is influenced by the combination of the DFE and the print engine.

Whilst attracting considerable interest for short run, multicolour printing – with each print varying from the next if required – early digital printing had a number of limitations that have now almost completely been addressed in the latest generations of digital presses. These limitations of the earlier digital machines included:

- difficulties in matching bright and special brand/house colours
- limitations in availability of substrates (especially of top-coated filmic substrates)
- poor performance in rub or scuff testing
- cost of labels
- press reliability/durability.

Against this, digital colour printing offers benefits to brand owners in terms of short-run proofing, extended proofing, test marketing and product trialling. Potential additional benefits, yet to be fully exploited, include the ability to produce colour labels with special brand protection, anti-copy or anti-counterfeiting features on each label or batch.

Whilst the early generations of colour digital presses were cost-effective for very short runs of say 5000–10 000 labels, runs over this size could be produced more economically by short-run mechanical printing presses. The new generations of higher-speed, longer-run digital colour presses are claimed to be cost-effective for up to 30 000–50 000 labels.

4.7 Print finishing techniques

4.7.1 Lacquering

Lacquering is similar to applying a coating on a press but carried out off-press on a machine equipped with a roller coater after the printed label sheets have dried. Coatings, which may be UV cured, may be used at varying thicknesses to suit the requirement or function of the label. Being a separate operation, lacquering increases the unit cost of the labels.

4.7.2 Bronzing

Bronzing is a means of creating a metallic appearance – usually gold – on wet-glued labels printed in sheets. A special adhesive or bronzing base is applied to the sheet in the areas to be bronzed, on a single colour, sheet-fed press. This machine incorporates an in-line bronzing application system which applies the bronze powder to the sheet. Special dusting devices distribute the powder evenly all over the sheet but it only adheres in the treated areas. The sheet is then cleaned to remove excess bronze powder and burnished to develop the bronze lustre. The process is relatively slow and expensive. It is primarily used to produce labels for high added-value products such as expensive wines and spirits and cosmetics.

4.7.3 Embossing

Embossing is undertaken between a male and female die. The female die is a depressed image, whilst the male die is prepared so as to push the label paper into the female embossing die to create a raised (embossed) area on the label design.

4.8 Label finishing

4.8.1 Introduction

Once printed, most labels have to be cut or punched to a specific size or shape as part of the label manufacturing process. Cutting, in particular, is one of the most important operations in the finishing of most labels and this may be carried out by straight cutting or by die-cutting. Other labels (leaflet or booklet labels) may have to be applied from a carrier web for subsequent automatic application, whilst some others may be bronzed, embossed or laminated. These operations are all part of the various label-finishing stages and techniques.

Wrap-around labels for cans, for example, are cut to a rectangular size on a guillotine so as to wrap around the body of the can; wine labels are cut to rectangular shapes to be applied to the front or back of glass bottles; beer-bottle front labels or champagne labels may be punch cut in oval or special shapes, whilst self-adhesive labels are mainly all die-cut to size and shape.

The methods and techniques required to produce finished labels ready for application to bottles, cans or packs are:

- guillotining to cut sizes
- punching
- die-cutting
- on-press slitting and sheeting (for some applications).

Special label-placement techniques may be used to apply the label to the container as well as techniques for folding, inserting or crimping.

Originally, labels were simple, mainly rectangles or squares, sometimes with rounded corners – followed later by circles and ovals. However, as the market developed further, the demand for unique shapes – which would aid recognition – increased and, with these new and challenging requirements, a whole new separate industry and technology servicing the label printer evolved.

A guide to some of the main label-finishing options is as follows.

4.8.2 Straight cutting

Many labels are printed in sheets on offset presses with, sometimes as many as, 100–150 labels on each sheet. These sheets of multiple labels must be straight cut, either for use as square or rectangular labels or before cutting out or punching to shape.

Straight cutting requires piles of up to 1000 printed and properly stacked label sheets to be fed into the back gauge of a guillotine, usually using air glides which float the paper into position. With the pile against one of the rigid sides of the guillotine, the pile is positioned for the first cut. Once positioned, a clamp descends on the pile to hold it in position whilst the cut is made with a guillotine knife. The clamp then rises ready for the pile to be moved to the next cutting position. All subsequent cuts on modern guillotines are carried out using programmed gauges, with the back gauge moving automatically to the correct position for the next cut. Tolerances for straight label cutting are fairly critical.

4.8.3 Die-cutting

Shaped designs of self-adhesive labels and some wet-glue and in-mould labels have to be die-cut as part of their manufacturing and finishing procedure. Depending on the type of label and the printing and/or die-cutting requirement, the operation may be performed using high or hollow dies (ram punching), flat dies, rotary dies or, most recently, with digital die-cutting.

High or hollow dies used for ram punching glue-applied labels to shape are made of cold-rolled steel which is forged and welded to create the required shape and height. The inside of the dies are parallel for about 25 mm, after which they flare out. In use, the stack of labels is either stationary with the die moving or the die is stationary and the stack is pressed against it. In either case, sufficient space must be allowed between each label to ensure clean cutting. Sharpness of the die is critical, as are finished cut-label tolerances.

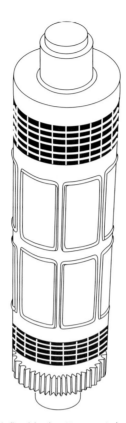

Figure 4.12 Magnetic cylinder with flexible die. (Source: Labels & Labelling Consultancy.)

Flat dies are most commonly produced by bending lengths of an accurately fashioned steel rule which has been finished to a cutting bevel along one edge. This rule is around 0.4 mm in thickness and nominally 12 mm in height. To form a cutter, the rule, once bent to shape using a special bending tool, is placed in a base into which the shape or shapes of the label(s) have been cut. In this way, the rule is supported during use on the press and retains a high degree of accuracy.

Rotary dies are engraved by electronic discharge from a cylinder of solid steel so as to leave the cutting edge standing proud around the cylinder circumference. An alternative method is to use thin steel plates that have the die configuration etched over the surface. They are then mounted for use by wrapping the thin steel around a magnetic cylinder (Fig. 4.12).

Each of the types of rotary die requires some form of final finishing following on from the machining or etching, which is undertaken using computer-guided equipment and which sets the seal of quality on the die.

Digital die-cutting of labels has been a relatively new development which has evolved from a laser-cutting technology used to cut out die-base boards prior to inserting the rule. As with variable imaging or digital printing, the required shape and size of label is programmed by a computer, with a laser beam and lens system used to direct the beam in cutting out the label shape. Both paper and filmic labels are being cut with a degree of success although the technology is still a somewhat expensive method of cutting labels to shape unless there are a large number of size changes and shapes required each day or week. However, if digital label printing is to fully develop its potential for printing self-adhesive labels in small quantities and on demand, then a digital method of label die-cutting on demand to any shape or size will also be required. The technology is available; it will be demand and economics that determine its eventual wider usage and success in the label industry.

4.8.4 Handling and storage

Most labels form a relatively expensive element of the total finished, labelled product. Not unnaturally, the label user organisations expect them to be in pristine condition and to perform well on the label-application line.

For wet-glue labels, customers normally expect labels to be packed in bundles with flat packing pieces, for example cardboard top and bottom. PE shrink films or PE bags should be used to protect paper-based labels from moisture, changes in relative humidity (RH) and to maintain hygiene. Storage conditions are quite critical in maintaining flatness and the correct moisture content. Both RH and temperature-storage conditions are likely to be specified by the end-user. Pre-conditioning of labels to a specified RH in the label user plant is also recommended prior to usage.

Handling and storage of self-adhesive labels also requires special attention. Like wet-glue labels, self-adhesive paper labels are also affected by temperature and humidity. Higher temperatures can cause the adhesive to soften and flow; lower temperatures may cause the label to begin de-laminating from the backing paper. Again, recommended temperature and RH conditions laid down by the laminate supplier or end-use customer should be followed.

A further problem with self-adhesive labels is that excessive pressure on a stack of labels may squeeze the adhesive out around the edges of the labels. Reels of labels should therefore be stored and packed flat (cheese fashion). Reel stocks can take on curvature the longer the reels are stored. The curvature will be worse the nearer one gets to the core. Because of the limited shelf life of self-adhesive labels and potential adhesive ageing, a recommended use-by-date should be followed.

4.9 Label application, labelling and overprinting

4.9.1 Introduction

Once labels have been printed and finished, whether in cut sheet label sizes, punched into shapes, in a roll on a backing liner (self-adhesive) or in reels for sleeve or wrap-around film labelling, they are despatched to the packaging and labelling facilities for application to bottles, cans, packs or products using a labelling or label-application machine – which may be hand-operated, semi-automatic or fully automatic. In high-speed labelling applications, incorporated into dedicated bottling or canning lines, de-palletising, washing, pasteurisation, inspection, container filling, capping, sealing, neck-foiling, labelling, dating/coding, carton or case filling, shrink wrapping and palletising are carried out in one in-line continuous operation.

The method of applying the label will vary according to the product to be labelled, the label type and the specific labelling requirement, i.e. front label only, front, back and neck, wrap-around body label, tamper-evident label, etc. Some of the key label-application methods are set out later.

4.9.2 Glue-applied label applicators

All glue-applied labels, whether paper, metallic foil or film materials, are affixed to containers by a labelling machine into which the bottles or cans are fed on a conveyor line. Glue is applied to the back of the label or in a strip to the label on the container, and then the labels are applied to the containers automatically and continuously at high speeds. Capable of handling glass and plastic bottles, metal and special-shaped containers (depending on the user requirement and system), glue-applied labellers can apply various combinations of body, shoulder, back, wrap-around, neck-around and deep-cone labels.

Modern wrap-around labelling systems have also been designed for the application of labels from the reel using low-cost paper labels as well as a variety of plastic film labels, including reverse-printed transparent film. In operation, glue is applied onto the leading edge of the film via mechanically driven glue rollers. The paper or film is cut to length and wrapped around the container and then glued on the trailing (overlap) edge to form a complete wrap. Glass, plastic and metal containers – square, cylindrical, and oval – can be handled at speeds up to 800 bottles per minute when using 2-applicator stations.

Because of the high speeds used in most glue-applied label lines, there are usually special requirements for the glues/adhesives and the labels to ensure optimum productivity. The most important label characteristics for glue-applied paper label line efficiency are size tolerance, moisture absorbency, grain direction, wet strength and label memory, stiffness and curl characteristics.

Glue-applied labelling machines may be straight through lines, which have outputs up to 24 000 containers per hour or high-speed rotary systems, which today can achieve outputs of up to 160 000 containers per hour.

4.9.3 Self-adhesive label applicators

All machines for applying self-adhesive labels need a means of peeling the silicone-coated backing paper from the web of die-cut labels and at the same time applying the individual die-cut labels to the product to be labelled.

Removal of the backing paper is achieved by passing the backing paper over a 'beak', pulling the backing paper backwards and leaving the label to be dispensed moving forwards into the correct position for applying onto the container or pack. It is then pressed into contact by rollers, brushes, air jets or tamper pads (Fig. 4.13).

Depending on the design of the labelling head and method of pressing the label to the pack or container, self-adhesive applicators can be used to apply front, back and neck labels to the body of a container, to the top or the bottom of a pack or bottle, to apply labels around corners, into recesses, onto delicate surfaces (using air pressure) and onto all sizes of containers – from small pharmaceutical bottles up to large drums or beer kegs. They are therefore the most versatile of all labelling machines, yet have one of the lowest capital purchase costs, the greatest flexibility in use and application, high labelling accuracy – even for the smallest of labels – and provide ease and simplicity of operation with high operational efficiency and short changeover times (15–30 min).

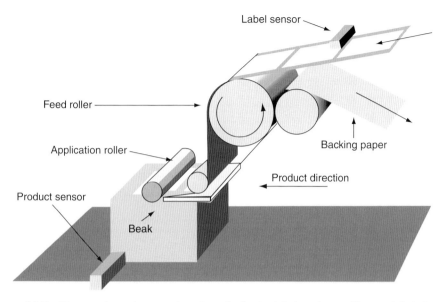

Figure 4.13 Diagram shows the operation of a self-adhesive label applicator. (Source: Labels & Labelling Consultancy.)

These key advantages largely outweigh the higher cost of the self-adhesive label itself and make self-adhesive labelling a cost-effective solution for many industries and particularly for the cosmetics, toiletries, healthcare and beauty, pharmaceutical, industrial products, food/supermarket and added-value drinks sectors.

Self-adhesive label applicators may be manually operated, semi-automatic or fully automatic in-line filling and labelling systems that incorporate conveyors, filling and capping, web feed and dispensing, photoelectric controls and backing waste removal. They may also incorporate VIP heads into the line to add variable text, date or batch codes, bar codes and price-weight information. A wide variety of applicator output speeds are available for almost any labelling requirement and, by combining applicator heads in one line, it is possible to label, say, up to a thousand bottles per minute (60 000 bottles per hour), using a 6-station applicator.

Self-adhesive label-application heads can also be installed as additional units on wet-glue and glue-applied labelling machines for applying tamper-evident labels, promotional labels, etc., or for applying neck labels to bottles decorated with wrap-around film labels.

New types of applicators are now available that can automatically separate and apply linerless reel-fed self-adhesive labels to bottles.

4.9.4 Shrink-sleeve label applicators

Shrink sleeves are applied on an application machine which takes the reels of tubular sleeving, opens it on a mandrel and feeds the opened tube to a rotary knife which then cuts it to the required label length. After placing over the bottle, the labels need to pass through a heated tunnel so that the label can be shrunk to the required bottle shape – even if this means a shrink of up to 30% or 40%. To prevent movement of the sleeves before entering the shrink tunnel, a pre-shrinking unit can be incorporated.

Rather than just a label-applicator line (as with glue-applied or self-adhesive labellers), shrink sleeving is a complete system, which includes pre-heating and post-curing. The more shrink required, say, for a taper neck, the longer the heated shrink tunnel required. In simple terms, shrink sleeving is not just about slipping a sleeve over a bottle but about the technology of differential shrinking. Speed is not a constraint, as two carousels can label up to 42 000 bottles per hour.

Complete body and tapered neck bottles can be labelled with one sleeve although this means that a whole bottle sleeve has to be produced even if only a front, back or neck label is required. Either pre-fill sleeving or post-fill sleeving is possible.

Key advantages of shrink-sleeve labelling systems include 360° full body and neck decoration, decoration of complex bottle shapes, provision of tamper-evidence and relatively low running costs.

4.9.5 Stretch-sleeve label applicators

Stretch sleeves are supplied to labelling lines in flat, collapsed form in a reel and are then passed through a buffer unit prior to cutting and application. To apply the sleeve, it is first opened up and guided by a mandrel and guide to a cutting wheel, where it is cut precisely to the desired length by servo-controlled cutting knives. Before the cutting process is complete, the sleeve is already positioned on the transfer element, which stretches the sleeve and pulls it down over the bottle to a pre-determined application height. The stretchable film then shrinks itself elastically to fit, thus offering a completely glue-free container-decoration process.

Generally seen as more relevant where easy removal of body labels on returnable bottles is required, stretch-sleeve labelling is primarily used for large-size bottles of carbonated drinks. Stretch-sleeve application is a glueless all-round 360° body labelling process which offers precise positioning and easy removal of labels for returnable bottles.

4.9.6 In-mould label applicators

Unlike all other methods of label application in which labels are applied on a packaging or decoration line after the container had been made, in-mould label application is undertaken as an integral part of the container-manufacturing operation. In this, the label is placed in the mould before the bottle is blown or the tub is formed by injection moulding or thermoforming, so as to become part of the container itself. The label is therefore part of the container wall and does not have the raised edge characteristic of other labelling methods.

In blow-moulding label application, pre-printed and cut-to-shape labels are placed in the moulding machine in stacks, from which they are picked up and placed into the mould during the mould-opening cycle and held in place by a vacuum. Accuracy of label placement in the mould is essential for good results. The plastic that forms the bottle comes out in the form of a tube at about 200°C (about 400°F) and as high air pressure is blown into the tube, it expands to conform to the bottle shape. The hot plastic fuses the heat-seal layer to the bottle.

In the labelling of injection-moulded containers, a PP label is used with PP-injected container. Because of the high temperatures and pressure involved in injection moulding, the label fuses to the container without the need for a heat-seal layer.

Undertaking in-mould label application requires special moulding equipment and moulds and the necessary modifications that will enable labels to be inserted and positioned accurately in the mould before moulding.

4.9.7 Modular label applicators

One of the more recent developments in application technology is the introduction of modular labelling systems in which the wet-glue, self-adhesive, shrink or wrap-around film labelling heads are all incorporated into one labelling line. This offers bottlers complete flexibility in their bottling plant for body, neck, front and back labelling using any one, or combination, of label types and application method.

Modular systems are seen as cost-effective to purchase, flexible in application and offer brand owners a wide label and bottle image choice. Modular systems are seen as one of the fastest growing label-application methods and are expected to continue finding a key role in bottle decoration and branding.

4.10 Label legislation, regulations and standards

In recent years, there has been an increasing range and variety of legislation, regulations, national or international standards, codes of practice, certification schemes, etc., which relate to labels and the labelling of a wide range of products and sectors – food, dangerous substances, cosmetics, textiles, floor coverings, crash helmets, aerosols, electrical appliances, medicines, toys, pet foods and materials handling.

Such legislation, standards or codes which have implications for labels or labelling may arise from Acts of Parliament, Statutory Instruments (SI), British Standards, etc., in the UK, from European Commission (EC) Directives or from international sources such as the ISO (International Organization for Standardization), IATA (the International Air Transport Association) or the world's maritime or rail organisations.

4.10.1 Acts of Parliament

The law-making body or legislature in most European countries is Parliament. An Act of Parliament (viz. Consumer Protection Act, Food and Drugs Act, Weights and Measures Act) is enforceable by all courts as the law of the land, unless and until it is repealed or amended by the Parliament.

4.10.2 EC regulations and directives

Apart from the legal requirements of national legislation relating to labels and label usage, it is also necessary to take account of regulations and directives emanating from the EC.

Directives are formulated by the European Commissioners in Brussels in the interests of harmonisation of trade and are implemented by SIs in individual countries. SIs are legally enforceable and affect many aspects of manufacturing and trading, including labelling. Such SIs include those for food labelling, cosmetic products and dangerous substances.

4.10.3 Standards

National standards institutions, such as the British Standards Institution (BSI), are the recognised bodies for the preparation and production of national standards. Standards are coordinated internationally by the ISO.

Standards are prepared under the guidance of representative committees and are widely circulated before they are authorised for publication. They include glossaries of terms, definitions, quantities, units and symbols, test methods, specification for quality, safety and performance, preferred sizes and types, codes of practice, etc. Examples include 'Recommendations for informative labelling of textile floor coverings' or 'Standards relating to the testing and performance of labels for use in maritime conditions' or those for 'Safety signs and colours'.

It is always advisable to check for possible legislative or standardisation requirements when designing or developing labels for new applications, requirements or markets.

4.11 Specifications, quality control and testing

4.11.1 Introduction

Almost all label buyers and label end-users today will expect that the labels they purchase and use will meet key materials, colour, print, label line and end-user performance criteria. These criteria will be established with the label printer and set out in the label specifications prior to the commencement of the job and before the ordering of the label materials, inks, varnishes, etc.

Many packaging and label end-user organisations undertake their own tests during the development of new labelling solutions to ensure that the specifications they draw up for the label printer and converter will meet:

- all the necessary brand identity and image requirements
- brand or house colour criteria
- label line, handling, distribution and end-usage needs in terms of rubbing, scuffing, durability
- any product or usage resistance demands
- any legislative requirements (such as 3 months immersion in sea water)
- safety, waste or environmental demands.

Once established in the specifications, the materials suppliers (inks, papers or films), label printers and converters and application companies will be expected to check, test and confirm that all criteria and specifications are met throughout the label order.

Test procedures that are accepted by brand owners, label buyers, packaging and printing companies, paper and film suppliers, ink manufacturers and the like have been established over a long period of time by organisations such as Smithers Pira, Technical Association for the Pulp, Paper and Converting Industries (TAPPI), American Society for Testing and Materials (ASTM International), Graphic Arts Technical Foundation (GATF) or by the relevant trade or industry associations, such as FINAT (the International Federation of Manufacturers and Converters of Self-Adhesive and Heat Seal Materials on Paper and other Substrates) or the Tag & Label Manufacturers Institute Inc (TLMI). Some of the more common test procedures or requirements are set out later.

4.11.2 Testing methods for self-adhesive labels

Testing of pressure-sensitive adhesives has evolved from the subjective evaluations used in the early days of the industry to the standardised tests now used throughout the world. Standard tests to assess the adhesive properties of a self-adhesive label are now published by FINAT (see test methods later) and the European Association for the Self Adhesive Tape Industry (Afera) in Europe and by TLMI and ASTM in America. Test methods to measure the three fundamental properties – 180° peel, shear and quick stick – are described, plus a test for adhesive coat weight, which is important not only for performance but also for commercial reasons.

4.11.2.1 Peel adhesion test method

Designed to quantify the performance or peelability of pressure-sensitive materials, peel adhesion is measured using a tensile tester, or similar machine, which is capable of peeling a laminate through an angle of 180° with a jaw-separation rate of 300 mm per minute with an accuracy of ±2%. Adhesion is measured 20 min and 24 h after application – the latter being considered as the ultimate adhesion.

4.11.2.2 Resistance to shear test method

Designed to measure the ability of an adhesive to withstand static forces applied in the same plane as the labelstock, resistance to shear is defined as the time required for a standard area of pressure-sensitive coated material (using at least three strips) to slide from a standard flat surface in a direction parallel to the surface. The resistance to shear is expressed as the average time taken (for the three strips) to shear from the test surface.

4.11.2.3 Quick-stick test methods

Designed to allow end-users to compare the 'initial grab' or 'tack' of different laminates, the quick-stick value is the force required to separate, at a specific speed, a loop of material (adhesive outermost) which has been brought into contact with a specified area of a standard surface. It is tested using a tensile tester, or similar machine, with a reversing facility and a vertical jaw-separation rate of 300 mm per minute with an accuracy of ±2%. It is an extremely useful test for those working with automatic labelling equipment where a 'grab' or 'tack' value is of particular importance.

4.11.2.4 Adhesive coat weight test method

Designed to determine the amount of dry adhesive material applied to the surface of a pressure-sensitive label construction, adhesive coat weight is expressed as the weight of dry adhesive on a standard area of material – in grams per square metre (g/m^2). It is tested using a template to cut samples, a circulated hot-air oven and accurate balance and a beaker of solvent.

4.11.3 Testing methods for wet-glue labels

A wide range of equipment and apparatus is available for carrying out various test procedures to establish individual wet-glue label properties. In practice, however, most bottling and packaging plants only need to apply a limited number of tests. The more important tests for wet-glue labels and label papers include the following.

4.11.3.1 Tear strength test method

Designed as a simple test for use by bottling plants and small printing companies, tear strength testing is normally carried out using a tensile testing machine and a 10 or 15 mm wide paper strip.

4.11.3.2 Water absorption capacity test method

Designed to ensure that labels bond quickly and positively, water absorption is measured by the $Cobb_{60}$ test (ISO 535 1991). Suitable labels should have an absorption capacity of between 7 and 11 g/m² after 60 s of water contact. Labels with a low absorption capacity may have problems due to the edges lifting after application. Labels with excessive water absorption capacity will tend to curl excessively. The standard time for water contact is 60 s, but when comparing more absorbent papers a shorter time may be preferred. In all cases, the time for water contact used in the test must be recorded.

4.11.3.3 Caustic soda resistance test method

Designed to test the resistance of the label to the effect of caustic soda, resistance can be tested by placing 120 cm² of label paper in a sealed measuring vessel containing 20 cL of 1.5% sodium hydroxide solution and shaking vigorously (around 30 times). A sample with adequate caustic soda resistance will not disintegrate, whilst there should be no contamination of the solution caused by 'pulping' of the fibres.

4.11.3.4 Paper weight test method

Designed to test the weight per unit area (basis weight) of a label paper, testing can be determined with sufficient accuracy by the use of pocket scales. Thickness or caliper can be measured using a standard micrometer or a special paper micrometer.

4.11.3.5 Bending stiffness test method

Designed to measure the flexural stiffness, i.e. bending strength of paper and labels, bending stiffness is measured using a bending stiffness tester, for example Thwing-Albert Tester, which provides a comparison of the bending strength of new labels with that of labels known to handle satisfactorily in labelling machines.

4.11.3.6 Digital printability testing

At the present time, there are no national or international standards for the printability testing of label papers for digital printing. To begin to address this issue, a Digital Label Printing Testing Taskforce has been established by FINAT, the worldwide label industry association and IGT Testing Systems, one of the world's leading companies in the field of printability testing, tack measuring equipment and related products.

This Taskforce will initially be looking at issues such as toner, dot spread, absorption and fastness and investigating whether existing commercial printing systems are suitable for reference testing or can be modified. It is also part of the programme to develop a terminology dictionary for uniform communications between the different stakeholders in the market – paper makers, laminate makers, ink manufacturers, press manufacturers, label printers and label users.

4.12 Waste and environmental issues

Waste and environmental issues have been well to the fore in the label and packaging sectors for several decades and have acquired some notoriety, partly due to the high level of visibility that post-consumer packaging and label waste 'litter' arouses.

The driving force behind the prominence of environmental issues for packaging and labelling is threefold: governments, commercial factors and consumers/consumer groups.

- *Governments* around the world, particularly in Western Europe and North America, have introduced measures ranging from container deposits, packaging levies, bans on certain types of packaging and mandatory recycling rates.
- *Commercial factors* and supply-chain issues are playing a role as companies respond to the environmental challenge. Some retail chains have eliminated or banned what they consider to be environmentally unacceptable types of packaging. They are also responding to pressures from customers and governments by introducing their own environmental scorecard schemes and/or chain-of-custody requirements.
- *Consumers and consumer groups* are showing ever more interest in the environmental credentials of the products that people buy and the companies that they buy them from.

Environmental legislation, bringing these various issues together, has been a key issue for the packaging and labelling industry from the late 1980s through the 1990s and into the current decades. Existing, and forthcoming, regulations are becoming important drivers for waste and environmental issues in the rigid plastics, metal, glass and paper-based packaging and labelling sectors.

Labels and labelling environmental and waste issues initially followed on from an EU Directive on Packaging and Packaging Waste, as well as from schemes and requirements introduced by individual countries around the world.

In the label field, increasing numbers of paper manufacturers and labelstock suppliers today have become certified to the ISO 14000 Environmental standard and additionally obtained European Eco-Label or Environmental Management and Audit System (EMAS) certification. Most also have the forest sustainability accreditation through the Forest Stewardship Council (FSC) and Programme for the Endorsement of Forest Certification (PEFC) schemes. Many label printers also have ISO 14000 or are working to Environmental Scorecard schemes introduced by the likes of Walmart, P&G or Marks & Spencer.

Certainly, the major label user organisations – the brand owners, retail groups, consumer products manufacturers, etc. – have been increasingly presenting the label industry supply chain with environment or sustainability assessment documents, with requirements for materials efficiency, for the use of responsibly sourced raw materials (made in an efficient, ethical and environmentally responsible way), chain of custody, (CoC), third party certified

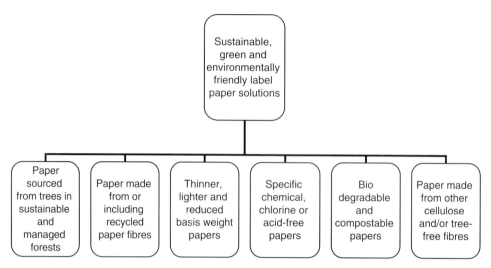

Figure 4.14 Impact of the environment on label papers. (Source: Labels & Labelling Consultancy.)

materials, supply-chain transparency about where and how materials have been produced, and on the handling and disposal of label materials waste.

Together, all these initiatives over the past few years have already led to lighter and thinner label materials, to sustainably managed forest paper products, increasing recycled paper content in label papers, non-tree fibre pulps, chemical-free label papers, biodegradable substrates, compostable labelstocks, corn starch and other bio-label films, recyclable grades, environmentally benign or recycling compatible adhesives, chain-of-custody certification and a whole host of label waste recovery, recyclable and re-processing solutions (Fig. 4.14).

Major initiatives have also been taking place in terms of the waste generated by self-adhesive backing (liner) materials. Collection, recovery and recycling schemes have now been established within the label industry – which aim to turn the liner waste into products as diverse as fuel pellets, compost, building materials, paper hand towels and decking – into packaging grades and, most recently, into new self-adhesive liner papers. Linerless label solutions have also been introduced.

Websites

www.afera.com
www.gain.net/PIA_GATF/non_index.html
www.finat.com
www.tappi.org
www.tlmi.com
www.igt.nl

5 Paper bags

Smith Anderson Group Ltd, Fife, UK, and
Welton Bibby & Baron Ltd, Radstock, Somerset, UK

5.1 Introduction

Paper bags, in all their different forms, make an essential contribution to industry and are a vital part of everyday shopping. Made from a renewable and natural resource, paper bags are a proven packaging medium – part of the High Street scene beyond living memory.

Paper offers strength, rigidity, breathability and versatility. Paper can be made in various colours, its surface can be glossy or matt, it can be printed in a wide variety of ways to produce an appropriate surface appearance and it is cost-effective.

This chapter traces the development and use of paper bags from the development of the early prototype machines right up to the high-value industry into which it has grown in the first years of the twenty-first century.

The earliest record of machines to make paper bags dates back to the 1850s. The leaders in the new emerging market appear to have been Bibby & Baron (now Weldon, Bibby & Baron in Radstock, Somerset, England) of Bury, Manchester, and Smith Anderson at Fettykil, Fife in Scotland. These companies are still major paper bag manufacturers today with the latter still being in family ownership. Prior to the 1850s, many products were probably wrapped in paper and secured with string.

There were major advances in the growth of the industry from 1850 until World War II during which period the basic shapes of bags and the machinery to make them were invented and developed. The companies named were prominent in these developments.

Paper bag shapes are flat, satchel (with side gussets) and – most importantly for the future of the industry – square-bottom, which were free-standing. Section 5.4 examines these shapes in more detail.

Printing techniques were also being developed. We again hear the name Bibby & Baron as the inventors of the flexographic printing process in Liverpool in 1890 (Wmich, 2012).

As retail shopping developed, large quantities of paper bags were required for the packaging of staple foods, such as dried fruit, flour, sugar, tea and coffee. Other products, which used millions, even billions, of paper bags, included coins, potato crisps, ice-lollies and gramophone records.

Handbook of Paper and Paperboard Packaging Technology, Second Edition. Edited by Mark J. Kirwan.
© 2013 John Wiley & Sons, Ltd. Published 2013 by John Wiley & Sons, Ltd.

Expansion of the industry was checked during World War II. Paper was scarce and a quota system continued for several years after the war was over.

Manufacture of paper bags continued its inexorable expansion in subsequent years. Paper bag makers increased production, new factories were opened and engineering companies were constructing more advanced machines to make and print paper bags. Coupled with this, a parallel industry emerged making machinery for the automatic opening, filling and sealing of paper bags in food-processing factories.

Today, bags are widely used for food service packaging (catering), food ingredients, pet foods, bread, potatoes, sterilisation packs, vacuum cleaners and the travel industry. They are also increasingly in evidence at point of sale in the High Street.

The paper bag displays resilience as a form of packaging, quickly adapting to constantly changing needs and fashions.

5.1.1 Paper bags and the environment

Paper is increasingly recognised by consumers and governments as not only a natural but also a renewable and recyclable resource from which to manufacture high-performance packaging.

Paper recycling schemes have become widespread since the 1990s, and there is increasing demand for packaging to be made from previously used materials. In addition, managed forestry in Europe and North America ensures that we have a secure future source of supply.

Legislation in the Republic of Ireland in 2002 imposed a levy on plastic carrier bags, and this had an immediate and visible effect of litter reduction. In department stores, the paper carrier bag has provided a popular and economic replacement. Other EU nations are considering similar action as they strive to comply with collection and recycling targets.

5.2 Types of paper bags and their uses

5.2.1 Types of paper bag

Section 5.1 gave details of the main shapes of paper bags, which have been developed over the past 150 years. This section examines these shapes in more detail with illustrations of the types of bag, their main applications and the materials used. The main types of paper bags are:

- flat and satchel
- strip window
- self-opening satchel, (SOS), bags – pre-packed
- self-opening satchel, (SOS), bags – point of sale
- self-opening satchel, (SOS), carrier bags – pre-packed
- self-opening satchel, (SOS), carrier bags – retail/point of sale.

5.2.2 Flat and satchel

5.2.2.1 *Flat bags*

This is the most basic form of paper bag (Fig. 5.1). It is two-dimensional, and its use is confined almost entirely to retail point of sale. Much favoured in the retail sector by

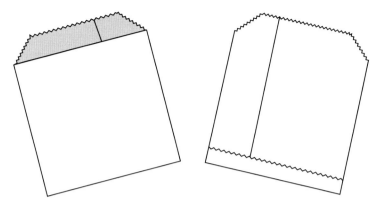

Figure 5.1 Flat bags.

Figure 5.2 Satchel bag.

greengrocers, confectioners, bakers, clothes shops, pharmacies and ironmongers, most of these bags are produced from white (bleached) or brown kraft.

5.2.2.2 Satchel bags – bags with side gussets

The construction of these bags is similar to the flat bag, but additional paper is folded on each side to form the gussets – hence the term 'satchel' (Fig. 5.2).

Satchel bags, when opened, have the advantage of being three-dimensional and provide greater ease of handling and filling compared with a flat bag. Their use is mainly at point of sale in the same retail sectors as flat bags. Although three-dimensional, satchel bags are not 'free-standing' and therefore have limited use in factories for pre-packed food and other

172 Handbook of Paper and Paperboard Packaging Technology

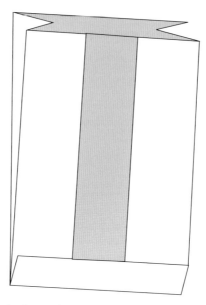

Figure 5.3 Micro-perforated polypropylene strip window bag as used to pack bread and baguettes.

products. Most satchel bags are produced from kraft paper but may have a coating or inner ply of a protective material if required for hot or greasy products, for example bakery and hot snacks.

5.2.2.3 Medical and hospital bags

An important and specialised application for flat and satchel-shaped paper bags is sterilisation packs for medical and hospital use. Bleached white kraft is the preferred material. Due to the stringent requirements of the sterilisation process, the bag and the paper must have special properties, which are described in Section 5.3.4.

5.2.3 Strip window bags

These bags are satchel shaped but merit a separate classification because of their special method of construction in which a paper reel is sealed to a reel of plastic film on the bag machine to produce a strip window. The plastic film, usually polypropylene (PP), or polyethylene (PE), may be micro-perforated to suit product packaging needs. Developed entirely for bread and baguettes, these bags (Fig. 5.3) are used in factory bakeries as well as in 'in-store' bakeries. The combination of paper and micro-perforated film ensures the bread remains fresh. Bags are supplied to large-scale factory bakeries in 'wicketted' bundles to meet the needs of automatic packing machines, as will be described in Section 5.4.5.

5.2.4 Self-opening satchel bags (SOS bags)

For a better appreciation of the uses of this shape of paper bag (Fig. 5.4), one needs to consider two entirely separate applications:

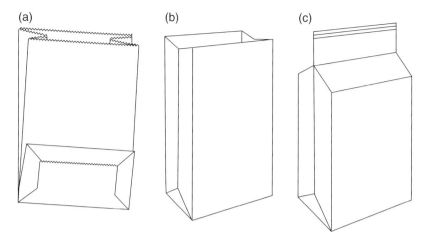

Figure 5.4 Self-opening satchel bags (SOS): (a) folded, (b) open and (c) filled and sealed.

- pre-packed products
- point of sale.

5.2.4.1 SOS bags for pre-packing

The introduction to this chapter emphasises the essential contribution that the SOS shape of bag has made to the manufacturing industry. Figure 5.4 shows the SOS bag in the (a) folded form as delivered from the bag machine, (b) fully open and, lastly, (c) filled and sealed. It shows that the SOS bag, when fully opened, is free-standing and consequently is ideally suited to being opened, filled and sealed on fully automatic machinery. Products packed in these bags comprise not only many basic foodstuffs, for example flour, sugar, cereals, tea, coffee, biscuits and confectionery, but also food mixes for both retail and catering requirements. Numerous other industries are served, including cat litter, small pet-food packs, agrochemicals, horticulture and DIY.

The base material used is generally bleached kraft – brown kraft to a lesser degree. SOS bags may be made in either single ply paper or, where necessary, multi-ply. However, a wide range of coatings, lacquers, laminations and protective lining materials may be used to supplement the kraft paper with two main objectives:

- On the outside of the bag, the aim may be to achieve the maximum visual impact. This requires a coating on the paper to achieve the best print surface: use of gloss or matt over lacquer or a layer of transparent film printed on the underside for optimum effect.
- Inside the bag, additions to the base paper are chosen according to the nature of the product and protection required, for example length of shelf life, grease, flavour and odour protection. This is achieved through grease-resistant or greaseproof papers, wet-strength and plastic-coated papers. A lining of paper or plastic film or foil is sometimes added to provide additional mechanical strength or product protection.

5.2.4.2 SOS bags for use at point of sale

The SOS bag is constructed in such a manner that it forms a rectangular base and can stand up unsupported. This feature, essential for use on automatic machinery in factories, is also of benefit when used 'in-store' and at point of sale. It provides 'stackability'. Significant markets in this sector include bags, often with windows of various shapes, for doughnuts and cookies in supermarkets, popcorn bags in cinemas, 'pick "n" mix' confectionery selection and bags for prescriptions awaiting collection in pharmacies. Bleached kraft is the most popular material in this sector. Whilst graphic design is important for instant visual impact and the material must be compatible with the product, choice of material is less critical than that described in Section 5.2.4.1. The bag is handed to the consumer very soon after packing. 'Shelf life' can be measured in minutes rather than weeks and months.

5.2.5 SOS carrier bags with or without handles

This section is also divided into two separate categories:

- pre-packed products
- point of sale.

5.2.5.1 SOS carrier bags for pre-packing

A large proportion of carrier bags are delivered to factories for pre-packed products. Due to the weight and bulk of the contents, handles are fitted for ease of carrying (Fig. 5.5). Amongst the many products packed in these larger bags, pet foods, cat litter, potatoes, charcoal for barbeques and horticultural products predominate.

The range of papers used and the reasons for the selection of these materials are largely as explained in Section 5.2.4. However, physical strength is an additional factor. Two-ply bags made from strong papers are necessary for many of these larger packs, which weigh up to 12.5 kg. Many pre-packed carriers have punched windows using netting or plastic film – particularly those for the potato market, which were developed by Welton Bibby & Baron.

5.2.5.2 SOS carriers for use at point of sale

This sector, more than any, provides the most public profile for the paper bag. An attractive paper carrier bag offers excellent publicity for retail shops – a walking advertisement in the High Street. Not only the retail trades but also manufacturers who wish to promote their products make use of this powerful advertising medium by distributing carrier bags to retail outlets. Exhibitions and shows offer an additional means of publicity.

The paper carrier bag is used extensively as a 'carry-out' or 'carry-home' pack for a wide range of ready meals and snacks for High Street fast-food outlets. They can be made with either twisted paper or flat tape handles.

A smaller and specialised market exists for luxury gift carrier bags. These bags, often handmade, use high-quality art papers with board attachments and coloured rope or cord handles. They are used as point-of-sale bags for luxury goods, such as perfumes and jewellery. In addition, they are sold empty, mainly in boutiques and greeting card shops for customers' use as presentation carriers.

Finally, mention should be made of the supermarket check-out bag ('shopper' or 'chuck' bag). These large capacity brown kraft SOS bags have fulfilled a demand in Europe, but

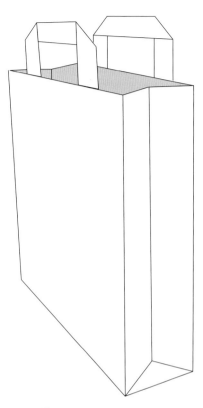

Figure 5.5 Block-bottomed SOS – self-opening satchel – carrier bag with handle.

mainly North America for transporting goods from a supermarket check-out to a car boot (trunk). They are often reused in the home as waste bags.

The papers used for this sector (apart from the 'chuck' bags) are usually higher weight and often coated papers to achieve better printing results, combined with sufficient physical strength to protect and carry the contents. Frequently, the contents are already in sealed containers to protect against moisture or grease penetration. If not, a wide range of coatings and protective layers can be added to the inside of the carrier bag.

5.3 Types of paper used

Section 5.2 refers to the different papers used in paper bags according to the application. This section examines these papers in more detail.

5.3.1 Kraft paper – the basic grades

'Kraft', the German word for strength, is the grade of paper most commonly used. Sulphite paper is used for flat and satchel bags but it can lack whiteness. Kraft paper is used in both unbleached form (brown kraft) and bleached form (white kraft). Within these two main categories, numerous qualities are available according to the mill process. Machine-glazed (MG), supercalendered (SC) and machine-finished (MF) krafts are all available according to the degree of smoothness required. Striped or ribbed krafts are used and are traditional for certain retail trades, particularly clothes and fashion shops. Coloured krafts can be obtained but minimum tonnages mean that a solid ink coverage is often a more practical proposition.

Other grades of paper used for special applications are as follows.

5.3.2 Grease-resistant and greaseproof papers

These are used widely for direct food contact and pet foods where protection from greasy and oily products is required.

5.3.3 Vacuum dust bag papers

Special grades of paper have been developed for paper bags used in vacuum cleaners. The important factors are porosity and filtration to ensure the air flows freely through the paper but not the dust.

5.3.4 Paper for medical use and sterilisation bags

A special government-approved paper has been developed for this purpose. Porosity, bacteria-proofness and wet-strength properties are all important factors to ensure that sterilisation bags withstand steam autoclaving and remain sterile.

5.3.5 Wet-strength kraft

This grade is often specified where exposure to damp or moisture is likely. It is an essential requirement for hospital bags which are steam sterilised.

5.3.6 Recycled kraft

Recycled kraft is used for carrier bags and other applications where a recycled paper is acceptable and is often specified by customers. It is not suitable for some medical or food grades where virgin kraft is used. There are many grades of recovered paper which can be used for recycling and some papers containing 100% recycled fibre have been approved for direct food contact.

5.3.7 Coated papers

Kraft papers can be coated with china clay to provide an extra smooth finish and achieve enhanced print standards. Extrusion coating of kraft papers with various grades of plastic film is used to both protect and enhance appearance.

5.3.8 Laminations

Kraft papers are used in conjunction with all plastic films and/or with aluminium foil to provide up to 4-ply laminations depending on the level of protection required from the bag. Aluminium is used particularly where odour, flavour retention and light protection are important factors.

5.3.9 Speciality papers

Many other grades have been used to produce bags with specialised requirements, some on an experimental basis. These include:

- water-soluble paper
- flame-retardant paper
- wax-impregnated paper
- bitumenised kraft
- creped kraft.

5.3.10 Weights of paper

Grams per square metre (g/m^2) is used for measuring the weight of paper. The range of weights used depends on the style of bag and machinery used. In general, lighter substances can be used to make flat and satchel bags and range from 35 to $80 \, g/m^2$. The range for SOS bags is between 45 and $130 \, g/m^2$ – the minimum weight is higher due to the manufacturing process of SOS bags. The total grammage of paper bags can, of course, be increased by making 2-ply bags. The grammage of some coated and processed papers needs to be increased to match the equivalent strength of a kraft paper. (Note: $lbs/3000 \, ft^2 = 1.627 \, g/m^2$.)

5.4 Principles of manufacture

5.4.1 Glue-seal bags

5.4.1.1 Flat and satchel bags

The bags are formed from a reel of the required material which is usually about 20 mm wider than the total width of the bag, i.e. face + gusset × 2. The 20 mm of extra material becomes the overlap to form the side seam along the bag.

The bag is shaped over a forming plate matching the size required. Guide wheels tuck in extra paper to form a gusset if a satchel bag is required. Paper from the reel is cut to the required length by a knife or a blade mounted on a rotating cylinder. Glue is applied to the web at two separate stations to seal the bottom seam and the side seam.

5.4.1.2 SOS bags

As can be seen from the illustrations in Section 5.2, the method of construction for SOS bags is more complex. The free-standing bag is formed by a series of folds, slits and creases.

Heavier paper is needed to withstand this process, and some machine speeds tend to be slower than for flat and satchel bags. The principles relating to reel width and glue application are the same as for flat and satchel bags. Fixed size high-speed SOS machines are widely used within the food service market sector which uses over 1 billion paper bags in the UK alone every year.

5.4.2 Heat-seal bags

As the construction depends on a heat seal, the reel or reels of paper must include a thermoplastic material to form the inside of the bag. This can be in the form of an extrusion coating or a separate web of heat-sealable film. Virtually all heat-seal bags are of SOS shape. The bottoms are formed by a series of folds, cuts and creases. Heat is applied to both the base of the bag and the sidewall by heated machine parts and a hot-air flow.

5.4.3 Printing on bag-making machines

Depending on the nature of the design required and the number of colours, the printing process is often included in the bag-making machinery. Paper from the reel is first fed through the printing unit before passing through to the bag-making process. Quick drying water-based inks are essential to avoid reduction in machine speeds. Printing can be offered from one colour to eight colours and include an ultraviolet (UV) lacquer.

5.4.4 Additional processes on bag-making machines

As paper from the reel passes through the bag-making machine, other processes and attachments can be included.

5.4.4.1 Punching

If windows or die-cut handles with card inserts are required in the bag, the paper web can have a punched cut-out. Unless the bag is totally lined with film, a narrower reel of film to suit the width of window is used. Bags can have a large number of very small holes punched, for example ventilation holes for bulbs.

5.4.4.2 Paper handles

Narrow reels of folded or twisted paper and reinforcing paper to secure the handles are fed on to the main web of paper to produce carrier bags.

5.4.4.3 Lacquers and adhesives

A strip of heat-sealable polyvinyl acetate (PVA) coating can be applied to the top of the inside of the bag. Hot-melt adhesive can be sprayed on to the outside of the bag. These provide customers with the means to heat-seal bags when filled.

5.4.4.4 Metal strips

Metal strips are a means of reclosure.

5.4.4.5 Reinforcement strips

A supplementary reel can be used to provide a reinforcing patch on the base of the bag.

5.4.5 Additional operations after bag making

Additional processes include the following:

- Wicketting of bread bags to be used on automatic machines in bakeries. This is a process whereby the bags are collated in any quantity on a U-shaped metal wicket wire utilising holes pre-punched through the lip of the bag. The purpose is to enable packing on automatic filling lines where the bags are mechanically selected and air inflated to enable the product to be inserted. This packing method is most suited to factory bakers of bread and farm-produce packers.
- 'Stringing' of flat and satchel bags for use in retail outlets. This consists of threading a string through one corner of a stack of bags, which can then hang on a hook. (The stringing operation is a single stitch carried out on a large sewing machine.)
- Fitting of 'tin-ties' around the top of the bag for reclosure.

5.5 Performance testing

5.5.1 Paper

Standard tests exist in the paper trade for measuring the strength and other properties of paper. These include burst, tear, tensile, stretch, smoothness, stiffness, thickness, porosity and substance. More specific tests are carried out on special grades, for example wet-strength and grease-resistant papers.

5.5.2 Paper bags

When the reels of paper have been converted into paper bags, performance tests can be conducted both by the bag maker and the bag user to ensure conformity with agreed standards. Many of these tests will be of a practical nature, such as handle-held multiple jog testing to mimic use of a bag whilst walking.

5.5.2.1 Hospital bags

These can be tested in a steam steriliser to ensure that they withstand the steam sterilisation cycle and remain sterile.

5.5.2.2 Dust bags

Performance tests are conducted in the vacuum cleaner. These tests are designed to ensure the maximum level of dust retention and maintain sufficient porosity to prevent strain on the motor.

5.5.2.3 Paper bags for food use

Food manufacturers are, of course, concerned that paper bags will preserve their product and provide the required shelf life. The moisture vapour transmission rate (MVTR) of the bag is an important factor in heat-sealed bags. In addition, accelerated shelf-life tests can be conducted on the bags.

5.5.2.4 Physical strength

Drop and tear tests have been devised to measure the strength and resistance of a paper bag to bursting. There are weight tests designed to gauge the strength of handles on carrier bags.

5.6 Printing methods and inks

5.6.1 Printing methods

5.6.1.1 Flexographic printing, off-line

Due to advances in flexographic printing in the last 15 years, 'off-line' or pre-printing by the flexo process has become the norm for multicolour illustrations with the additional use of UV lacquers. The use of the common, or central, impression (CI) cylinder for half-tone printing has resulted in ever-finer screens being printed on suitably smooth paper surfaces. It now requires a trained eye to detect the difference between flexo and photogravure print.

5.6.1.2 Flexographic printing, in-line

The ability to print the paper reel on the bag machine from rubber plates was viewed as a major development in developing paper bags (see the chronicle of events in Section 5.7.1). This in-line printing is used extensively for printing bags but is generally restricted to four colours. As aniline dye inks were used in the early days, the process was initially known as the aniline printing process. Inks today are based on polyamide resins. The plates were originally based on rubber, but this has been replaced with photopolymer plates.

5.6.1.3 Photogravure

This is an intaglio process using etched cells as distinct from the relief printing of flexography. Gravure printing is synonymous with excellence in print quality and definition of very fine print. It has the drawback of high origination costs, which means it is only viable for long print runs. Its use has become more limited in recent years due to the advances in the quality of flexographic print, which in commercial terms is now similar to photogravure.

5.6.1.4 Silkscreen

This stencil printing process is slow and expensive but achieves impressive results. Silkscreen may be used for printing or over-printing on paper bags after manufacture usually for the packaging of luxury items.

5.6.2 Inks

The nature and viscosity of printing inks depend, of course, on the process being used. In addition, printing inks which are used on paper bags for food and hospital products must conform to strict trade standards.

Special 'indicator' inks are used for paper bags for ethylene oxide (EtO) gas and steam sterilisation to show that the packs have passed through the sterilisation process.

5.7 Conclusion

5.7.1 Development of the paper bag industry

The introduction to this chapter referred to the age of the paper bag industry; listed in the following is a chronicle of some major events in the development of the industry. It is not comprehensive but serves to demonstrate the direction in which the industry was evolving.

1630	The first recorded reference to grocery bags
1858	Bibby & Baron of Bury, England, installs the first machine to make paper bags
1870	Margaret Knight founded the Eastern Paper Bag Company (USA). Just before that, she had been an employee in a paper bag factory when she invented a new machine part to make square bottoms for paper bags
1872	Luther Crowell also patented a machine that manufactured paper bags
1872	Robinsons of Bristol, England, acquires a New York patent for making 'satchel' shape bags, i.e. bags folded with a side gusset
1880/1890	Smith Anderson (Fife, Scotland) having made paper since 1853 build factory incorporating the first automated bag-producing machines
1890	First aniline press built by Bibby and Baron in Liverpool – but inks subject to smearing/bleeding
1902	Hele Paper Mills provided the idea of converting printed reels of paper into bags. This cheaper process was soon seen as a powerful advertising medium leading to rapid expansion of the bag industry
1902	A Bristol, England, factory has 17 paper bag machines, 400 workers are still engaged in making bags by hand
1905	Holweg built an aniline press as a tail unit on a bag machine
1930	A Bristol factory, opened in 1912, now produces 25 million bags a week
1932	Welton Bag Company opens factory to make paper bags and carrier bags at Radstock, England

During the 1930s, a European patent was registered for printing paper from rubber stereos. The ultimate aim of printing the reel of paper on the bag-making machine had now been achieved. The process was called Aniline or Densatone but is now generally known as flexographic printing. Since the 1990s, flexographic printing has developed in quality terms to resemble that of photogravure.

5.7.2 The future

The paper bag industry has shown that it is adaptable and responsive to challenges. Paper, the raw material, is a natural and renewable resource. Paper bags have an assured future in the forefront of all the available packaging materials. Having regard for the current packaging legislation trends in Europe, paper bags are well placed for a period of sustained growth.

Reference

Wmich, 2012, *Introduction*, Department of Paper Engineering, Chemical Engineering & Imaging, Western Michigan University, Kalamazoo, Michigan, visit http://www.wmich.edu/ppse/flexo/

Websites

http://www.smithandersonpackaging.co.uk – for further information about paper bag manufacture
http://www.welton.co.uk – for further information about paper bag manufacture
http://www.wmich.edu/pci/flexo – for further information about flexographic printing
http://www.hgweber.com – details about bag-making machinery

6 Composite cans

Catherine Romaine Henderson

Communication Consultant, Greer, SC, USA
Chapter commissioned by Sonoco, Hartsville, SC, USA

6.1 Introduction

By utilising the best combination of materials, composite can construction ensures optimum presentation and mechanical strength as well as hermetic protection. This choice of materials and production techniques for can components – body, top and bottom – offers packaging solutions that are flexible and cost-effective for mass consumption and luxury products, as well as for numerous industrial applications. Whilst dry-food packaging is the most common application for today's composite cans, more and more retailers are seeing its value as a customisable option for vendable and non-food products.

In the latter part of the first decade of the twenty-first century, economic conditions worldwide caused consumers, manufacturers and consumer products companies to re-evaluate priorities. Consumers wanted to spend less whilst getting more value, packaging manufacturers wanted to curtail costs and consumer products companies wanted to satisfy consumer demands. In 2010, the *Packaging Digest* magazine reported that the issues facing packaging over the next few years will include: the economy, sustainability and private branding (*Packaging Digest*, 2010). The consensus is that the economy will remain unpredictable. Downturns will drive private label brands to increase as retailers try to improve margins. And sustainability is here to stay. In fact, many companies are using sustainability efforts as a way to differentiate themselves in the marketplace.

Sustainability is a term that comes up repeatedly in the packaging industry today. Currently, efforts are underway to develop standards for sustainability worldwide. There is work in progress by the International Organization for Standardization (ISO) and the European Union's Packaging Directive. The Sustainability Consortium is also working globally to develop science and tools that address the environmental and social implications of products throughout the life cycle. (See Tables 6.1 and 6.2 for information on sustainability packaging issues.)

In most cases, a sustainable composite can is one that delivers a clear environmental advantage over the package it replaces either through the use of more sustainable materials, source reduction, less energy, water or raw material to produce or results in fewer carbon

Table 6.1 Sustainable packaging issues

Reduced energy use	51%
Reduced waste	50%
Increased recycled content	33%
Increased pallet/truck capacities	31%
Reduced transportation costs	29%
Reduced primary packaging	24%
Reduced secondary packaging	17%

Source: Food Engineering, Packaging trends survey (2010).

Table 6.2 Top issues facing the global packaging industry

Sustainability	Cost	Innovation	Availability of material, quality and performance issues
39%	33%	16%	13%

Source: Survey: Sustainability, cost challenge global packaging industry (Du Pont, 2011).

emissions. For example, in 2009, Kraft Foods replaced the steel cans used for its Maxwell House brand of coffee with high-performance composite paperboard cans. The new package features a 30% reduction in weight, uses 50% recycled content (including 25% post-consumer) and eliminates approximately 8.5 million pounds of packaging per year. Kraft Foods also reports that compared with metal cans, paperboard cans require less energy to manufacture and result in fewer greenhouse gas emissions (http://sustainablegallery.brandpackaging.com/Reduced/#id=4&num=5).

Many manufacturers rely on sustainability as a key selling point, either in the use of recycled material, incorporation of post-consumer recycled content, Sustainable Forestry Initiative (SFI) certification or a combination of all of the above. This consumer-driven requirement is becoming a primary consideration in packaging choice. In January 2011, Pira International, a well-renowned research firm, reported that 'Consumer awareness of environmental issues and new material developments are the most important drivers in the development of sustainable packaging'.[1]

More than 3 years after the initial global economic downturn, consumers are continuing to demand more value (bulk and big value packages), convenience (easy opening, resealable packages) and sustainable options (new designs, less material, improved recyclability). In addition, packages that meet these criteria whilst providing shelf differentiation (shape, on-the-go sizes) are more successful in the marketplace.

In August 2010, SustainableBusiness.com reported that the sustainable packaging industry is expected to exceed $142 billion (US) by 2015. The reasons given for this surge in growth include increased awareness, government initiatives and regulations as well as a way for companies to cut costs and reduce packaging waste. Additionally, consumer demand for more sustainable packaging choices is driving growth in key markets including cosmetics

---

[1] Pira International conducted a survey of key players across the global packaging value chain. Seventy-nine per cent of respondents rated 'Increased exposure of consumers to environmental issues' and 'Advances in materials technology' as either a growth drive or major growth driver.
Source: http://www.pira-international.com/consumer-awareness-of-environmental-issues-and-materials-technology-to-drive-sustainable-packaging-d.aspx

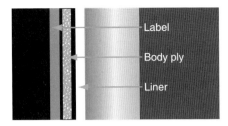

Figure 6.1 Composite can construction. (Source: SONOCO.)

and personal care, food and beverage, food service and shipping markets, healthcare, and others (SustainableBusiness.com News, 2010).

The bottom line is that the composite can is still one of the most versatile packages in the marketplace. The composite can is produced at a relatively low cost and in most cases is a sustainable option. It also provides some general performance characteristics, such as containing the product and allowing it to be readily and easily dispensed. It also delivers adequate shelf life for the type of food it contains.

In May 2011, DuPont conducted an online survey of more than 500 packaging professionals worldwide to identify the top issues facing the global packaging industry. The survey consisted of open- and closed-ended questions. As the results in Table 6.2 show, nearly 40% of respondents said that sustainability issues have top priority closely followed by cost implications. Respondents expressed interest in bio-based and compostable materials for packaging (Du Pont, 2011).

6.2 Composite can (container)

6.2.1 Definition

A composite can is a convolute-wound, spiral-wound, linear-draw or single-wrap rigid body with one or both end closures permanently affixed. Whilst paper is the primary component of the canister body, the total construction of the package involves several layers of material, often including aluminium foil and plastic (Fig. 6.1).

6.2.2 Manufacturing methods

There are several processes which can be used to manufacture composite cans.

6.2.2.1 Convolute winding

The convolute manufacturing method involves winding multiple layers of a single-ply of material around a rotating mandrel to form a round (or non-round) can body. The body material is coming in at a right angle to the mandrel, with all of the paper going in the same direction (Fig. 6.2a). A cutting operation sizes individual cans to customer specifications.

6.2.2.2 Spiral winding

The plies of a spiral can are wound around a stationary mandrel in a helical pattern, bonded with an adhesive in high volume, continuous production (Fig. 6.3). Individual

186 Handbook of Paper and Paperboard Packaging Technology

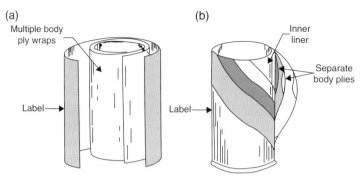

Figure 6.2 Construction of sidewalls of composite containers: (a) convolute construction, and (b) spirally wound construction. (Courtesy of the Packaging Society.)

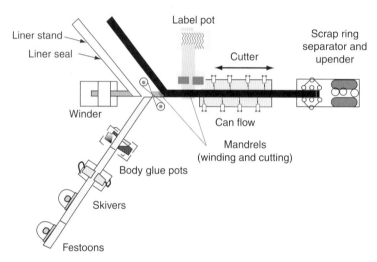

Figure 6.3 Basic can technology. (Source: SONOCO.)

sizes are then cut to customer specifications. The spiral can is divided into five distinct parts (Fig. 6.2b):

1. an inner liner
2. paperboard body, separate plies (typically two-ply)
3. label
4. top closure
5. bottom closure.

6.2.2.3 Linear draw

Multiple plies are pulled/drawn around a stationary mandrel in the same direction of the mandrel (Fig. 6.4).

Composite cans   **187**

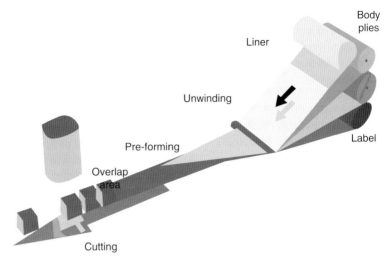

Figure 6.4 Multiple plies wound around a stationary mandrel (linear draw). (Source: SONOCO.)

Figure 6.5 Single-wrap can. (Source: SONOCO.)

6.2.2.4 *Single wrap*

In this manufacturing process, composite cans are made one at a time. An already-sized, pre-printed, unglued blank is formed around a mandrel. Once the overlap seam is heat sealed, the can is positioned to receive the bottom-end closure which is heat sealed and clamped to the body. The top is then heated and curled. The single-wrap can is available as straight wall or tapered to nest, or stack, one inside another (Fig. 6.5).

6.3 Historical background

The composite can has a long history that dates back more than 100 years when packaging was in its infancy and packages were designed to hold and transport lightweight, easy-to-hold dry products. In its earliest beginnings, it was an offshoot of the paper tube. In the late

1800s, gunpowder, oatmeal and salt were commonly packaged in paper tubes with crimped ends. At some point, near the turn of the twentieth century, paper end plugs were added, and composite canisters were born.

Whilst the early construction of these packages involved nothing more than unlined paper bodies with crimped paper or metal ends, interest in the composite can began to grow. Most commonly seen on the pharmacist's shelf, composite cans had become the package of choice for items like Epsom salts, sulphur and other powdered drugs.

During World War II, when the USA was trying to limit the use of steel and direct all available materials to the war effort, the composite can was used as a paint can. And by the mid-twentieth century, the composite can market had experienced rapid expansion. At this point, the product was offered with lined or sprayed can bodies, coupled with the appearance of opening and pouring features. It was also during this tremendous growth period that manufacturing and distribution processes were becoming more complex. As production became more sophisticated, so did the demands placed on manufacturers. New nationwide distribution channels led to high-volume, high-speed manufacturing. Package design began to focus on product protection, ease of handling and user convenience. The resulting versatility of the composite can made it the package of choice for many products, such as cleanser, caulks and frozen berries.

It was also during the 1950s that makers of refrigerated dough began seeking an inexpensive, convenient package for their products. The composite can was selected because of its low cost, its ability to hold internal pressure and an opening feature that did not require the use of a can opener.

Due to the success of the composite can in the dough market, numerous technological advances were being implemented. These advances included the development of high-speed winding and cutting equipment, the use of improved liners, like aluminium foil, for enhanced product protection and special metal-end designs coupled with better seaming techniques.

Over the next decade, the composite can migrated from specialty markets to the more high-volume commodity-oriented segments. In the 1960s, the first commercial shipments of frozen citrus concentrate packaged in composite cans took place in Florida. Within a few short years, the composite can achieved package of choice status for 6 and 12 oz, a position it still maintains in the twenty-first century.

Paralleling the growth of the concentrated juice market was the petroleum industry. During the 1960s and through the 1980s, the composite can was the package of choice for the quart-size container of motor oil. This success, coupled with some technical experimentation in coffee and solid shortenings, led to further successes over the next several decades.

In the 1970s, Procter & Gamble made headlines when the company introduced a brand new product in a brand new package. The Pringles® potato crisp, a uniquely shaped chip, was the first large-scale snack food to be packaged in a hermetically sealed composite can.

Major manufacturing companies seized the opportunity to develop the snack-food business and began, rapidly, establishing production systems and technical support groups to satisfy the needs of this emerging market. The success of the nitrogen-flushed, hermetically sealed composite can is well documented and created numerous opportunities to use the composite can as a package for processed meat snacks, powdered infant formula, peanuts and specialty nuts, noodles, coffee and other food items.

Table 6.3 Leading brands packed in composite cans

Segment	Brand names	Time in market (years)
Snacks	Pringles/Planters	40
Refrigerated dough	Pillsbury	60
Concentrates	Minute Maid, Seneca	60
Adhesives/sealants	DAP, DOW, GE	≥40
Cleansers	Ajax/Comet	60

Table 6.4 Products packed in composite cans

Foods	Misc. non-food	Beverage	Adhesives/Sealants
Refrigerated dough	Petroleum products	Frozen concentrates	Caulks
Snacks/nuts/chips	Agricultural products	Coffee	Adhesives
Frozen fruits	Household products	Powdered milk products	Sealants
Dried fruit	Pet products	Powdered beverages	
Salt/spices	Dish detergent	Powdered infant formula	
Solid/liquid shortening	Cleansers		
Confectionery			
Nutraceutical products			
Cookies/crackers			
Powdered foods			
Dried meats			
Cereal			

6.4 Early applications

Many of the earliest composite can applications, as mentioned in Section 6.3, are still in use today. Table 6.3 illustrates some of the world's most famous brands and the length of time they have been marketed in composite cans.

6.5 Applications today by market segmentation

The packaging industry, projected to reach US $733.9 billion by 2014 (http://www.plasteurope.com/news/detail.asp?id=217866), is one of the world's largest and most diverse manufacturing sectors.

More than half a decade after its introduction as a commodity package, the composite can still maintains an admirable position in the marketplace. Pira International reports that paper and paperboard is still the single largest sector amongst packaging materials (*Market Intelligence Guide for Packaging*, p. 6).

To summarise, the major markets for the composite can in the packaging industry include foods, non-food, beverage and adhesive/sealant products. Further segmentation of each market is listed in Table 6.4.

6.6 Designs available

Whilst the earliest composite cans were available with limited design options, technology has enabled a total redesign of the package. The composite can allows industrial and consumer companies to create packaging based on product attributes and customer preferences with different sizes and shapes with numerous opening-feature options.

6.6.1 Shape

Shaped containers are an ideal and proven way to establish product differentiation on the cluttered shelves of drug, grocery, specialty and retail stores. The traditionally round composite canister is now available in unique shapes – rectangular, triangular and oval. The ability to print high-impact graphics is an added bonus to other can features like stackability and delivering eye-catching billboarding opportunities.

6.6.2 Size

- *Diameter.* In the case of a round can, the size is determined by the diameter. There is a difference between the way this is expressed in the USA compared with Europe.
  - In the USA, the dimension used is the diameter of the metal can end which is applied to the tube, and in Europe the dimension used is the internal diameter of the tube itself. Popular diameters in the USA expressed in inches and sixteenths of an inch are 202, 211, 300, 401, 502 and 603.
  - In Europe popular diameters are expressed in mm (US measurements in brackets): 65 (211), 73 (300), 84 (307), 93.7 (313), 99 (401) and 155 (603). (The 73, 84 and 93.7 mm are also produced with other styles of closure, which are within the tube diameter. The 93.7 mm is used for powdered milk and drinking chocolate and the 73 and 84 mm are used for gravy powders and have paper bases.)
- *Height.* A wide variety of heights is available.

6.6.3 Consumer preferences

Manufacturers of consumer products are taking consumer preferences into consideration when determining composite can characteristics. For example, many conventional canned goods now offer an easy open (EZO), closure feature to eliminate the need for any type of opening device, for example can opener, twist key. Certain demographics are also considered. For example in the USA, where the population is ageing, many can features are designed to make handling and opening easier.

6.6.4 Clubstore/institutional

With the explosion of clubstore and institutional buying opportunities for the general public, larger cans and combination packs are becoming more prevalent in the marketplace. Often multi-packs are bundled together with shrink-wrap packaging or labelling. These larger packages are often considered value buys for consumers with less disposable income due to economic pressures. In addition, many companies are opting for value-size packages for those who do not want to buy in bulk. These larger-scale packages give consumers the value they seek and the option to buy in a grocery store.

6.6.5 Other features

Because of the composite can's versatility, there are numerous ways to enhance the package based on particular customers' needs:

- *Combination packs*. As manufacturers of consumer products continue to consolidate, more combo-packs are appearing on the shelf. One of the fastest growing markets is the pet-care industry. It is very common to see cans of pet treats sold in combination with other pet-care products. In this type of scenario, multiple packaging options are combined for promotions.
- *Pressure-sensitive coupon application*. To increase shelf appeal and sales, many manufacturers apply a coupon directly to the end or body of the composite can prior to shipment. These immediately redeemable in-store money-off coupons or recipe ideas often influence consumers' purchasing decisions.
- *Insert-inside overcap*. Similar to the pressure-sensitive coupon application, an insert is attached to the inside of the overcap and is not visible to the consumer prior to opening the product. In many instances, a coupon is included for cents off on subsequent purchases.
- *Flip-lid and scoop*. For consumer convenience, H.J. Heinz introduced its Nurture Growing Baby Follow-on Milk product, available in the UK, in a sustainable composite can that features a custom overcap (flip-lid) and scoop to measure the right amount of formula.

6.6.6 Opening/closing systems

In addition to strength and versatility, the composite canister is also known for its numerous options for opening and closing systems. Consumers prefer easy opening and dispensing features that provide resealability to maximise freshness. Paper, aluminium, steel, plastic or membrane closures are fitted on cans by single or double seaming, gluing, pressure inserting or heat sealing.

Closure choice primarily depends on the product to be packaged, as well as the ease of use, protection needed, dispensing requirements, the opening and re-opening ability and the necessary hermetic properties. Choices include:

- glued or heat-sealed paper bottoms and/or lids
- steel and aluminium bottoms and/or EZO lids with single or double seaming
- lever system with or without pilfer-proof membrane
- seamed metal ring with a foil membrane and plastic overcap
- plastic lids and/or bottom with pour spouts
- rolled edge with flat membrane and plastic overcap
- sealed recessed foil membrane with pull-tab and plastic lid
- glued-in or sealed plastic closure with hinged lid and pilfer-proof membrane
- paper lid with or without flat membrane
- two-stage lid, that is removing a membrane closure and using the plastic over cap with a living hinge for resealability.

Table 6.5 Features and benefits of various opening ends for composite cans

Steel end options	
Type	Full panel
Features	Full panel easy open ring-pull end
	Food and non-food applications
	Double reduced steel available in select diameters
	Safety fold available in select diameters
Benefits	Easy open
	Lithography, embossing available
	Style, design options
	Recloseable with overcap
	Vacuum or nitrogen flush capable
Aluminium end options	
Features	Full or partial-pour panel with easy open ring-pull end
Benefits	Easy open
	Folded edge or hot melt bead provides cut protection
	Recloseable with overcap
	Vacuum or nitrogen flush capable
Flexible membrane options	
Features	Peelable, flexible multi-layer or foil membrane heat sealed to a tinplate ring or tinplate steel ring
	Valve available for pressure applications
	Ring-pull option for some diameters
	Levelling feature
Benefits	Easy open
	Printing, embossing available
	Recloseable with overcap or plastic plug
	Pressure, vacuum or nitrogen flush capable
Plastic end options	
Features	Resealable moulded plastic overcap
	Available in colours to complement label
	Product protection through resealability
Benefits	Easy dispensing
	Preserves freshness and integrity of product
	Increases shelf life
	Provides tamper resistance, deters pilferage
	Spill proof
	Dual function as opening and reclosing feature

6.6.6.1 Top-end closures

Top-end closures are often, but not always, differentiated from bottom-end closures by the presence of an opening feature. Various types of opening features are available in the market today due to the wide range of products packaged in composite cans. In today's economy, consumer preferences dictate the types of opening features which are most frequently used. Some of these opening features are also used with shaped and non-round composite cans. Details of various designs of opening are given in the following text. For their features and benefits, see Table 6.5.

Steel ends

- full panel
- partial pour
- Mira strip (peelable and ring-pull options).

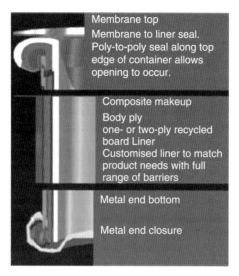

Figure 6.6 Cross section of typical composite can construction with metal end. (Source: SONOCO.)

Aluminium EZO ends

- full panel
- partial pour.

Flexible membrane

- paper or aluminium foil
- recessed membrane with plastic plug
- flat membrane applied to curl-top can
- combination membrane with steel reel for retort applications.

Plastic

- full-panel EZO end
- rotor-tops
- shake and pour
- overcaps
- flip-top
- hinged.

6.6.6.2 Bottom-end closures

Bottom-end closures are primarily steel (Fig. 6.6), although plastic, paper (Fig. 6.7) and aluminium are used in some applications.

The strength of the steel end is generally correlated with basis weight and temper. Commonly used base weights range from 55 to 107 lb per basis box, with temper ranging from T-3 to T-5.

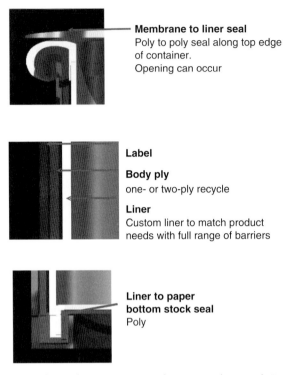

Figure 6.7 Cross section of typical composite can with paper or plastic end. (Source: SONOCO.)

Coatings may also be applied to the end closure. This is done to protect the raw material from attack by the packaged product or other environmental elements. Examples of coatings include tinplate (the most common), vinyl, epoxy and phenolic. If maximum protection is needed, a sealing compound is applied to the end. The compound serves as a gasket to seal the can and protect the contents. The chosen compound is based on the properties of the product being packaged.

In the latter part of 2008, reports and concerns about bisphenol A (BPA) on tin plate ends (those used on composite cans and other packages) escalated to near-panic levels. For many years, the chemical has been under fire from the scientific community and is considered to be a carcinogen. In spite of studies and conclusions from the U.S. Food and Drug Administration (FDA) and the European Food Safety Authority, which independently arrived at the same conclusion – BPA in food packaging is safe – efforts to eliminate it from metal ends continue. As recently as April 2011, reports indicate that BPA poses no danger to humans at any age, but the controversy still exists. Today, BPA-free options are readily available in the marketplace (http://www.packagingdigest.com/article/343256-FDA_says_BPA_is_safe_for_food_and_beverage_packaging.php).

6.7 Materials and methods of construction

As mentioned in Section 6.2, the composite can is made from several layers of material (Fig. 6.8) and is sealed with top and bottom ends. Material selection and the final can composition are based on customer needs and end-user preferences.

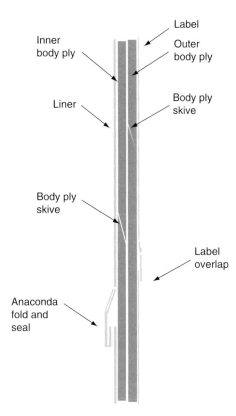

Figure 6.8 Cross section of composite can body. (Source: SONOCO.)

6.7.1 The liner

The function of the liner is to provide product protection. It must be impervious to attack by the product and at the same time resist transmission of its own components into the product. The liner must also maintain its integrity under impact from external forces, particularly during shipping. In some applications, such as salty snacks, it must also resist abrasion and puncture from within. In some cases, the liner must act as a barrier to oxygen and moisture, whilst in others, it serves only to prevent product leakage.

Depending on the product to be packaged, the liner is structured with two or more layers of material. Materials used on the product contact surface include polypropylene, polyethylene (PE), polyethylene terephthalate (PET), Surlyn®, metalised PET (MPET) and others. In order to maintain a complete barrier with spiral winding, the edge of one spiral winding overlaps the preceding winding, the edge is turned through 180° and heat sealed, face to face, with the preceding winding in the overlapped area. This is known as an 'Anaconda fold and seal' (see Fig. 6.8). Aluminium foil is used under the contact surface where high barrier and product protection are both required. A kraft (paper) layer completes the liner structure. It serves as a carrier through the converting process that produces the liner structure.

The ratings in terms of 'high', 'medium' and 'low' for moisture vapour transmission rate and oxygen permeability are shown in Table 6.6, and examples of typical liner structures, their barrier ratings and typical uses are shown in Table 6.7.

Table 6.6 Ratings for various levels of functional barrier

	Approximate barrier parameters	
	MVTR* g/24 h/container 38°C/90% RH	O_2 permeability cc/24 h/container 25°C/50% RH
High	<0.02	<0.01
Medium	<1.00	<0.50
Low	<5.00	<1.00

*Moisture vapour transmission rate.

Table 6.7 Barrier ratings for typical liner structures

Liner structures commonly used	Barrier properties	Typical uses
Kraft/foil/polypropylene	High	Hermetic food (coffee, nuts, snacks, shortening, peanut butter, infant formula, etc.) solvent- and latex-based paints
Kraft/foil/Surlyn®	High	Hermetically sealed packs for food, including nuts and snacks
Kraft/foil/high density polyethylene (HDPE)	High	Solid shortening, powder beverage, snacks
Kraft/MPET/Surlyn®	High	Hermetic snack foods
Kraft/MPET/HDPE	High	Coffee, nuts, snacks, solvent- and latex-based paints
Kraft/PE/foil/Coex HD	High	Refrigerated dough
Kraft/foil/vinyl slipcoat	Medium	Refrigerated dough, adhesives, caulk, powder beverage, non-hermetic food
Kraft/PE/PET/Surlyn®	Medium	Pet food, paper-bottom can
Kraft/LDPE/OPP	Medium	Solid shortening
Kraft/LLDPE/White HDPE	Low	Hard to hold frozen concentrate
Kraft/Surlyn®/White HDPE	Low	Hard to hold frozen concentrate
Kraft/White HDPE	Low	Frozen citrus concentrate
Kraft/HDPE	Low	Motor oil

6.7.2 The paperboard body

The paperboard body of the composite can is its source of strength (Fig. 6.6). The body must provide axial strength sufficient to withstand multi-pallet stacking of finished goods and sidewall strength necessary to absorb impact during distribution.

There are generally two types of paperboard used in composite cans: unbleached (brown) kraft and recycled. Both provide similar axial and sidewall strength, but kraft may resist more shearing because of its longer fibres. Recycled board is an affordable alternative, and due to regulations in some geographic locations it is a required component of packaging. In fact, most composite cans in the market today use recycled paperboard. Decades of research in paper science have improved the properties of recycled paper grades to match kraft paper, whilst also providing some cost savings.

In order to achieve a neat joining between adjacent spirally wound layers, the edges are 'skived'. This means that the edges are cut at an angle so that the adjacent edges mesh neatly together. This also adds strength to the composite can.

6.7.3 Labels

The label is the outer layer of the composite can that meets a variety of needs. It serves as a billboard for product information, nutrition information and attention-grabbing graphics, as well as providing additional barrier protection.

6.7.4 Nitrogen flushing

The materials and methods of constructing a composite can depend on its end-use and desired performance characteristics. When composite cans were being developed for the snack industry, the concept involved creating a hermetically sealed container that could replace the vacuum-packed steel can. This process – known as nitrogen flushing – was developed to accomplish this, and it is still in use today to extend the shelf life of certain products, such as nuts, snacks and chips (potato crisps).

The most common way to remove oxygen is called flushing. It works on the principle of flooding the container with a non-reactive gas, such as nitrogen, just before it is filled.

The type of gas flush is generally determined by the nature of the product. For example, powders have a tendency to pack together, that is particles compact together, making it more desirable to purge in several stages: firstly, to purge the can prior to filling; secondly, to purge the product in the filler; and finally, after filling, prior to seaming, to remove oxygen from the head space. For snacks, it may be only necessary to flush the filled can prior to seaming.

6.8 Printing and labelling options

6.8.1 Introduction

The two most common label materials are aluminium foil and paper. Foil is most often used for aesthetics and barrier properties when packaging products such as cleanser and some food stuffs. An example is the Pillsbury® refrigerated dough canister. On paper, there is more of a matt, less glossy finish, such as the Pringles® Potato Crisps canister.

A thermoset or thermoplastic coating is applied over the foil or paper for print protection and added barrier properties, particularly moisture. These coatings also provide shelf appeal, delivering a high-gloss appearance.

Ensuring proper registration on graphics is a complex process. Both aluminium foil and paper provide good offset litho, flexographic and rotogravure print surfaces. Paper is less expensive, but aluminium foil adds additional barrier protection.

It is important to remember that a package is a communication vehicle, which plays an important role in the purchasing process – both as a billboard for the brand and as a bearer of product information. Therefore, the impact the label has on the effectiveness of the composite can is paramount to the end-user, the consumer. It must be eye-catching and appealing.

There are several printing options for composite can labels. In the following sections, we will focus on the most common printing options: flexographic, rotogravure and offset printing.

6.8.2 Flexographic

Flexographic printing is an efficient, cost-effective and versatile printing method widely used for packaging products and envelopes. It provides high print quality on a wide variety of absorbent and non-absorbent substrates, including flexible materials, label stock, corrugated substrate, rigid plastics, envelopes, tissue paper and newsprint. It is still the fastest growing analogue print method on a global scale (http://www.gc3summit.org/LinkClick.aspx?fileticket=UmjNPklnuSo%3D&tabid=275).

Flexography is a relief printing process, in that the image is pressed onto the label material. This printing process is so named for its flexible rubber plates, which are also the key to the method's versatility and increasing popularity (see EFTA, European Flexographic Technical Association, www.efta.co.uk). The increasing use of flexography also increases the demands on the print quality of the end product.

Flexography accounts for nearly 60% of the $440 billion (USD) printed packaging market, due to:

- its ability to print on practically any substrate, absorbent or non-absorbent, including very thin extensible films
- the use of fast-drying inks, be they solvent-based, water-based or ultraviolet (UV)-curable
- the development of photopolymer plastic plates with good half-tone reproduction and low dot gain
- quick make-ready and changeover times making it profitable for short-run printing and its cost-effectiveness for many applications.

Other areas of growth are expected to occur in developing areas where grocery stores and retail outlets will replace outdoor markets. As urbanisation increases, so will the demand for flexographic printing.

6.8.3 Rotogravure

Once an art form that has been totally digitised, rotogravure is a printing technique characterised by high print quality and large quantity runs – hundreds of thousands or even many millions. Tiny ink volumes are transferred from the steel-based gravure printing cylinder to printing dots on the label substrate. Millions of printing dots show up to the human eye as letters, text and images.

Material knowledge, papermaking competence and production facilities are the secrets to ensuring superior printing performance. Gravure is a quality printing process producing excellent quality and constant reproductions throughout the entire print run. The secret of the gravure process lies in the cylinder. Increased robotics and total plant computer control systems will bring many further advantages for gravure.

Gravure's environmentally friendly printing process allows the use of:

- papers with higher recycled-fibre content
- highly effective solvent-recovery installations
- industry-leading methods to save paper, ink and energy
- less residual ink solvent content in gravure products.

In short, rotogravure is a very simple printing process that can produce millions of perfect copies at enormous speed. It also produces superb colours and good gloss on relatively low-quality paper. Because the gravure process can guarantee consistent reproduction of high-quality graphics, it suits the packaging market where buyers are greatly influenced by the quality of the printed image. The ideal substrates are generally smooth in finishing (clay-coated paper, films and foils) since effective ink transfer depends on thorough cell contact with the substrate to be printed. Presses for packaging gravure printing must meet different requirements due to the variety of substrate colours and finishing processes. Running at lower speeds enables the processing of difficult materials and drying of special inks. Packaging presses can combine in-line finishing processes including laminating, cutting, creasing, embossing, etc. (see ERA, European Rotogravure Association, www.era.eu.org)

6.8.4 Lithography (litho/offset) printing

Lithography is an 'offset' printing technique. Ink is not applied directly from the printing plate (or cylinder) to the substrate as it is in gravure or flexography. Ink is applied to the printing plate to form the image to be printed and then transferred or offset to a rubber blanket. The image on the blanket is then transferred to the substrate to produce the printed product.

Lithography is based on the principle that oil and water do not mix (hydrophilic and hydrophobic process). Lithographic plates undergo chemical treatment that renders the image area of the plate oleophilic (oil-loving) and therefore ink-receptive and the non-image area hydrophilic (water-loving). Since the ink and water essentially do not mix, the fountain solution prevents ink from migrating to the non-image areas of the plate.

The structure of the label substrate has a strong influence on print quality and ink drying, making this a field of high priority for research and development. Properties such as the base paper's formation and topography and the surface's porosity and smoothness must be optimised for the end-use and printing method.

6.8.5 Labelling options

- Spiral labels – roll-fed
  - printed using rotogravure or flexographic printing processes
- Convolute (strip) labels – roll-fed
  - more cost-effective
  - printed using rotogravure or flexographic printing processes
- Convolute (strip) labels – sheet-fed
  - can only be printed offset/lithographic
- Single wrap
  - pre-printed on the body blanks
  - liner, body and graphics are all together as one piece – wrapped to make the body
- Post-filling labelling techniques
- No label, no printed surface
  - customer applies labels.

Whilst many improvements can be readily made, there has to be a balance between cost and market demand. For the future, efforts will be focused to enhance label appearance with a more glossy finish. The differences between rotogravure and flexographic processes are narrowing as printers have made tremendous strides to improve the quality of the latter. Rotogravure printers are working toward reductions in pre-press costs (see Future Trends for information on the film label).

6.9 Environment and waste management issues

6.9.1 Introduction

During the latter part of the twentieth century, the composite can received extensive scrutiny in terms of meeting current environmental initiatives underway at that time. The package's body plies are made from recovered and recycled fibre and can have a post-consumer recovered waste content of over 50%. In many communities, this qualifies the composite can for placement in the material flow streams of curbside recycling programmes. In addition to the body, the metal ends of the can are also recyclable.

Today, the composite canister continues to fit into the majority of environmental initiatives demanded by the marketplace and continues to be recognised for its environmentally friendly attributes. These attributes include the introduction and use of a paper bottom end, which can increase the amount of recycled post-consumer content up to 70%, reduction of material into the waste stream and other efforts to enhance the can's ability to be recycled.

6.9.2 Local recycling considerations

When selecting packages, manufacturers and co-packers take a close look at local and regional recycling laws. In the USA, some states require that certain types of packaging are recyclable and use a minimum of recycled content whilst still meeting FDA food contact standards. The wide availability of paper and plastic recycling resources in the USA make composite cans an excellent package for a variety of products. Whilst there is a preference for foil composites in Europe, there is also a good deal of aluminium can packaging due to its easy recyclability.

In the USA, a marked focus on sustainability often translates to recyclability. In 2009, global composite can manufacturer Sonoco introduced a completely recyclable composite can as part of its True Blue™ line of sustainable packaging. The company's Linearpak® shaped package contains an average of 55% recycled content (50% post-consumer). The low-barrier version of this can is completely recyclable in the mixed paper stream of most local and curbside recycling programmes. The paperboard in the package itself is made from 100% recycled paper.

6.10 Future trends in design and application

6.10.1 Introduction

The ongoing development of new materials, sizes, shapes and technologies indicates that market applications for the composite canister will continue to grow. In 2008, the World Packaging Organisation published a position paper for the global packaging industry entitled, *Market Trends and Developments*. Packaging engineers are focusing on ways to increase performance and convenience by:

- enhancing existing features and materials
- identifying ways to further reduce costs associated with materials and processes
- redesigning closure systems
- developing sustainable packaging solutions.

These concerted efforts to improve shelf life, satisfy environmental requirements and improve performance lead to new technologies. Some of the most intriguing areas under consideration include:

- high barrier materials
- active packaging
- intelligent packaging
- nanotechnology.

(http://www.worldpackaging.org/uploads/paperpublished/2_pdf.pdf, p. 5.)

6.10.2 Increase barrier performance of paper-bottom canisters

Creating a truly hermetic barrier performance for paper-bottom cans would allow them to be used for more difficult to package products like coffee and powdered infant formulas.

6.10.3 Totally repulpable can

Efforts are directed at removing any tin or foil from the composite can in order to make a totally recyclable package. This includes coatings, linings, labels and ends.

6.10.4 Non-paper-backed liner

Research in this innovation is focusing on material and source reduction whilst providing some barrier protection.

6.10.5 Film label

With the demand for better print quality rising, a film label would allow for improved shelf appeal through special printing techniques (e.g. reverse printing) and provide some water resistance.

6.10.6 Killer paper

In January 2011, the American Chemical Society reported significant strides in the development of 'killer paper', a material in testing for use in food packaging. The paper is coated with the bacteria-fighting nanoparticles of silver and is proving to prevent product spoilage. Should the product prove successful, it can replace other, more traditional methods of preservation including radiation, heat treatment and low-temperature storage.

6.11 Glossary of composite can-related terms

Anaconda seal – In order to maintain a complete barrier where the edge of one spiral winding overlaps the preceding winding, the edge is turned through 180° and heat sealed, face to face, with the preceding winding in the overlapped area.

Body stock – A term used to identify the inner plies of a composite can body – excludes the liner or label.

Caliper – Thickness of a sheet of paper or paperboard measured under certain specifically stated conditions, expressed in thousandths of an inch. Units are called 'mils' when referring to paper and 'points' when referring to paperboard.

Canboard – A generic name for the board used in composite cans. It is generally a medium-strength board with good surface smoothness.

Can dimensions – Can diameter and height measurements are expressed in inches and sixteenths of an inch. The standard 12-oz juice can measures 211 × 414, which translates to 2 11/16 in. in diameter by 4 14/16 in. in finished can height. The diameter of the can is given first. In Europe, dimensions are measured in millimetres (mm), which refer to the internal diameter of the can.

Composite can – A term used to describe a can made from more than one type of material. The tops and bottoms can be made of metal, plastic, paper or a combination of materials.

Convolute can – A laminated fibre can be made by winding material around a form with the material fed at right angles to the axis of the form.

Finished can height – The overall height of a finished can with one end seamed on, often referred to as 'one end on height'.

Gas retention – The measured ability of a can to retain gas during a specified time.

Gassing – The act of removing air from a can and replacing it with a non-reactive gas, such as nitrogen or carbon dioxide.

Hermetic can – Air or gas-tight canister.

Hermetic seal – The bonding of two parts together so that it is air- or gas-tight.

Kraft – A term meaning strength applied to pulp, paper or paperboard having a minimum of 85% virgin wood fibres and produced by the sulphate process.

Label – The outermost ply of a composite can. This layer may be printed. Even if it is not, it is still called the label.

Liner – The innermost ply of a composite can, constituting the can body's principal barrier to moisture or gas transmission.

Linerboard – Solid bleached or unbleached kraft paperboard, either Fourdrinier or cylinder, combination paperboard produced from a furnish containing less than 80% virgin kraft wood pulp.

Locking plastic overcap – An advancement over the traditional plastic overcap, the plastic lid and metal ring combination is designed to flex and lock with an audible sound. The lid actually attaches inside the rim of the metal ring.

Mira strip tape – Plastic strip adhered to the top of a can-opening feature, which is used with crimp seams (such as frozen concentrate cans).

Nitrogen flushing – During the filling process, the can is filled with a nitrogen gas before inserting the product. The nitrogen gas displaces oxygen allowing for longer shelf life.

Overcap – A cap fitting over another closure that will allow resealing once the can is opened. An overcap must be slightly larger than the end on the can for a proper fit.

Paperboard – A broad classification of materials made from fibrous matter on board machines, encompassing linerboard and corrugating medium. Most commonly made from wood pulp or paper stock.

Ply – The layers of paper used to form the can body. The paper on a can is usually described by the number of plies wound into the can: two-ply versus one-ply can.

Re-closure – A term used to identify or classify the type of end, plug or cap which can be replaced after opening a container.

Recycled stock paper – A paper made from pulp consisting of reclaimed paper waste materials.
Shelf life – The length of time a composite can, or material in a container, remains in a saleable or acceptable condition under specified conditions of storage.
Spiral can – Any composite can which has been formed by winding material around a mandrel at an angle to the axis (less than 90°).
Self-opening can – A method of construction that allows a refrigerated dough can to open when the label is removed, the body ply joint is exposed, allowing the liner heat seal to release and dough pressure pops the can to open.
Skive – In order to achieve a neat joining between adjacent spirally wound layers, the edges are 'skived'. This means that the edges are cut at an angle so that the adjacent edges mesh neatly together. This also increases the strength of the composite can.
Valved membrane end – Designed for coffee, this one-way release valve allows for packing and sealing immediately after roasting. This eliminates the need for extended hold times for degassing, and at the same time maximises flavour and aroma. A vacuum is no longer necessary. Because the end is a peelable foil membrane, a can opener is not needed.
Wicking – The tendency of liquid to be absorbed by osmosis through a sheet of paper.

References

SFI is an independent, non-profit organisation responsible for maintaining, overseeing and improving a sustainable forestry certification programme that is internationally recognised and is the largest single forest standard in the world. SFI program components include *FI forest certification*, SFI chain-of-custody, SFI certified sourcing and SFI labels, all of which are audited by independent, third-party certification entities (http://www.sfiprogram.org).

Du Pont, 2011, *Du Pont Packaging News, Survey: Sustainability, Cost Challenge Global Packaging Industry*, 12 May, visit http://www2.dupont.com/Packaging_Resins/en_US/whats_new/article_20110512_global_survey.html

Food Engineering, 2010, *Food Engineering Packaging Trends Survey*, Kevin J. Higgins, 01/09/2010, visit http://www.foodengineeringmag.com/Articles/Feature_Article/BNP_GUID_9-5-2006_A_10000000000000891700 (*Note*: this is the result of an exclusively web based survey sent to readers by e-mail in April 2010).

Kalkowski, J., 2010, *Packaging Digest*, 1 January 2010, visit http://www.packagingdigest.com/article/442112-Packaging_trends_for_2010.php

SustainableBusiness.com News, 2010, *Sustainable Packaging Industry to Reach $142B by 2015*, 13 August, visit http://www.sustainablebusiness.com/index.cfm/go/news.display/id/20855

Websites

EFTA, European Flexographic Technical Association, www.efta.co.uk;
ERA, European Rotogravure Association, www.era.eu.org;
Printer's National Environmental Assistance Center for technical notes on the main printing processes, www.pneac.org/printprocesses/lithography/moreinfo2.cfm;
Sonoco, www.sonoco.com

7 Fibre drums

Fibrestar Drums Ltd., Cheshire, UK

7.1 Introduction

A fibre drum is a cylindrical container with a sidewall made of paper or paperboard having ends and components made of similar or other materials such as metal, plastics, plywood or composite materials. The sidewall of drums used for industrial applications is made up of several layers of paper laminated together by convolute winding.

Fibre drums are used globally and offer a strong, cost-effective means for the packaging of solid, granular, powder, paste, semi-liquid and liquid products. They are widely used by the chemical, pharmaceutical and food industries as well as for special applications in other industries, such as the packing and dispensing of wire, cable and metal foil, adhesives, rolled sheet materials, dyestuffs and colourants. A US survey indicated that fibre drums comprised 30% of the industrial containers (drums) market (Hanlon et al., 1998).

A comprehensive range of open-top fibre drums is produced in sizes from 10 to 270 L capacity (2–60 imp. gal) and in a wide range of designs with respect to diameter, height, cross section, type of closure and lid/base construction (Fig. 7.1).

Fibre drums are strong and protect their contents during transportation and under compression whilst in storage. In general terms, they are capable of holding up to 250 kg (550 lb) of dry or semi-liquid product, but with today's health and safety legislation, typical pack weights are 25 kg (55 lb) or 50 kg (110 lb).

Fibre drums were originally introduced as an alternative to the metal drum and therefore had a circular cross section. In recent years, drums with a square cross section with rounded corners have become available (Fig. 7.2).

The square cross-section drums enable customers to make use of the space-saving attributes when such a pack is palletised for storage and distribution.

Fibre drums are made primarily, and in some designs exclusively, from fibrous materials, namely paper and paperboard, as in Fig. 7.3. Figure 7.4 shows a fibre drum with a strengthened fibre top rim or chimb. This all-fibre drum has a metal-closure band and options with respect to a plastic lid and base collar, all components being recyclable. Several other designs, however, incorporate metal components, plastic liners and coatings to meet specific application needs.

Handbook of Paper and Paperboard Packaging Technology, Second Edition. Edited by Mark J. Kirwan.
© 2013 John Wiley & Sons, Ltd. Published 2013 by John Wiley & Sons, Ltd.

206 Handbook of Paper and Paperboard Packaging Technology

Figure 7.1 Various designs of fibre drum.

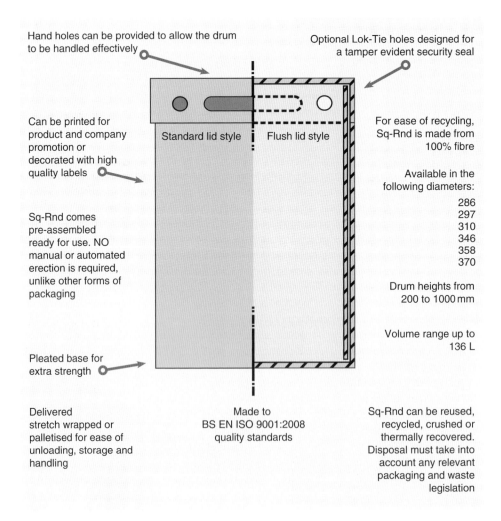

Figure 7.2 Square-round all-fibre drum with rounded corners.

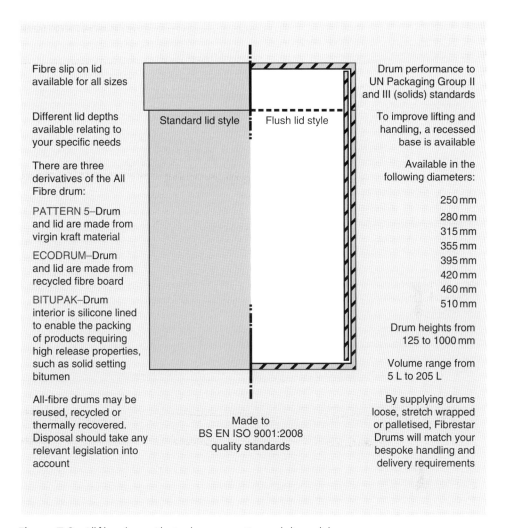

Figure 7.3 All-fibre drum with circular cross section and slip on lid.

One of the most important characteristics of the fibre drum is that manufacturers are able to supply users with a bespoke, fully customised drum, to match their packaging requirements exactly. Fibre drums are therefore not over specified and hence follow the European Packaging Waste Directive for optimised packaging. They are lighter than metal drums, and additionally, fibre drums are easily recovered and their component parts recycled.

7.2 Raw material

The sidewalls are constructed using virgin unbleached kraft or a recycled paper alternative, for example coreboard. A typical grammage would be 280 g/m^2. The strength of the drum is strongly influenced by the type of paper and the number of plies wrapped around the mandrel. End boards used in the base construction typically start at 1275 g/m^2 and can be as high as 1800 g/m^2.

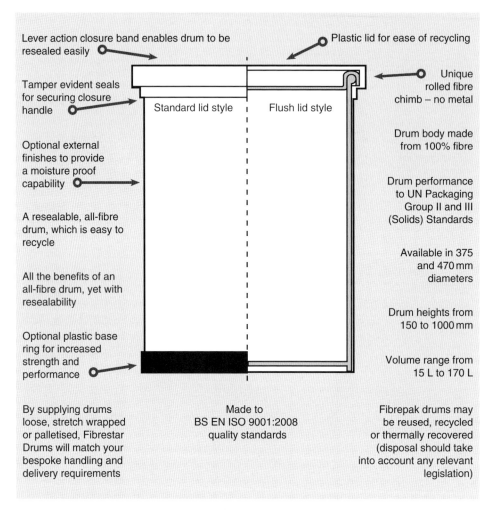

Figure 7.4 All-fibre drum with strengthened fibre top rim or chimb.

Laminates based on these materials with, for example, aluminium foil, polyethylene (PE) or other materials are used for additional functional properties, such as a moisture barrier or a silicone coating for product release, (see Fig. 7.5).

The additional barrier material may also be 'sandwiched' within the sidewall construction.

7.3 Production

7.3.1 Sidewall

The usual method for constructing the sidewall is by convolute or straight winding of paper around a mandrel, or forming tool, from a reel, i.e. continuous length. Adhesive is applied to the paper by an applicator roll rotating in an adhesive tray prior to wrapping around the

Figure 7.5 Drum with fibre body, steel chimb on bottom, slip on lid.

mandrel. A number of plies are built up on the mandrel, for example six plies. The finished cylinder is then removed from the mandrel. In this convolute winding, the diameter of the cylinder is the same as the diameter of the mandrel on which it was wound and its length is the same as the width of the paper from which it was wound.

After removal from the mandrel, the cylinder is left to stand for from 1 to 8 h to allow the adhesive to dry. It is then cut to the required drum length, any waste generated being recovered for recycling.

Winding onto the mandrel in this way means that the cross direction of the paper becomes parallel to the axis of the newly wound cylinder with the machine direction running around the circumference. This is an advantage as far as the strength of the drum is concerned. There will be enough plies of paper to give the required vertical stacking strength, and the body of the drum will have a greater resistance to sideways impacts.

Paper laminates with additional functional properties are applied to the cylinder in sheet form. Depending on the required position of the barrier material in the drum construction, the material may either be wrapped around the mandrel prior to or after the winding of the sidewall.

It should be noted that there are other methods of winding paper around a mandrel, but they do not have the strength required for industrial drums. The use of these methods, such as spiral and single-wall winding, is confined to the manufacture of small retail size drums, decorative drums and hat boxes.

The industry standard, and most commonly used, adhesive is sodium silicate. This adhesive is cost effective and efficient in the winding process and contributes to the strength of the drum.

For drums requiring moisture-proof capability, polyvinyl acetate (PVA) is used. There is also some use of polyvinyl alcohol and dextrin. The main requirements for the adhesive are good wet tack and a quick setting speed. It is important to use an adhesive with a high solids content to ensure that the amount of water added, as a consequence of using the adhesive, is kept to a minimum.

7.3.2 Drum base

The fibreboard discs are die cut from larger parent sheets; this is usually done in-house for low-volume requirements, and higher-volume requirements are bought in from specialist suppliers. For the all-fibre drum, the base is fixed in place by sandwiching it between a turned-over flange and an outer thinner disc.

A galvanised steel chimb on the bottom rim of the sidewall may be used to apply a plastic (PE) or steel base by crimping. This will make it easier for one person to move the drum by rolling it along on the base rim.

Caulking is used to seal the chimb 'joint', hence producing a liquid-carrying capability if required. A typical caulking compound is a moisture-curing polyurethane, which is applied as an extruded bead into the joint before it is crimped.

The chimb adds strength and enables the drum to meet the United Nations (UN) performance standards. The incorporation of a chimb to the edges of both the top and bottom ends is a global standard produced by all manufacturers in one form or another.

7.3.3 Lid

Fibre drums are known as 'open' top drums. This means that they have a lid which matches the diameter of the drum. The lid can be made from wood, plastic, steel or fibreboard. The lid is held in place with a metal band, normally steel (Fig. 7.6). (Readers should note that there is another type of drum, having a different design, known as a 'tight-head' drum where the top is permanently fixed to the body. These drums are made only from plastic or steel. Tight-head drums are used to carry liquid or semi-liquid products, with two 2-in. diameter apertures in the top which are closed by bungs.)

Depending on the global manufacturing location, there is a preference for the lid material. In the UK, the most popular choice is plastic, both high-density polyethylene (HDPE) and low-density polyethylene (LDPE). Steel lids are used in high-performance specifications. Within the USA, the choice of the lid material is split approximately, evenly between plastic

Fibre drums **211**

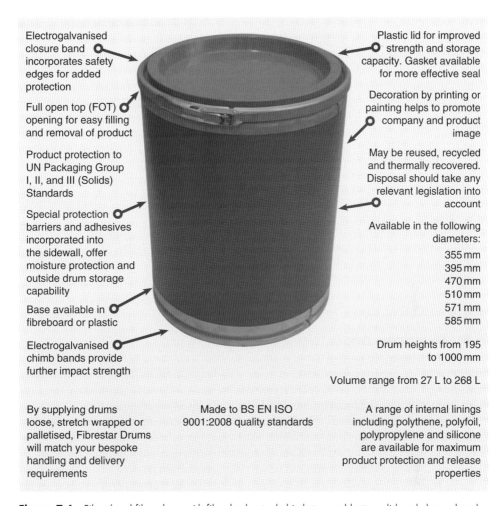

Figure 7.6 Fibre/steel fibre drum with fibre body, steel chimb top and bottom, lid and closure band.

and fibreboard lids. Steel is used where required. In mainland Europe, a wider variety of lid is used, including plywood, steel, fibreboard and plastic.

Where a plastic lid is used, as in Fig. 7.6, a 2″ aperture bung can be incorporated, which can take a bung closure. This is more commonly used in the USA and is used for filling, using a lance or pipe, and dispensing liquid and semi-liquid products, such as PVA adhesives. An option for easily dispensing the product with the drum on its side is to fit a large tap in the bung housing.

Steel lids can be either varnished, galvanised or hot-dip galvanised for corrosion/weather protection.

In order to assist the stacking of drums, and the stability of a stack of drums, it is usual for the lids to be designed in such a way that they locate within the base chimbs of the drums on the next layer. For drum specifications that require an air-tight seal, a gasket is incorporated into the plastic or steel lid. Gasket material varies between manufacturers, but the most common is a polyurethane foam gasket flowed into the lid, or a rubber gasket ring bonded to the lid.

The choice of the lid material comes down to cost and performance requirements, all the materials quoted being accepted around the world.

There are several ways of securing the lid to a drum:

- For the design of drum shown in Fig. 7.6, with a metal chimb at both ends, for example Fibrestar 'Leverpak', the lid is secured in place by a metal-closure band or closure ring. The ring is normally made of steel and galvanised for anti-corrosion protection. The locking-handle design varies around the world. There are two basic designs: one incorporates a locking latch whilst the other incorporates a one-piece handle with the provision for a security seal.
- The majority of closure-band designs allow for the use of a tamper-evident security seal. There are many designs of standard industry seals, but one of the most popular is known as the 'fir tree' or 'Christmas tree' seal.
- For drum styles that have slip on lids, such as those shown in Figs 7.3 and 7.4, the end-user would usually tape it in place with wide self-adhesive tape.
- The square cross-section all-fibre drum, Fig. 7.2, has a provision for securing the lid by means of plastic tie straps, which are inserted through a set of holes in its corners.

7.4 Performance

Fibre drums are used for the carriage of dry powders, granules, pastes and semi-liquid products and other materials. As already noted, the main industrial sectors that use fibre drums are the chemical, pharmaceutical and foodstuff sectors.

Fibre drums which conform with UN Packaging Group I, II or III Standards are often used to transport hazardous products in dry/solid form. Fibre drums cannot be used to transport or pack hazardous liquids. Tests are clearly laid down in the UN *Orange Book*, which defines the drum specifications that can be used for specific products.

Every country has its own national testing authority to undertake and/or validate testing carried out in other testing facilities. This testing may be carried out in facilities owned by drum manufacturers, provided such facilities have been approved by the national testing authority. The national testing authority issues certificates when test data are submitted to it in the required manner.

The tests themselves are clearly laid down by the UN. They incorporate a variety of drop tests at different angles of drum impact, along with a stacking test, at 23°C, 50% RH.

Stacking performance, due to the very nature of the fibre drum being a bespoke designed container for a specific, vertically stacked application, means that the stacking performance or capability will vary as between the various sizes, design and specification of drum. As a guideline, a drum for packing 25 kg (55 lb) of product could be designed to allow a maximum of 10-high stacking; whereas a drum for packing 200 kg (440 lb) of product could be designed to allow a maximum of 3-high stacking.

Handling of fibre drums can be undertaken manually, but above 50 kg (110 lb) handling would typically be by means of vacuum lifting, mechanical drum-handling equipment or forklift truck. Fibre drums can be handled by all these methods.

When fibre drums are used to pack moisture-sensitive and flavour- or aroma-sensitive, semi-liquid and liquid products, the interior of the drum is integrally lined with a PE-lined paper or a PE aluminium foil lined paper, thus stopping the product from penetrating the

Figure 7.7 Inserted plastic bag.

fibre sidewall body. Such barrier lining protects the product from losing moisture, flavour or aroma and prevents contaminating odours, flavours and aromas from affecting the product. The base joint is also normally sealed with a caulking compound to prevent the liquid from leaking through the joint, whilst the lid will include a gasket. When required for a liquid, or semi-liquid product, a PE strip is heat sealed over the lap joint on the inside of the straight-wound drum to ensure that the product is not absorbed by the otherwise exposed (raw) edge of paper.

The use of an integral barrier material plus caulking allows drums to be used without the need for a separate loose polythene liner where the product requires that type of protection. Hot-molten products which solidify on cooling, such as bitumen, can be packed in fibre drums which have silicone-coated paper as the inside liner of the drum.

Removal of the product is usually facilitated by cutting away the container or by the use of special drum-heating equipment.

Fibre drums which are required to be suitable for external storage – the exterior surface of the sidewall body is likely to be exposed to the natural elements such as rain, snow, frost and sun – will incorporate a weatherproof barrier or coating and the lid will include a gasket for improved sealing. Fibre drums can be externally varnished to provide moisture protection.

The vast majority of fibre drum users – when packing, for example dry powders – would use a separate plastic bag, usually PE, as the first level of protection (Fig. 7.7). Other higher specification co-extrusions can also be used. The bags can be inserted by the drum manufacturer. Bags can be fixed to the drum base by heat sealing or with the use of a semi-permanent adhesive. This has several advantages:

- drums are ready to use when delivered to the end-user
- end-user stock control and labour costs are lower
- fixed bags speed up product discharge and prevent them, accidentally, entering reactors and blenders
- semi-permanently fixed bags are easily removed from the drum for reuse or disposal.

If the highest barrier/moisture protection is required, the drum would, depending on the product, have a PE/aluminium foil lining or PE barrier built into the drum sidewall, with a gasket seal added to the lid.

Figure 7.8 Drums in air-cleaning facility.

Fibre drums incorporating a bag with high barrier properties can be used for aseptic, hot-fill packaging.

Fibre drums with recessed ends provide edge protection for reels of materials such as plastic film, aluminium foil and rolls of plastic sheeting and floor coverings.

The wire-manufacturing industry uses a special design of fibre drum, which incorporates an additional core secured into the base of the drum. There is a choice of design. The core may be glued around a stapled-in locating disc or it may be glued into a hole formed by a 'donut'-shaped base disc. Wire and cable products are packed in the void between the inner core and the drum sidewall. Drums such as these are used for long-duration feeding of welding wire and other similar products, thereby reducing the number of changeovers for manufacturers.

Cleanliness of the drums is an important feature and some customers, such as those in the pharmaceutical/fine chemicals sector, require packs that have passed through an air-wash system which removes fibrous debris from the inside of the drum, as shown in Fig. 7.8.

7.5 Decoration, stacking and handling

The standard brown/beige fibre drum can be roller coated either with a colour or clear varnish to provide a wipe-clean surface, corporate image ('house' or brand colour) and, as noted earlier, a moisture-resistant surface. Water-based inks and varnishes can also be used.

The body can also be silk screen printed, normally up to a maximum of three colours. Printing can include product/user instructions, safety and hazard-warning data. If the design required and volume required are cost effective, a pre-printed outer wrap can be applied when the cylinder is manufactured, hence allowing multicolour complex designs to be added to the fibre drum.

Labels can also be applied to the drum body if required, and these labels, including the adhesive, are usually application specific, for example for internal or external storage. Where

Figure 7.9 Printed and labelled drums and drums with snap on handles integrated into the lid for ease of handling.

the label is to be applied by the filler, the fibre drum manufacturer can print a label location guide on the drum surface to enable the filler to position the labels correctly (see Fig. 7.9).

Pigmented plastic lids are available to complement drum sidewall decoration.

7.6 Waste management

Fibre drums can either be reused, the component materials recovered and recycled, or disposed of in energy-to-waste systems. Internationally agreed identification code data and a recycling logo can be applied to drum sidewalls and bases which indicate their composition (SEFFI, 2012). This information is used as a guide to recycling. Fibre drum manufacturers can provide information and support on all environmental and waste-management issues. This can include advice with respect to drum recycling companies and equipment to assist recycling.

Fibre drums can be used to pack waste materials which are sent for incineration – the drum is easily incinerated and has a lower cost than steel or plastic drums.

7.7 Summary of the advantages of fibre drums

Fibre drums are bespoke designed packs, matching customers' specific requirements:

- They are cost effective compared with metal alternatives.
- They are available globally in a wide range of diameters, heights and styles.

- The fibre is naturally renewable, and both the fibre and the other components are easily recoverable and recyclable.
- They can incorporate linings and barriers, increasing their performance and range of applications.
- They are approved for the packing and transportation of hazardous solids.
- They are used predominantly by the chemical, pharmaceutical and food industries.

7.8 Specifications and standards

A British Standard for fibreboard drums, BS 1596:1992, provides a minimum specification, a series of definitions and the normal size/capacity range. BS 1133-7.4: 1989 Packaging Code describing paper and board wrappers and containers including fibreboard drums is also a current standard.

European standard BS EN 12710:2000 covers the construction requirements of fibre drums in the 15–250 L range.

Hazardous products which are allowed to be packed in fibre drums are listed in the UN *Orange Book*, with its proper title being *Recommendations on the Transport of Dangerous Goods: Model Regulations*.

References

Hanlon, J.F., Kelsey, R.J., and Forcino, H.E., *Handbook of Package Engineering*, 3rd edn., CRC Press, Boca Raton, Florida, p. 201.

SEFFI, 2012, *Materials Identification Codes/Symbols, for Drum Sidewalls, Base Materials and Components*, visit European Association of Fibre Drum Manufacturers, visit http://www.seffi.org./regulatory

Websites

SEFFI (European Fibre Drum Association), www.seffi.org
Industrial Packaging Association, www.theipa.co.uk
Fibrestar Drums, www.fibrestar.co.uk

8 Multiwall paper sacks[1]

Mondi Industrial Bags, Vienna, Austria

8.1 Introduction

Multiwall paper sacks are concentric tubes of one to six layers (or plies) of paper with a choice in the type of end closure. Many different designs of paper sack will be described in this chapter. The designs differ mainly in respect of whether the sack is to be filled through an open mouth or a valve, which in turn depends on the product and the volume to be handled. Valve sacks are closed manually or automatically and can be equipped with valve sealing for better closure, for example by sealing with polyethylene (PE) coating. Open mouth sacks can equally be sealed manually or automatically with various methods, including sealing through hotmelt, sewing, etc.

Other key features of the specifications are the number of plies, the quality of paper and other materials used by way of paper coatings, impregnations, laminations or whether separate liners are incorporated depending on the product, protection and performance required.

Paper sacks were developed in the 1830s and became a major type of packaging from the 1920s. Their traditional uses from the early days were for building materials, chiefly cement, foods such as flour, dried milk, sugar and potatoes, animal feed, chemicals and fertilisers. Today, it is claimed that over 2000 different products are packed in paper sacks in the USA. Advantages for paper sacks include easy bulk palletisation, stacking and handling, and the fact that plain paper sacks, as for instance are used for cement, are permeable to air allowing the products to 'breathe'.

Paper sacks have been used to pack up to 50 kg of products, but this weight has been reduced in some regions and industries to ease handling and meet health and safety requirements. Furthermore with a wide use of paper sacks in the retail sector today, where there are customer convenience needs, the weights range from 25 down to 10 and even 2.5 kg. Products packed for the retail trade include cement, other building materials, human food,

---

[1] This chapter is adapted from the chapter under the authorship of ETAPS (The Environmental and Technical Association for the Paper Sack Industry) published in the first edition.

Handbook of Paper and Paperboard Packaging Technology, Second Edition. Edited by Mark J. Kirwan.
© 2013 John Wiley & Sons, Ltd. Published 2013 by John Wiley & Sons, Ltd.

Table 8.1 End-uses of paper sacks (Europe)

European end-uses for 2010

Product end-use	Quantity used in 2010 (in million units)	%
Building materials	3416	63.9
Mineral products	95	2.3
Food products	592	11.2
Animal feed	392	7.5
Chemicals and fertilisers	343	8.1
Seeds	98	1.8
Miscellaneous	277	5.2
Total	5345	100.0

Note: Eurosac members comprise about 70% of the market.
Source: Eurosac.

pet food, pet litter, animal feed and gardening products. The retail trade also requires a higher quality of printing and features such as carrying handles, sealable valves or biodegradability of the packaging.

Over seven billion paper sacks are used per annum (pa) in Europe and over four billion pa in the USA. The typical end-use range is shown in Table 8.1.

Multiwall sacks can be biodegradable. An example of this is the Terra Bag® made by Mondi Industrial Bags. This valved paper sack can have a plastic bag for better product moisture protection which is made from biodegradable film so that the entire bag is biodegradable. It has been certified according to EN 13432, the European standard for packaging recoverable through composting and biodegradation. This specification won the Eurosac Grand Prix of the year 2010 (Mondi, 2012).

8.2 Sack designs

There is a wide range of sack designs from which users may choose to meet their requirements. When making the choice, account will need to be taken of the properties of the product to be packaged, the requirement of the filling, closing and distribution systems, and the needs of the final user.

8.2.1 Types of sacks

The first basic division of paper sack types is *open mouth* and *valve* designs. Each of these may be subdivided into *pasted* or *sewn* closure types and into further subdivisions by the sack body being either *gusseted* or *flat*. Further minor variations can also arise from the inclusion of certain other design features.

Not all the possible combinations of design features are practicable due to the restrictions in manufacturing machinery or sack design geometry.

A schematic range of multiwall sack designs are shown in Figs 8.1 and 8.2. Each design will now be discussed briefly outlining the important features, advantages, limitations, etc.

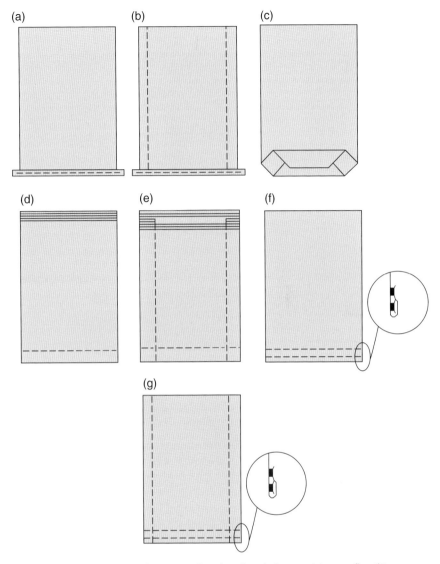

Figure 8.1 A schematic range of open mouth multiwall sack designs: (a) sewn, flat; (b) sewn, gusseted; (c) pasted, flat; (d) pinch closed, flat; (e) pinch closed, gusseted; (f) pasted, double-folded, flat; and (g) pasted, double-folded, gusseted. (From Kirwan, 2005.)

8.2.1.1 Open mouth sacks

There are four basic open mouth sack designs depending on the type of closure types, namely sewn, pasted, pinched and double-folded closures (Fig. 8.1).

8.2.1.1.1 Open mouth, sewn, flat sack

This is the simplest form of multiwall paper sack (Fig. 8.1a). It may be directly compared with the traditional jute and cotton sacks that have been used for centuries to pack granular and

220 Handbook of Paper and Paperboard Packaging Technology

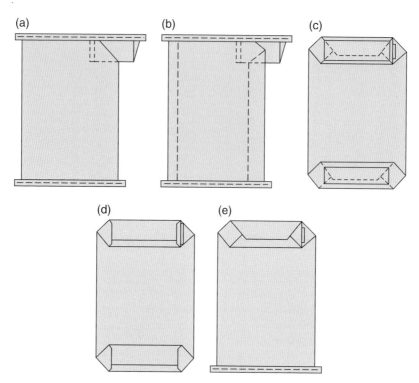

Figure 8.2 A schematic range of valved multiwall sack designs: (a) sewn, flat; (b) sewn, gusseted; (c) pasted, flush-cut, flat; (d) pasted, stepped-end, flat; and (e) pasted and sewn, flat. (From Kirwan, 2005.)

powdered products. The name 'pillow sack' is sometimes used to describe this design, which has an inherent disadvantage in that, when filled, corners tend to jut out. These can give rise to difficulties after palletising, by snagging against objects to cause tearing and leakage.

There are certain considerations that may dictate the use of these sacks. For instance, it is possible to include a layflat seamless polyethylene film tube as the innermost ply. Such a sack may be heat sealed within the line of the sewing to give a hermetic closure. These layflat film tubes may also be made longer than the paper plies by including a Z-fold section in the tube length. This enables it to be pulled out, filled and closed as a separate operation prior to the closure of the paper plies.

8.2.1.1.2 Open mouth, sewn, gusseted sack

By having gussets inserted at the sides of the sack, the user is ensured of a rectangular block shape after filling (Fig. 8.1b). The choice of face width, gusset depth and sack length will be governed by factors mainly associated with the volume of contents and palletising requirements.

Because of the gusset folds, it is not possible to include an internal polyethylene layflat film liner contiguous with the paper plies. In certain circumstances, it is feasible to include an edge-folded layflat tube of polyethylene film, which is inserted into one side fold of each gusset. It is also possible for a layflat film liner to be inserted and fixed to the inner paper ply with hot melt adhesive. A Z-fold may also be included in this construction, as with the open mouth, sewn, flat sack.

The most popular gusset sizes are 50, 75, 100, 125 and 200 mm. Larger gussets than these are possible but the filled shape of the sack becomes less rectangular.

8.2.1.1.3 Open mouth, pasted, flat sack

The pasted bottom closure (Fig. 8.1c) will automatically give filled sacks a rectangular end, and a sewn closure at the top will allow them to be either butted or overlapped when stacked onto pallets. A layflat polyethylene film liner can be incorporated into this sack during manufacture, which may be made longer than the sack plies by the inclusion of a Z-fold.

The open mouth, pasted, flat sack may be designed as a baler bag. Here the sack is a preformed wrapper for packing single items, quantities of small containers, trays of eggs, etc. Baler bags can be made with bottom widths up to approximately 350 mm. The top of a baler bag is generally folded down and sealed by either tape or adhesive.

8.2.1.1.4 Open mouth, pinch closed, flat sack

The pinch closed sack (Fig. 8.1d) is one with an envelope-type closure, as the design uses an extended flap at each end, which is folded over and glued down. This allows a fully sealed barrier sack to be made using a thermoplastic-coated aluminium foil to paper laminate as the innermost ply. This may be sealed at the side seam and at each closure to encapsulate the contents. It is also possible to produce an open mouth bag with a rectangular pasted bottom on one end and an extended flap at the opposite end.

Pre-applied hotmelt adhesive may be employed on the open mouth flap of the sack for the user to close by reactivating with hot air, for example Mondi Hot Lock Bag®.

Performance figures for pinch closed sacks compared with the same size of sewn sacks show the design makes a stronger sack, which effectively means that they may be of lighter construction to give equal performance.

If a barrier ply is present next to the sack contents, it is generally important for the sack closure to include an effective seal of the barrier ply. With pinch closed sacks, a heat seal may be included above the fold line of the flap so that, after closure, no stress is put on the heat seal by the contents. By using a layflat tube of polyethylene film for the innermost ply, which is heat sealed at both closures, the product can be very effectively encapsulated.

There is a variation to this design in which the open mouth flap is trimmed away during the final stage of manufacture to allow the user to make a normal sewn closure. A tear string may be incorporated into the closed flap, during manufacture, as an easy-opening device.

8.2.1.1.5 Open mouth, pinch closed, gusseted sack

The inclusion of gussets overcomes the problem of sharp corners and the filled sack can assume an almost perfect rectangular block shape (Fig. 8.1e). As with sewn sacks, the gussets prevent the incorporation of a contiguous polyethylene film tube but a side-folded layflat film tube inserted into one side of each gusset is possible. The design variation, in which the top flap is trimmed off to allow closure by sewing, is also available, as is the easy-opening tear string.

8.2.1.1.6 Open mouth, pasted, double-folded, flat sack

A pasted bottom closure may be achieved on a flush-cut multiwall paper tube by simply folding the bottom edge over twice and gluing (Fig. 8.1f). To ensure a strong closure and to resist any tendency for the bottom to unfold, the outer ply is slit on the edges to form a flap which is adhered to the body of the sack above the folding. These sacks are also known as 'double fold' and 'roll bottom' sacks. The user closure is normally made by sewing. A layflat polyethylene film may be incorporated as the innermost ply at the tubing stage of manufacture.

8.2.1.1.7 Open mouth, pasted, double-folded, gusseted sack

The presence of gussets results in a rectangular shape after filling, and the packer/filler closure is normally made by sewing (Fig. 8.1 g). A contiguous layflat tube cannot be incorporated into the construction, but it is possible to include an edge-folded layflat polyethylene film tube into one side of each gusset.

8.2.1.2 Valve sacks

A valve sack is closed at both ends during manufacture and includes an opening for filling purposes in one corner. Valve sack designs may be divided in a similar manner to open mouth sacks – first into sewn and pasted types and then into the gusseted and flat varieties (Fig. 8.2). There is a further subdivision of the valve, pasted, flat sacks into those made from either stepped-end or flush-cut multiwall tubes.

8.2.1.2.1 Valve, sewn, flat sack

The valve of this sack can be inserted only by a complex manual operation during manufacture and the filled sack has the disadvantage of protruding corners (Fig. 8.2a). The valve pasted, flat sack has superseded it almost totally.

8.2.1.2.2 Valve, sewn, gusseted sack

This design of sack also requires complex manual corner folding for the valve insertion stage of manufacture (Fig. 8.2b). Although the presence of the gussets overcomes the protruding corners problem, the demand for this type of sack is negligible and it has largely been replaced by valve pasted sack designs.

8.2.1.2.3 Valve pasted, flush-cut, flat sack

In the manufacture of paper sacks, the paper is drawn from the reels and formed into a flattened tube which is then separated into the required length needed for the sacks. With flush-cut sacks, the preformed tube is cut into lengths with a guillotine or chop knife (Fig. 8.2c).

The use of a flush-cut tube is the simplest way to achieve a folded and pasted closure. Sacks made in this way are usually given an additional paper-capping strip on each end to strengthen the closure and resist sifting. Alternatively, longitudinal slits may be made in the ends, down to the diagonal fold, to form rectangular flaps for pasting down.

8.2.1.2.4 Valve pasted, stepped-end, flat sack

The multi-ply tubes for stepped-end sacks are made in similar manner to those for flush-cut sacks, but the individual plies are perforated before the tube is formed (Fig. 8.2d). The tube is separated into sack lengths by pulling apart the perforations. By this means, the ply ends may be stepped relative to each other, and by perforating in a different pattern for each ply, it is possible to substantially increase the area that is available for folding in and pasting. The stepping also allows the individual ends of the plies to be pasted and incorporated directly into the closure. Stepped-end sacks do not generally require caps and may be made from multiwall tubes of two to six plies.

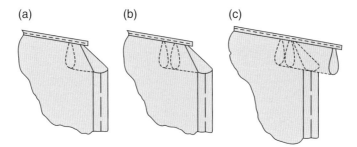

Figure 8.3 Valve designs for sewn sacks: (a) plain sewn valve; (b) internal sewn valve; and (c) exterior sewn valve. (From Kirwan, 2005.)

8.2.1.2.5 Valve pasted and sewn, flat sack

This design of sack enables a pasted valve to be used for filling and allows the final customer easier access to contents through the sewing line (Fig. 8.2e). Carrying handles may also be included with the sewing line.

8.2.2 Valve design

Valves used in paper sacks are held closed after filling, either by the pressure of the contents of the sack, by the folding down of an external sleeve or heat or ultrasonic sealing of a layer applied to the valve during sack production. These are termed 'internal' and 'external' valves, respectively, and either design may be incorporated into sewn or pasted types of sacks. There are many possible combinations of the various (roughly 150) valve design features, especially with pasted valves, and it is not possible to illustrate all of these in this guide. (Users are advised to consult paper-sack suppliers for more specific information.)

8.2.2.1 Valve designs for sewn sacks

Sewn sack valves are made by folding in one corner of the sack prior to sewing the closure. An extended corner or 'notch' may be employed in the multi-ply tube to allow a greater length of paper to be folded in to form the valve.

The simplest valve is made without the use of any additional paper, but generally a folded paper patch is inserted and sewn in as the internal or external component. Figure 8.3 shows the three basic sewn valves.

8.2.2.2 Valve designs for pasted sacks

The simplest type of filling valve is for one corner of a capped flush-cut pasted sack to be left unglued. Such a design is weak and a strengthening patch of paper is usually included under the corner fold.

An improved type of valve is achieved by the inclusion of a flattened tube of paper or plastic film into one corner to form either an internal or an external valve. In addition to the tubular component of the valve, it is possible to insert other paper or plastic components to produce a whole variety of designs to improve the efficiency of the valve or to tailor the valve design to the filling and closure requirements.

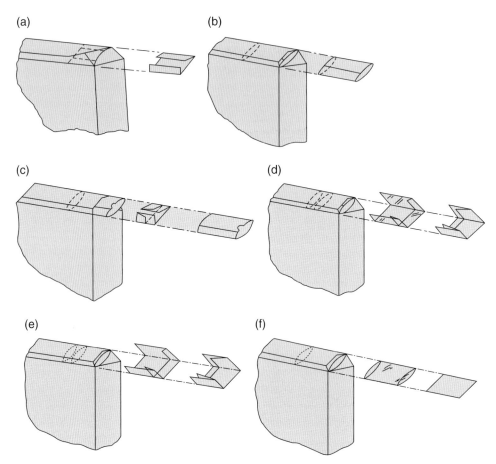

Figure 8.4 Valve designs for pasted sacks: (a) patch valve, (b) internal sleeve valve, (c) external sleeve valve, (d) small valve in larger bottom, (e) polyethylene film valve and (f) polyethylene tubular sleeve valve. (From Kirwan, 2005.)

It is not possible to illustrate all the combinations of design features, but Fig. 8.4 shows the six basic types of valve design used in pasted sacks.

8.2.2.2.1 Patch valve
A single ply of folded paper is positioned in the valve opening to give strength and rigidity (Fig. 8.4a).

8.2.2.2.2 Internal sleeve valve
A tubular paper sleeve is positioned in the valve opening and protruding into the sack (Fig. 8.4b).

8.2.2.2.3 External sleeve valve
This type comprises a tubular sleeve extending out of the valve opening, often with an internal pocket formed by a folded paper patch (Fig. 8.4c). This sleeve can be supplied with a thumb notch to facilitate easy opening.

8.2.2.2.4 Small valve in larger bottom

This is similar in formation to the external sleeve valve but includes a preformed valve sleeve smaller than the bottom width (Fig. 8.4d).

8.2.2.2.5 Polyethylene film valve

This valve is comprised of two sheets, one paper and one polyethylene, generally slightly offset from each other (Fig. 8.4e). They are inserted together, folded and positioned in the sack aperture to form an internal valve.

8.2.2.2.6 Polyethylene tubular sleeve valve

A paper strip is attached to a tubular polyethylene sleeve and positioned in the sack aperture (Fig. 8.4f). The width of the sleeve may be less than or equal to the width of the bottom.

8.2.3 Sewn closures

Sewing can be seen on open mouthed sacks in Figs 8.1 and 8.2 and on valved sacks in Figs 8.2a, b and e. There is a choice for the user of the type of sewing and the materials used in sewn closures. Stitching can be with either one or two sewing threads and may be in combination with other ancillary crepe paper tape and cords.

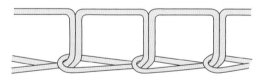

Figure 8.5 Single sewing or chain stitch. (From Kirwan, 2005.)

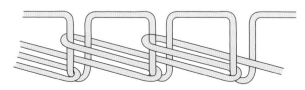

Figure 8.6 Double sewing. (From Kirwan, 2005.)

8.2.3.1 *Single sewing or chain stitch*

Single sewing, Fig. 8.5, can be readily unravelled and is a feature that may be used effectively as an easy-opening device by incorporating a ripcord.

8.2.3.2 *Double sewing*

With double sewing, Fig. 8.6, a second thread loops through and round the stitches on the underside during the sewing process and is very effective in increasing resistance to unravelling. This stitch is not used for easy opening.

8.2.3.3 Sewn closure constructions

In addition to sewing threads, a number of other materials may be used in forming the sewn closure.

Crepe tape may be folded over the end of the flattened multi-ply tube and sewn through. This has the effect of cushioning the stitching and assisting in preventing the sifting out of fine powdery contents. Different colours of crepe tape may be used for product identification.

In addition to the crepe tape, sewing may be made through a jute string generally called a filler cord, a soft cotton string called a filter cord, or a double-folded paper tape. These cushion and reinforce the stitches and also tend to block up the sewing holes and resist sifting.

An additional crepe tape may be applied over the stitching and adhered with a quick-setting latex adhesive to fully seal the sewing holes. Latexed closures are used as a means of preventing the entry of contamination and sifting of contents.

For users who require a carry-home pack, it is possible to incorporate a carrying handle into the manufacturer's sewing line. One design uses a thick polypropylene string, in a similar manner to the jute filler cord, which is left unsewn at the centre portion of the face. An alternative is a pre-made plastic handle, generally of low-density polyethylene (LDPE), which is inserted under the sewing tape and sewn in.

8.3 Sack materials

The materials used in multiwall paper sacks are most easily divided into those of the sack body and the ancillary materials employed elsewhere.

8.3.1 Sack body material

The construction of a multiwall paper sack is based on the use of sack kraft – a strong, durable and versatile paper that can be coated, laminated with other materials, modified to give it additional strength properties, printed, glued and sewn in the converting operations to produce a product of high strength and quality.

8.3.1.1 Sack krafts

The word 'kraft' means 'strong' in a number of languages and sack kraft is so named as it is one of the strongest papers made. It is manufactured using wood-pulp fibres made by the sulphate papermaking process. This uses sodium sulphate as the main chemical, which produces strong, undegraded cellulose fibres.

The sheet is formed by flowing a suspension of wood-pulp fibres in water onto a moving wire mesh. Most of the water drains away through the wire to leave a consolidated layer of intermeshed fibres. This is then pressed by mechanical presses, dried by steam-heated rolls and then wound into reels.

This mode of manufacture imparts different properties to the paper in the direction of movement, compared to the properties across the sheet. For example, sack kraft has higher tensile and lower stretch properties in the machine direction (MD) compared with those in the cross direction (CD). This arises from the random lattice-work orientation of the fibres as they form a sheet on the moving wire in the papermaking process. Both the tensile and the stretch properties contribute to strength, and the overall effect is high sheet toughness in both directions.

Paper produced from cellulose fibres will readily absorb water unless the sheet is treated by 'sizing' with an additive to inhibit ingress of water into the dry sheet. All standard sack kraft is sized on the paper machine, generally with rosin. The degree of sizing is measured by the $Cobb_{60}$ test (ISO 535: 1991).

Sizing does not impart strength-when-wet properties to the sheet because it only slows down the rate at which the paper absorbs water. Wet strength is achieved by the addition of a liquid resin during the papermaking process. This locks the fibres together at their points of contact so that they are not readily pulled apart when the sheet is soaking wet. Normal wet-strength treatment will allow approximately 20–30% of the strength of the dry sheet to be retained after the sheet has been thoroughly soaked in water.

The natural colour of sack kraft varies from dark to light brown, a shade which is suitable for many applications. By bleaching the wood-pulp fibres during manufacture, a pure white sack kraft can be made. When used on the outside of a paper sack, the white appearance enhances presentation and provides an improved surface for printing. In addition to the natural and bleached colours, a range of coloured sack krafts can be made by the addition of dyes during the papermaking process or by surface-colouring the paper after it has been manufactured. The range of colours available is normally restricted to a number of standard shades with any one manufacturer, and users should consult sack suppliers to obtain details and samples.

8.3.1.2 *Extensible sack krafts*

The term 'extensible' is used to describe those sack krafts which have been given enhanced MD stretch properties, either in the papermaking process or as a subsequent operation. This increase of stretch is generally associated with a slightly lower corresponding tensile, but the overall effect is for most extensible krafts to be tougher than normal sack kraft. All are available in normal and wet-strength grades, either natural or bleached.

8.3.1.2.1 Low-stretch creped (LSC) krafts
These are sack krafts that have been subjected to a wet-creping process, usually on the papermaking machine, to give a greater MD stretch. The creping process permanently wrinkles the sheet so it is rougher in appearance, more porous and more flexible than normal kraft.

8.3.1.2.2 Microcreped krafts
These sheets are mechanically crimped, or compacted, with a barely visible creping in the MD during papermaking, to give greater MD stretch. Typical of a number of systems for producing such paper is the 'Clupak' process.

8.3.1.3 *Coated sack krafts*

Certain commodities packaged in paper sacks require some form of barrier protection to restrict either the ingress or the egress of vapours or liquids. For this purpose, sack krafts can be coated with a range of materials to give strong, flexible barrier sheets for inclusion in paper-sack constructions.

Protection of the packaged material from the water vapour in the atmosphere is a common user requirement, and the barrier generally used is polyethylene-coated sack kraft. This is readily available in a range of grades to suit most requirements.

8.3.1.4 Laminated sack krafts

For greater strength or barrier protection of sensitive commodities, laminates of sack kraft with foil, film, woven or non-woven plastics maybe utilised. These materials are relatively more expensive than normal and coated sack krafts, and the choice would depend on the barrier requirements and/or the hazard level of the distribution system.

8.3.1.5 Non-paper materials

A number of non-paper materials, such as plastic film, layflat plastic-film tubing or woven or non-woven plastics, may be included in paper-sack constructions where high barrier or high strength is a prime requirement. The use of thermoplastic film can make it possible to include a heat seal in the closure of some types of sack. There are limitations to the use of these materials in the manufacture of certain types of paper sacks. For example, it may not be possible to include some of the reinforced and high-strength materials into stepped-end sacks, as these are not compatible with the normal manufacturing process.

8.3.1.6 Special purpose sack krafts

Although the majority of user requirements can be met by paper-sack constructions using combinations of materials described in the previous subsections, there is occasionally a need for a special material for a specific requirement. A number of such special purpose materials which meet particular requirements are discussed in Section 8.3.1.7. Any such need should be discussed with sack manufacturers for advice on the most appropriate material that can be used in specific situations.

8.3.1.7 Summary of sack body materials

The following notes give information on most sack body materials in general use but are not intended to be fully comprehensive. Where applicable, abbreviations in general use have been included in brackets after the name of the material. There is no international agreement on names and abbreviations, and where there is more than one alternative in general use, all are included.

8.3.1.7.1 Natural sack kraft (NK or KS)
Generally in grammages (substances) of 70, 80, 90 and 100 g/m^2, 110 g/m^2 unbleached and 120 bleached are used for one-ply sacks. Higher, lower and intermediate grammages are obtainable but not usually held as stock by sack manufacturers. It may be used generally throughout a sack in all plies.

8.3.1.7.2 Wet-strength kraft (WIS or WS)
The most generally used grammages are 70, 80 and 100 g/m^2 but other grammages are also available, although not as normal stock items. Widely used for the outer plies of sacks which may be exposed to rain.

8.3.1.7.3 Bleached kraft (BK or BlK or FB)
Available mainly in 80 and 90 g/m² grades, but other grammages are available if required. Mainly used as the outside ply to enhance presentation.

8.3.1.7.4 Wet-strength bleached kraft (B/WS or FBWS)
The general trade standard is 80 g/m² but higher and lower grammages can be obtained if required. Used for the outside plies of sacks which are liable to become exposed to rain.

8.3.1.7.5 Coloured krafts
Generally available in the 80 g/m² grade with or without wet-strength additive. The colour required should be written in full to prevent confusion with paper grade abbreviations. Bleached or natural kraft may be used as the base paper, each of which will give a distinctly different shade to the imparted colour.

8.3.1.7.6 Low-stretch crepe (LSC)
Available as natural or bleached, and generally as the 80 g/m² grade, with or without wet-strength additive. The normal level of creping is between 5% and 9% for MD stretch. The creping process also imparts higher air porosity to the sheet.

8.3.1.7.7 Microcreped fully extensible kraft (EK or ExtK)
Such krafts are defined as having an MD stretch greater than 7%. It is a tough paper that can be very similar in appearance to standard sack kraft as the microcreping is barely apparent. It is available in both natural and bleached grades with or without wet strength and in the same range of grammages as normal kraft.

8.3.1.7.8 Microcreped semi-extensible kraft (S/EK or SExtK)
The shortened name in general usage for this material is 'semi-extensible kraft'. It is defined as having the MD stretch within the range 4–7%. The same grades are available as for fully extensible kraft. The stiffness of semi-extensible kraft allows good runnability on converting machinery, similar to normal sack kraft, and high field performance in the finished sacks, making it an ideal material for many applications.

8.3.1.7.9 Polyethylene-coated kraft (polykraft or PEK)
LDPE has reasonable moisture and water vapour barrier properties, but its grease and odour barrier properties are not good. It is generally available with 80 g/m² natural kraft as the base paper and 15, 23, or 34 g/m² of coated polyethylene. The most popular and the most general purpose coating weight is 23 g/m². A coating weight of 34 g/m² is one which is relatively free from pinholes and film rupture caused by surface fibres. The 15 gsm coating weight is not a good barrier to water vapour but does resist the ingress of liquid moisture and may also be used as a means of preventing paper fibre contamination of the packaged product.

Coating weights above 34 g/m² cause the paper to curl and give serious problems in the converting operation and are not normally recommended.

8.3.1.7.10 Silicone-coated kraft
The silicone coating is not a barrier ply for liquids or vapours. It is a release coating used on the inner surface of the innermost ply to enable sack kraft to be easily stripped away from sticky or mastic contents.

8.3.1.7.11 Waxed kraft

The wax coating will partially or totally impregnate the sheet. It is used only to a small extent, mainly as a grease or odour barrier. Although it has some moisture and moisture vapour barrier properties, these are inferior to those of other materials available.

8.3.1.7.12 Polyvinylidene dichloride coated kraft

Polyvinylidene dichloride is an excellent barrier for moisture, moisture vapour, grease, solvents, oil and odours. It is appreciably more costly than polyethylene-coated kraft and its use is justified where a good odour or grease barrier is a requirement.

8.3.1.7.13 Foil laminated kraft

Aluminium foil is virtually a complete barrier for odours, grease and water vapour. In paper sacks, it is generally used at thicknesses between 15 and 25 μm as a laminate with sack kraft. At 25 μm thickness, it is virtually pinhole-free. Below this thickness, the effect of the presence of pinholes is virtually eliminated by the use of a plastic coating, usually polyethylene or ionomer. Either of these two materials may also be used to laminate the foil to the sack kraft. The plastic coating will also allow a sack closure to include a heat seal with certain designs of sacks.

8.3.1.7.14 Bitumen laminates

Coatings of bitumen on sack kraft and laminates of two krafts or LSC kraft sheets, using bitumen as a laminant, have been used in paper-sack manufacture since the early years. It is a general purpose barrier against moisture and water vapour but has been largely superseded by polyethylene-coated kraft. It is not a high-performance barrier because a totally continuous coating of bitumen is not easy to achieve in practice. This barrier is also not resistant to creasing, and the fold lines in a paper sack are weak areas of the barrier.

8.3.1.7.15 Vegetable parchment

This is a traditional grease-resistant material, but its use has been largely superseded by plastic film or coated kraft.

8.3.1.7.16 Scrim and cloth laminates

Scrim laminates, comprising an open mesh woven cloth adhered to a single layer of sack kraft or sandwiched between two layers of sack kraft, will give a very high resistance to sack burstage and tear propagation. The degree of toughness of the laminate will be governed by the quality of the scrim cloth used; very open mesh cloth mainly giving resistance to tear propagation, and closer mesh, resistance to burstage.

8.3.1.7.17 Woven plastics/sack kraft laminates

Laminates of sack kraft and woven plastics are very tough and resistant to tear, burst and puncture. Sack kraft contributes stiffness to the laminate and the sack.

8.3.2 Ancillary materials

In addition to the materials comprising the plies of the sack, other ancillary materials are also included, being either necessary to the construction or optional, at the request of the user.

8.3.2.1 Sewing tapes

The most important requirements for sewing tapes are softness and bulk to assist the blocking of the sewing holes, and strength, flexibility and extensibility to suit the intermittent action of the sewing head. Sewing tapes are most generally made from crepe paper, with a creping level of the order of 10–12%.

Overtaping the stitching in the same operation as sewing is generally made using a wider version of the same crepe tape in conjunction with a quick-setting latex or hotmelt adhesive. Overtaping with a polyethylene-coated kraft tape, either creped or non-creped, may also be used in conjunction with a heat-sealing unit. This is most generally undertaken as an additional operation.

8.3.2.2 Sewing threads

Sewing threads are either made of natural cotton, synthetic filaments or a mixture of the two. Each thread is comprised of a number of strands or plies. Three and four ply threads are used for sacks containing up to five plies of paper and five-ply threads for sack constructions of six plies.

8.3.2.3 Filler (filter) cords

A cord of jute, double-folded crepe tape or fibrillated plastic string may be used in the sewing line to cushion the stitches and block the sewing holes to resist the sifting of fine powders, etc. Jute and cotton cords may be applied on one or both sides of the sewing line.

A single cord or double-folded crepe tape may be used in conjunction with single sewing for use as an easy-opening device.

8.3.2.4 Plastic handles

Injection-moulded LDPE carrying handles may be included in the manufacturer's sewn closure. These are inserted under the tape and sewn through during the closing operation and are available in a range of colours.

A simple carrying handle may be made using a heavy-duty polypropylene string.

It would also be possible to produce a handle with the usage of woven material for the bottom patch. The woven material is based on polypropylene and strong enough for up to 25 kg sack weight.

8.3.2.5 Adhesives

Starch and modified starch-based adhesives are generally used for the side seams, cross pasting and end closures, generally with a wet-strength additive included. When special coatings or laminates are employed in the sack construction, synthetic emulsion adhesives may be necessary. Hotmelt adhesives may also be used with certain designs of sacks as a quick-acting adhesive or as a pre-applied hotmelt adhesive which can be reactivated for the user's closure.

8.3.2.6 *Printing inks*

Paper sacks are generally printed with flexographic water-based printing inks which are formulations of pigments, resins and other additives for colour fastness, printing opacity, etc. Lead pigments are no longer used in these formulations. Over-varnishes are also available to give an enhanced finish.

8.3.2.7 *Slip-resistant agents*

There are a number of proprietary materials which may be applied to the outside of paper sacks to impart resistance to slipping during transportation. These are generally colloidal suspensions of a finely divided solid material, such as silica, in water. High application levels may also give difficulties with the sliding of sacks in chutes.

8.4 Testing and test methods

In present-day papermaking, automatic controls are incorporated to check sheet properties and adjust the machine settings to ensure consistent and high quality. In addition, all finished paper coming from the machine is sampled and tested on a frequent and regular basis for quality control and quality assurance purposes. Sack manufacturers may also test the incoming consignments of sack kraft and the other materials used in their production.

Specialist equipment is required for paper testing, which must be undertaken in an atmosphere which is controlled for temperature and humidity. Paper testing is not normally undertaken by users, but a sufficient knowledge of testing and quality is needed when changing specification levels or establishing new sack specifications.

Testing encompasses the two broad areas of the materials and the finished sacks, each of which requires the application of quite different techniques and principles.

8.4.1 Sack materials

A multiwall paper sack is a flexible form of packaging and is required to resist the stresses arising from the contents as well as those of the filling, handling and distribution systems. Testing should measure properties which best relate to sack performance in these areas.

Sack kraft is made from natural fibres which will readily take up or lose moisture, depending on the ambient temperature and humidity of the air. If there is no control of this, the absorbed moisture can vary widely and will have an appreciable effect on the physical properties and test results. For this reason, it is necessary to keep sack kraft for 24 h in a controlled standard atmosphere of temperature and relative humidity (RH) before any physical testing is undertaken.

In the UK and Ireland, the standard atmosphere used in the paper-sack industry is 23°C and 50% RH, which gives an equilibrium moisture content of approximately 9.5% to sack kraft.

Sack kraft, like all other paper materials, is composed of a mass of intermeshed fibres and any testing will show a degree of variation of properties both across and along the sheet, i.e. in the CD and MD. Because of this, it is always necessary to undertake a number of replicate tests on each sample of paper and sufficient samples should be taken from each consignment, for meaningful average results to be obtained.

There are a number of ISO and CEN Standards on sampling, conditioning and testing, details of which are given in Table 8.5.

8.4.1.1 Strength tests

The more important of the tests for strength are those which measure the resistance to breaking or the energy needed to cause rupture.

8.4.1.1.1 Tensile strength

The tensile strength of paper is the maximum pulling force that a narrow strip will stand before it breaks. Generally sack kraft has a higher tensile strength in the MD compared with the CD.

It is measured by an instrument in which a strip of sack kraft is clamped between two jaws which are gradually moved apart until the strip breaks. The machine records the final force before breakage as well as the amount of stretch that has occurred.

8.4.1.1.2 Stretch

This is determined at the same time as the tensile strength and is a measure of the elongation that will occur before a sheet is ruptured. The product of the tensile and stretch is an indication of the energy-absorbing ability or toughness of the sheet and may be used for comparison between materials.

8.4.1.1.3 Tensile energy absorption (TEA)

As the name implies, this is a test which gives a direct measurement of the energy needed to rupture a sheet. A special type of tensile tester is used, which not only measures the tensile and stretch but will integrate the two results and work out the total energy the paper strip absorbs before rupture occurs.

It is a test which is regarded as giving a better relationship to performance of finished sacks than all other tests and may therefore be used as an index when adjusting sack constructions or drawing up specifications. It is normally performed on two sets of sample strips, cut in the MD and in the CD of the sheet. The average between the two directions is normally used for comparing two or more materials.

8.4.1.1.4 Wet strength

Wet strength is defined as the level of strength that is retained by a paper sheet after it has become thoroughly wet, i.e. after a sample has been immersed in water for at least half an hour.

Traditionally the pneumatic burst test was used to establish the wet strength of sack kraft, but tensile testers based on load cells are now used more generally for this purpose.

Normal sack kraft will possess little or no strength after immersing for half an hour in water, but wet-strength sack kraft should retain an agreed percentage of its dry strength after soaking. Approximately 30% wet-strength retention is the level most generally used within the paper-sack industry.

8.4.1.1.5 Burst strength

The burst test is one that has largely gone out of favour, mainly because of the extreme variability of the results. In the test, a sample of paper is clamped above a rubber diaphragm covering an orifice of approximately 30 mm diameter. Air or hydraulic pressure is gradually increased under the diaphragm which expands until the paper sheet becomes ruptured. Hydraulic burst testers are not sufficiently sensitive for sack kraft testing, and pneumatic instruments are more generally employed.

Although it is regarded as an unsuitable test for assessing paper quality, the test may be used as a quick check for wet strength by burst testing a conditioned sample and another that has been soaked in water for half an hour. It may also be used as a quick check for confirming weakness in a paper sheet, such as that caused by over-creasing, abrasion, etc.

8.4.1.1.6 Tear strength

The tear strength is a measure of the resistance to tear propagation after puncture or rupture has occurred. In the test, a partially slit rectangular sample is held by two adjacent clamps, on either side of the slit, and the force to tear the sample fully apart is measured.

There is an inverse relationship between tear strength and tensile strength over which the paper-maker can exert a large degree of control. To obtain maximum tensile strength would result in a paper which possesses a relatively low tear strength. Good tear strength and good tensile strength are important in sack kraft, and the paper-maker will aim for a reasonable balance between the two.

8.4.1.2 Other physical properties/tests

There are a number of generally accepted testing methods for assessing other physical properties in addition to those for strength.

8.4.1.2.1 Grammage

The term 'grammage' is replacing the traditional terms of 'substance' and 'basis weight'. It is the weight in grams of 1 m^2 of paper, i.e. g/m^2, after conditioning in the standard atmosphere of 23°C and 50% RH for 24 h. The cutting of the sample to a suitably sized area for weighing must be done after the paper is conditioned.

In North America, the term 'basis weight' relates to the weight in pounds (lb) of a ream of paper, usually 500 sheets, with each sheet measuring 24×36 in. This amounts to 3000 ft^2 so that the most common basis weights for paper sacks are between 40 and 60 lb (equivalent to roughly 65–100 g/m^2).

8.4.1.2.2 Thickness (caliper)

Thickness is not normally measured on sack kraft as it has very little relevance to quality, performance or defining the grade. When required, it is measured by means of a dead-weight caliper micrometer and the result quoted in microns (thousandths of a millimetre) or thousandths of an inch.

8.4.1.2.3 Moisture content

The moisture content of paper is the loss in weight of a sample after oven-drying at a constant temperature of 105°C. A sheet of paper will rapidly lose or gain moisture when exposed to the atmosphere, and samples for moisture testing must be taken and placed immediately into a weighed container which should then be tightly closed.

Sack kraft, when freshly made, will have a moisture content of between 6% and 9%, and after conditioning for 24 h in the standard atmosphere of 23°C and 50% RH, the moisture content will be approximately 9.5%.

8.4.1.2.4 Air permeability (porosity)
The rate at which air can pass through sack kraft is important for the filling operation, particularly for valve sacks and for any sack to be filled with an aerated powdered product. This air permeability may be measured in a number of ways, the most general method being the Gurley test (ISO 5636/5 2003). In this test, the time is measured for 100 mL of air, under a fixed pressure head, to pass through a 6.5 cm^2 area of sample. The porosity of most sack krafts will generally be between 5 and 25 s.

8.4.1.2.5 Water absorption
It is important for sack kraft to resist the absorption of water into the sheet in order to maintain as much of the dry strength as possible. Sack kraft is treated, or sized, during manufacture for this purpose. The degree of sizing is normally measured by the Cobb$_{60}$ test (ISO 535 1991), which determines the weight of water absorbed by a given area of the paper surface after being held under 1 cm of water for 1 min. The Cobb value for most sack krafts is normally between 20 and 30 g/m^2.

8.4.1.2.6 Friction
Friction is the resistance to movement of two surfaces in contact, to either start or continue movement over one another. Sack krafts are normally fairly resistant to slippage, and it is generally some form of surface contamination that causes low friction to occur.

There are a number of methods for measuring friction. In one, a weighted sample of paper is pulled over another and resistance to movement is measured. Another method is for two samples to be held together on a level surface by a weight and the surface gradually inclined until slippage between the samples occurs. The angle with the horizontal when slippage first starts is a measure of the surface friction.

The measurement of friction is a most unreliable test as there are many factors which can affect the result. Strict care is necessary when taking and preparing samples, as even the touching of either surface by an operator is sufficient to appreciably affect the results.

8.4.1.2.7 Water vapour transmission rate (WVTR)
The rate at which water vapour can migrate through a barrier is dependent on the type of barrier material, its thickness and factors such as temperature and the humidity levels on either side of the barrier layer. It is measured by containing a quantity of a chemical that readily absorbs water in a metal can under a sealed-in sample of the barrier. The can is then left in a controlled atmosphere of temperature and humidity and weighed daily until the rate of increase in weight reaches a stable level. An atmosphere of 75% RH and 25°C is most generally used for testing barrier materials used in paper sacks.

The WVTR is always quoted in grams per square metre per 24 h together with the thickness of the barrier and the testing conditions. It is a time-consuming test which tends to give results with a degree of variability. A number of quicker, alternative non-gravimetric methods have been published, but with little information on comparison with this method.

8.4.2 Sack testing
There are two types of tests which may be undertaken on finished sacks, and these concern the quality of the finished sacks and sack performance.

8.4.2.1 *Quality of finished sacks*

To assess a consignment, sacks should be checked against the sack specification, which includes:

- dimensions
- sack construction, including quality of materials and number of plies
- printing, including print design, layout, register and colour
- supplementary features such as the use and positioning of adhesives.

All manufacture dimensions are subject to the normally accepted trade tolerances, which are given in Table 8.6. Repeat measurements should always be made from sacks sampled throughout the consignment.

One of the main criteria for the quality of the materials is that of grammage, which must be measured only after conditioning in the manner detailed in Section 8.4.1.2.

The type of material is generally visually apparent but checks, such as those for stretch or wet strength, may need to be made on certain materials.

The printing, especially with first consignments, should be compared for position and colour against the agreed design and layout specification.

8.4.2.2 *Performance tests*

There are two types of performance tests that may be undertaken on finished sacks, namely drop testing and field trials.

8.4.2.2.1 Drop testing

Drop testing consists of dropping filled sacks, generally until breakage occurs. By this means, different constructions or materials may be compared for performance in the same size of sacks. The sacks can be filled with either the normal or dummy contents. Drops may be onto the faces (flat drops) or the ends (butt drops) or a sequence of different types of drops. Drop testing is not an absolute assessment of sack strength. It must always be comparative, i.e. between two or more sets of sacks, one of which should be a control set of a sack construction that has given satisfactory performance.

Ideally, sacks used for drop testing should be conditioned immediately prior to the testing and, if possible, the drop testing should be undertaken in a conditioned atmosphere. This latter aspect is normally not possible, and it is therefore necessary that care to be exercised when planning testing to ensure that it is not undertaken in extreme conditions of temperature and humidity.

The main factors which influence drop test results are:

- weight of the contents
- type of the contents
- sack construction
- sack type
- degree of filling.

But dropping, consisting of dropping the filled sacks repeatedly onto one end until burstage occurs, will test the girth of the sack but will put no strain on the end closures. The results will indicate differences in body strength only.

Flat dropping will stress the whole sack, i.e. the body and the end closures. Such testing may be used to show differences in performance between different types of sacks as well as different constructions.

There are three ways in which sacks may be compared by drop tests.

8.4.2.2.1.1 Static drop height method
Here a height is chosen at which a set of sacks will survive a reasonable number of drops before bursting. The set under test is dropped from the same height and the average number of drops to breakage determined.

8.4.2.2.1.2 Elevator drop height method
The first drop is made from a height of 0.85 m, and the height of each subsequent drop is increased by 150 mm until breakage occurs. If sacks reach the maximum height of the drop tester without breaking, usually about 4 m, drops are continued at the maximum height until burstage occurs. In practice, this approach is better than that of the static drop height method, which can be extremely protracted if the sack constructions are appreciably different in strength.

8.4.2.2.1.3 Drop sequence method
A drop test sequence, comprising drops from a predetermined height onto face, butt and side in sequence, has been proposed as an international means of establishing a minimum standard for sacks, and other forms of packaging, for containing goods classified as dangerous. Users should be aware that whilst this test establishes a minimum standard, it does not establish that a particular construction is suitable for any specific distribution system. The hazards and journey lengths in distribution systems vary widely and a user should always think in terms of a preliminary limited field trial, if there is no prior experience of a particular distribution system. ISO 7965 1984 currently applies.

8.4.2.2.1.4 Field trials
Field trials are advisable with distribution systems involving multiple handling stages. Such a trial need only consist of a small initial consignment which is checked at despatch and upon arrival at the destination. Field trials in which consignments are monitored at various points within a distribution system can require a large number of observers and can be expensive to undertake.

This latter type of trial is only justified for complex or lengthy distribution systems in which initial trial consignments have shown unacceptable levels of performance or damage. The final analysis of the findings of such a trial should be aimed at the determination of sources of any damage and whether the need is for improvement in sack strength or the elimination of particular distribution hazards.

8.5 Weighing, filling and closing systems

The choice of filling and closing equipment is governed by a number of factors which include the nature of the product, the desired speed of operation and the type of sack.

Flow characteristics and bulk density of the product may determine the easiest method by which a sack can be filled. The particle size may also determine whether a satisfactory pack can only be achieved by either an open mouth or a valve sack system.

High-speed, fully automatic packing lines, in which the whole output of a mill or factory is funnelled through one packing line, will ensure labour is kept to a minimum. On the other hand, a number of simple packing lines will give flexibility in operation where various grades of one product are produced in the same factory.

The various systems available are most easily subdivided into those for open mouth sacks and those for valve sacks.

8.5.1 Open mouth sacks

Open mouth sacks are widely used for animal feed, seed, milk powder, etc. The equipment available for the filling, closing, conveying and palletising of these sacks may be chosen from a range of automatic, semi-automatic and manual units.

8.5.1.1 Weighing

All sacks containing products for sale must be filled within defined limits of weight accuracy. This can be done by gross or net weighing. The gross weighing is a slow operation and is usually linked to a simple packing line where operators are involved in both the weighing and the sewing operation. The net weigher can be used with either manual or automatic installations.

With gross weighing, the product being packed is weighed as the sack is filled, the flow of material being stopped once the correct weight is obtained. In net weighing, the product is pre-weighed into a weigh bucket after which it is discharged into the sack. Both types require a feeding system suited to the flow characteristics of the product and a device for reducing and finally cutting off the flow at the correct weight.

Weighers for free-flowing, granular products usually rely on either gravity or belt feeders. For other products vibratory, screw or fluidised feeders may be required.

The control systems for gross weighers are normally manual or semi-automatic. With semi-automatic systems, the flow of product is cut off just below the correct weight and the operator controls the final amount of material to enter the sack. Outputs of up to four sacks per minute are possible according to the nature of the product and the weight to be packed. The advantages of gross weighers are their relatively low cost, reduced headroom requirement and a low risk of inaccurate weighing due to the build-up of difficult materials on the machine components.

Net weighers, which always operate automatically, offer much higher speeds than gross weighers due to the separation of the weighing and filling operation. This means that whilst one sack is being filled, the next weighment is being prepared in the weigh bucket of the machine. The system also enables two or more weighers to deliver in turn through a common hopper to the filling spout where very fast packing is required. Typical outputs of free-flowing products, from a single weigher, range from 8 to 14 discharges per minute.

8.5.1.2 Sack applicators

The operation of placing and clamping empty open mouth sacks onto a filling spout may be manual or automatic. By using a net weigher, this can be linked to an automatic sack applicator where sacks are withdrawn from a magazine, opened by suction cups and lifted by

Table 8.2 Filling/weighing systems

Products				Weighments per minute		
Type of feeder	General description	Typical examples	Unit weight (kg)		Gross	Net
Gravity	Free-flowing, granular	Pellets, cubes	25		3–4	9–12
		Plastic granules	25		3–4	10–14
		Grain	50		3–4	6–8
		Fertiliser	50		2	10–12
		Sugar	50		2	8–10
Belt	Free-flowing, granular and coarse powder	Meals	25		3	8–12
		Pellets, cubes	25		4	10–14
Screw	Fine powders	Flour	50		3	6–8
		Fine chemicals	25		3	5–6
Fluidising	Fine powders	Flour	50		–	8–10
		Fine chemical	25		–	6–8
Vibrator	Coarse with lumps	Solid fuels	50		–	4
		Jumbo nuts	25		–	6
		Coarse flakes	25		–	5

pick-up arms onto the filling spout. Sack-detecting switches are used on the spout so that the weigher will only discharge if the sacks are correctly positioned.

8.5.1.3 *Filling*

The filling spout can be either a bird-beak type, which opens inside the sack, or an elliptical body, where the gripping jaws secure the sack from the outside. The elliptical units allow a dust-tight seal to be achieved during the filling operation. When the sack is correctly placed on the spout, the filling cycle is automatically started, the weigher discharges and the filled sack is released after a timed interval.

To assist light and aerated products to settle during filling, the sack may be mechanically agitated by a 'posser'. Two types are commonly used, one causing the filling spout to oscillate vertically and the other applying a vibration or oscillation to the base of the sack.

The transfer of the filled sacks to a closing unit is achieved by using a conveyor which receives the sacks directly from the filling spout. This avoids manual handling and distortion of the sack mouth.

8.5.1.4 *Summary of weighing equipment for open mouth sack filling*

Table 8.2 summarises the details of the filling/weighing systems available for open mouth sacks. The list of typical products is by no means complete and is provided for guidance only.

8.5.1.5 *Closing*

Open mouth sacks may be closed by sewing, bunch tying or, with certain types, by folding and gluing.

8.5.1.5.1 Sewing

Most open mouth paper sacks are closed by sewing. This gives a neat secure closure with a minimum requirement of ullage, or free space, at the top of the sack. Stitching equipment range from hand-held or suspended portable sewing machines to high-speed, automatic heavy-duty machines used in conjunction with conveyors. Either single or double thread stitching may be carried out, with or without kraft paper string or jute filler cord and may be sewn through crepe tape folded over the sack top.

Portable machines, which normally apply a single-thread stitch, are suitable for low or intermittent outputs, for example up to 2 sacks per minute. These should not be used on sacks with more than four plies. For low outputs in conditions requiring robust equipment, a heavy-duty pillar-mounted sewing machine may be used with a bogie on rails to carry the sacks.

The most widely used sack closing equipment comprises a heavy-duty sewing head mounted on a pillar over a conveyor. The sewing head is fitted with automatic knife-and-switch mechanism so that it starts when a sack is entered, sews, cuts the trailing sewing chain and stops again as the sack clears.

Sewing and conveyor speeds range from 6 to 15 m/min. At higher speeds, one sewing machine can close up to 20 sacks per minute.

Given adequate weighing and filling capacity, under typical conditions, one operator can fill and sew up to six sacks per minute. Two operators, one filling and one sewing, can handle from 12 to 16 sacks per minute. With sacks of heavy construction or if sewn-through-tape closure is in use, lower rates will apply or an extra operator may be required to fold the sack top before sewing. Table 8.3 shows the output of sacks per minute with respect to the type of equipment used.

For high-speed packing lines serving continuous production processes, double-head sewing machine units are recommended. If it is necessary to replace the closing materials or carry out adjustments, the operator simply rotates the unit to bring the reserve head into action.

Automatic sewing equipment eliminates the operation of forming and entering the sack top into the sewing machine. The sack may be fed forward by vertically mounted shaping rollers on either side of the conveyor or from a sack top stretching and forming system. Converging belts will then enter the sack into the sewing head. Guide boards, or a troughed conveyor, ensure that the sack remains upright during the closing operation. For tape closure, the sack top is trimmed flush before entering the sewing unit.

Automatic open mouth sack packing systems are available as either a combination of sack applicators and stitching units or self-contained integral filling and closing machines. These installations not only eliminate manual labour normally associated with sack packaging but have built-in safeguards against malfunctions, enabling the operator to perform other duties in the vicinity. The relative rigidity and handling qualities of a paper sack make it most suitable for use with automatic equipment.

Equipment is available to close sacks of specialised construction. For instance, sacks containing a sealable inner ply for containing finely powdered, toxic or hygroscopic materials may be closed by heat sealing and sewing. Successively sewing and heat sealing the inner liner below the sewing line achieves a closure. Any of these operations may be achieved by the inclusion of individual units in the closing line or by an integral machine which combines all operations. Other extras, such as a unit to clean the surfaces of the sack mouth, may also be incorporated in the packing line.

Table 8.3 Sewing closure – equipment options and outputs

Type of equipment			Sacks per minute with number of operators		
Scale	Filling	Sewing	1 operator	2 operators	3 operators
Platform	Hand spout	Portable	1–2		
Semi-auto gross	Hand spout	Pillar and bogie	2–3		
Standard net	Auto spout	Pillar and conveyor (low speed)	4–6	8–10	
High-speed net	Auto spout	Pillar and conveyor (high speed)	5–8	10–15	
Twin net	Auto spout	Pillar and conveyor (high speed)		12–16	18–24
High-speed net	Auto spout	Automatic	8–14		
High-speed net	Auto sack placer and spout	Pillar and conveyor (high spout speed)	8–14		
High-speed net	Auto sack automatic placer and spout		8–14*		
High-speed net	Integrated auto filling and sewing machine		8–14*		

*One operator in attendance but free to perform other duties in vicinity.

8.5.1.5.2 Closures other than sewing

Pinch closed sacks, which contain a pre-applied line of hotmelt adhesive at the mouth flap, are closed by a unit which reactivates the adhesive and then folds the flap over the mouth to form an instant closure. These units are built around a conveyor for moving the sacks and may also include a heat-sealing position and compression rollers to ensure complete adhesion and consolidation of the bonds.

There is also a unit to close open mouth sacks with the double-folded closure. It has a built-in conveyor and uses hotmelt adhesive which is applied to the top of the filled sack during the folding operation to give instant adhesion to the closure.

8.5.2 Valve sacks

By eliminating the need for the sewing operation, valve sacks can give faster packing speeds, together with automatic filling and weighing systems.

8.5.2.1 *Applicators*

A number of designs of sack applicators are available, which automatically place valve sacks onto filling spouts. These generally rely on suction cups for handling the sacks and for opening the valves. Whilst most valve sack placers are designed to work with single or twin

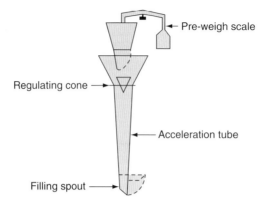

Figure 8.7 Gravity packers. (From Kirwan, 2005.)

spout packers, they are also offered to serve up to four to six closely spaced filling spouts. In addition, applicators are also available for high-speed rotary packers. The majority of designs make use of magazines for holding the empty sacks which may be replenished with fresh sacks either manually or automatically. Other designs are available, which present the sacks to the applicator by means of pre-made rolls of shingled sacks.

8.5.2.2 Weighing and filling

For the filling of valve sacks, several different types of packing equipment are available, the choice depending on the nature of the product. The method by which the product is accelerated to fill the sacks serves to classify the various packing machines. These may be further divided into those which operate below a pre-weigher and those which weigh the product in the sack and arrange for the flow of material to be cut off at the correct weight.

The filling machines cover a wide range of outputs from simple units to high-speed multiple installations. With the systems which automatically weigh and discharge the filled sacks, an operator has only to place sacks onto the filling spout or into the magazine of an applicator.

8.5.2.2.1 Gravity packers

The simplest type of valve packer employs gravity filling and is suitable for free-flowing, granular products (Fig. 8.7). For low outputs, the equipment takes the form of an inclined chute terminating in a filling tube above a sack chair, all mounted on a platform scale. Weight is controlled manually or by means of an automatic cut-off gate at the base of the overhead feed hopper.

A high output 'free fall' gravity packer operates in conjunction with a pre-weighing scale and consists of a vertical accelerating tube and filling spout, with automatic sack clamps. These packers require considerable headroom but may be grouped close together for operation by one person. They are particularly recommended for dense and free-flowing products.

8.5.2.2.2 Belt packers

The high-speed packing of granular material of any density is best undertaken with a grooved wheel or belt packer (Fig. 8.8). A pre-weighed amount of the product is fed

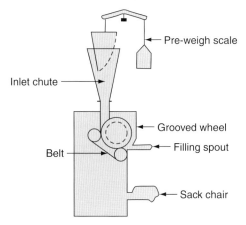

Figure 8.8 Belt packers. (From Kirwan, 2005.)

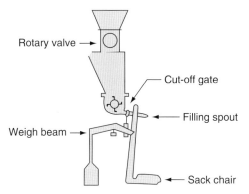

Figure 8.9 Impeller packers. (From Kirwan, 2005.)

vertically into the annular space between a continuously running belt in contact with the grooved wheel. This travels through 90° and delivers the product horizontally at high speed, through a throat and filling spout, into the sack. During filling, the chair supporting the sack may oscillate to settle the product. The chair may be tipped manually or automatically to discharge the filled sack. Grooved wheel packers are available in single or twin spout versions and may be grouped for one operator.

Some products, such as beet sugar pulp nuts, although free-flowing, would bridge if discharged directly into the narrow feed chute of a belt packer. This is overcome by inserting a belt feeder between the pre-weigher and the packer, which regulates the delivery rate of the material.

8.5.2.2.3 Impeller packers

For packing a range of ground rock-type products, the impeller packer is recommended (Fig. 8.9). A continuously running impeller feeds the product, through a cut-off valve gate and a flexible connection, to the filling spout and into the sack. A saddle on a weigh beam supports this, and, once the correct weight is obtained, the cut-off gate is actuated and the sack discharged. Impeller packers are available with weighing systems based on load cell controls.

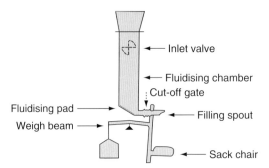

Figure 8.10 Fluidising packers. (From Kirwan, 2005.)

They are also supplied with up to four spouts for high-speed packing of cement and similar products. For low outputs, a simple impeller packer is available in which the filling spout and chair assembly is mounted on a platform scale, the cut-off gate being manually controlled.

8.5.2.2.4 Fluidising packers (air packers)

Fine and very fine powders may be made to flow readily by passing air through them. With valve packers based on this property, the fluidising is achieved by admitting low-pressure air through a porous pad in the base of a chamber. The fluidised product flows out by gravity, through a cut-off gate and the filling spout, into the sack. In the 'pressure flow' version of this machine, applying air pressure above the product in the chamber increases the packing rate. The control of weighing is similar to that of an impeller packer.

Fluidising packers can handle a wider range of products than any other type but are most suitable for material with a sufficiently high proportion of fine powder to hold the fluidising air (Fig. 8.10). They are particularly recommended for fine powders with difficult handling characteristics. By increasing the fluidising air considerably, some granular products may also be packed. Since the product delivered by a fluidising packer will be aerated, provision for the escape of entrained air may be needed. A vent may be incorporated in the filling spout, and if the sacks include a barrier ply, some perforations or other intelligent de-aeration constructions may also be necessary.

8.5.2.2.5 Screw packers

For powder products which do not flow freely, a screw packer may be used, either in connection with a pre-weigh scale or as a hand-controlled unit mounted on a platform scale (Fig. 8.11). The material is delivered from a small hopper fitted with an agitator, through a screw feed in the filling tube. Provision to assist settling of the product can be made during filling with net weighing.

8.5.2.3 Rotary packing system

For industries which pack one unvarying product at a constant high rate, rotary packing machines have been developed. These consist of a number of individual packing units which are mounted on a slowly rotating structure. Empty sacks are placed either manually or automatically on each of the filling spouts at one point. Filling takes place as they travel round to the automatic discharge point. The advantage of these machines is that the output is no

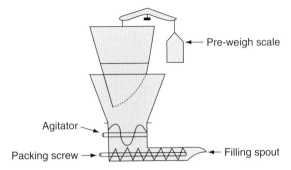

Figure 8.11 Screw packers. (From Kirwan, 2005.)

longer governed by the time taken to complete the weighing and filling of each sack. Rotary packers are available for both valve and open mouth sacks. They are used mainly for filling cement, plaster or chemicals into valve sacks at rates up to 135 tonnes per hour. A fully automated rotary system requires sacks to be presented to the spouts as flat as possible and free from sticking together, using one of three types of magazine as follows:

- horizontal sack-holding magazine with capacity of 600 sacks
- vertical magazine with storage capacity of up to 6000 sacks
- roll system (sacks on reel), which serves as a magazine holding approximately 4500 sacks shingled together to form a reel with a maximum diameter of 160 cm held in place by two plastic tapes.

In machines for filling open mouth sacks, it is more usual to employ two or three fixed pre-weighing scales, each serving several filling spouts. After filling, the sacks are led off tangentially onto a conventional sewing line or for heat sealing. If the operation is automatic, only one person is required to place empty sacks into a cassette magazine. If the operation is manual, up to three operators could be involved.

8.5.2.4 Output levels of valve sack systems

Table 8.4 summarises the output levels of available valve sack packing systems for various products. The list of products is not intended to be comprehensive and is provided for guidance only.

If sacks with tuck-in sleeves are used with valve packers having automatic discharge, an arresting device may be provided to hold the sack temporarily at a convenient angle.

8.5.3 Sack identification

It is often necessary to add information or a batch identification mark to sacks during the packing operation. This may be done by simple printing devices or by attaching pre-printed tickets.

On open mouth sacks, printing may be added by a rotating contact roller mounted beside the sewing head, applying a code or batch number close to the sewing line. Alternatively, for any type of sack, the printing may be done after filling by an ink-jet printer or a roller device as it passes along a conveyor. Special printing devices are available for valve sacks which

Table 8.4 Valve sack output levels for various products and packing systems

Type of packer	Products — General description	Typical examples	kg	1	2	3	4
Gravity with platform scale (semi-auto weigh)	Granular, free-flowing	Plastic pellets Dry sand	25 50	2 2	4 4		
Gravity free fall with pre-weigh scale	Granular, free-flowing	Plastic pellets Granular fertiliser	25 50	5–6 6	12–14 12–16	 —	 —
Belt, grooved wheel with pre-weigh scale	Granular, free-flowing	Granular fertiliser Granular sugar Plastic pellets Provender pellets Grain D.V. salt	50 50 25 25 50 50	6–7 4–5 6–7 6–7 5 6–7	14–15 8–10 14–15 14–15 10 14–15	20 — 20 20 15 20	20 — 20 20 20 20
Belt, grooved wheel with pre-weigh scale and belt feeder	Large cubes Light fine flake	Beet pulp nuts Cattle cubes Provender meals Soya meal Shredded beet Pulp	50 25 25 50 40	3 4–5 4–5 4 2½	6 10 8–10 8 5	9 15 12–15 12 7½	12 20 16–20 16 10
Impeller on platform scale	Pure ground rock	Portland cement Gypsum Hydrated lime China clay Fly ash	50 50 25 25 25	2–3 2 1–2 1–2 2			
Impeller with automatic weighing and sack discharge	Pure ground rock	Portland cement Gypsum Hydrated lime China clay Fly ash	50 50 25 25 25	5–6 3–4 3 3 4–5	10–12 7 6 6 8–10	15–18 10 10 9 12–15	20–24 14 14 12 16–20
Fluidising pressure flow	Powder Powder/fine-granule mixture	PVC resin Carbon black Starch Refractory powders Flour	25 25 50 50 50	5–6 3–5 4–5 5–6 4–5	12 8–10 8–10 12 8–10		
Screw with screw feeder, mounted on platform scale, hand control	Fine and very fine, non-free-flowing, powder	Resin powder Starch	25 50	1 1			
Screw with pre-weigh scale	Fine and very fine, non-free-flowing, powder	Resin powder Starch Flour Provender meal	25 50 50 25	2 1½ 1–2 2–3	4 3 4–5 4–6		

are either incorporated into the clamp which secures them to the packing machine spout or are an addition to certain sack applicators.

If printed tickets are required, these may be manually or automatically inserted into the sewing line of open mouth sacks. Automatic ticket dispensers either take the tickets from a reel or will feed them from a magazine. Self-adhesive tickets may also be applied to either valve–valve or open mouth sacks after they have been filled.

8.5.4 Sack flattening and shaping

To gain the maximum advantage from paper sacks as a means of storing and transporting materials, it is often desirable to flatten or shape them after filling. This provides firm, uniform packs, offering substantial savings in storage and transport costs, together with safer stacking or palletising.

The choice of equipment depends on the nature of the product and the type of sack. For light- and medium-density free-flowing products, a twin-roller sack-shaping machine, which serves to distribute the product evenly along the length of the sack, is adequate. For denser products which pack more solidly, the compression flattener, in which the whole length of the sack is subjected to uniform pressure as it passes between belts, is recommended. If the product is sensitive to excessive pressure or is of a coarse lumpy nature, a vibratory flattener should be used.

There is generally a greater benefit to be obtained from flattening sewn open mouth sacks than pasted valve sacks. Open mouth sacks should always be fed bottom first into roller or belt flatteners.

The flattening of sacks with barrier piles containing highly aerated products must be done very slowly, to avoid burstage, even if they are specially perforated. It is preferable to reduce aeration as much as possible by mechanically settling or 'possing' during filling and to allow time for further de-aeration before sealing the sack.

8.5.5 Baling systems

Baler packing equipment ranges from simple devices, which aid manual loading of baler sacks, to fully automatic power-operated machines. The method of functioning will depend on whether the goods to be packed are *fully compressible*, such as woollen articles, *semi-compressible*, such as small bags of flour, or *rigid*, such as boxes or cans.

A typical hand baler consists of a short chute, shaped to receive and clamp the mouth of the sack and support it in an inclined position. Articles are simply pushed through the chute into the sack. In another version, suitable for rigid articles only, the goods are loaded into a horizontal open trough, over which the baler sack is drawn. On releasing a catch, the trough tilts forward so that the goods and sack slide clear.

Semi-automatic baling machines include a receiving hopper into which the items are manually loaded. The sack is manually applied to the 'duck bill'-type jaws which grip it tightly. On closing the hopper, the charge is first subjected to alignment or pressure, by inward movement of one side of the hopper, and then pushed through a throat into the sack by a longitudinal ram. At the end of its travel, the ram pushes the loaded sack off the jaws. If the items are compressible, the side plate applies sufficient pressure to reduce the volume of the charge slightly; if they are rigid, it merely serves to line them up firmly.

In fully automatic baling machines, automatic collation of the articles forming the complete pack may take place directly in the hopper from above, or by forming the load alongside the machine and transferring it sideways into the hopper. The placement of the baler sacks on the sack jaws may be by hand or by mechanical means using suction cups.

Baling machines can be incorporated into continuous production lines, the completed packs being discharged through a sealing station to a conveyor.

8.6 Standards and manufacturing tolerances

The paper-sack industry participates actively on national and international standards committees, ISO, etc. These committees are concerned with the elaboration of standards dealing with terminology, dimensions, capacities, marking and test methods in the fields of packaging and unit loads.

8.6.1 Standards

In some instances, there are separate national standards but there is an increasing tendency towards common standards either internationally, (ISO), or across Europe, (CEN). Although many standards are not mandatory, they may become so.

Table 8.5 includes relevant standards for multiwall paper sacks which may be of interest to both manufacturers and users of paper sacks.

Table 8.6 indicates manufacturing tolerances.

In recent years, much attention has been paid to the suitability of paper sacks and other containers for the packaging and distribution of food products for human consumption. The question of possible migration of hazardous substances from paper, board and plastic film is under active consideration at national and international levels. User of packaging are concerned about the product safety of their packaging materials and in the case of food manufacturers and retailers are concerned about conditions of package manufacture and distribution.

United Nations approved specifications are becoming common in relation to the packaging of hazardous chemicals. National governments have regulations relating to the product safety of packaging materials. For example in the UK, the Food Safety Act 1990 is relevant as well as the following statutory instruments: SI 1523:1987 and SI 3145:1992. In the USA the US Food and Drug Administration, (F & DA) is reponsible for overseeing all aspects of product safety, including issues related to packaging.

8.6.2 Manufacturing tolerances

A paper sack is a pre-made form of packaging manufactured to the user's requirements of size and construction. During the manufacture, a small degree of variation can occur, which good manufacturing practice will keep to an absolute minimum. Table 8.6 summarises the tolerances which apply to sack manufacture and which are generally accepted throughout the industry.

Table 8.5 List of standards (ISO and EN) relative to multiwall paper sacks

International or European Procedure	Title
ISO 2206 1987	Complete filled transport packages
ISO 1924:2:2008	Determination of tensile properties of paper/board Part 2: Constant rate of elongation method (20 mm/min)
ISO 1924:3:2005	Determination of tensile properties of paper/board Part 3: Constant rate of elongation method (100 mm/min)
ISO 3676 1983	Plan dimension of unit loads
ISO 6780 2003	Pallets for intercontinental materials handling through transit specifications for principal dimensions and tolerances
ISO 6591/1 1984	Packaging – sacks – description and method of measurement
ISO 7023 1983	Packaging – sacks – method of sampling empty sacks for testing
ISO 7965:1 1984	Packaging – sacks – drop test
ISO 8351-1 1994	Packaging – method of specification for sacks Part 1 – Paper sacks
ISO 8367-1 1993	Packaging – dimensional tolerances for general purpose sacks Part 1 – Paper sacks
EN 277 1989	Sacks for transport of food aid (polypropylene fabric)
pr EN 1086 1993	Sacks for transport of food aid – selection recommendations for type of sack and liner (provisional recommendations) Sacks for transport of food aid (paper sacks)
ISO 535 1991	Sizing properties of paper – determination of water absorptiveness (Cobb method)
ISO 3689:1989	Determination of bursting strength of paper after immersion in water
ISO 3781:2011	Determination of tensile strength after immersion in water
ISO 2758:2001	Methods of determining bursting strength of paper and board
ISO 187:1990	Standard atmosphere for conditioning and testing and procedure for monitoring the atmosphere and conditioning of samples
ISO 536:1976	Method of determining grammage (basis weight) of paper and board
ISO 1974:2012	Method of determining internal tearing resistance of paper and board
ISO 5636:1 1984	Determination of air permeance of paper and board – General method
ISO 5636:3 1992	Determination of air permeance of paper and board – Bendtsen method
ISO 5636:4 2005	Determination of air permeance of paper and board – Sheffield method
ISO 5636:5 2003	Determination of air permeance of paper and board – Gurley method
ISO 6599-1:1983	Method of conditioning paper sacks for testing
ISO 2875:2000	Method of water spray test for filled transport packages and unit loads
ISO 291:2008	Standard atmospheres for conditioning and testing
ISO 3205:1976	Preferred temperatures for testing
ISO 1133:1;2011	Determination of melt flow rate of thermoplastics Part 1 Standard method
EN:648	Paper and board intended to come into contact with foodstuffs – determination of the fastness of fluorescent whitened paper and board
EN 27965-1:1992	Packaging – Sacks – Drop test – Part 1: Paper sacks (ISO 7965-1:1984)
EN 1086:1995	Sacks for the transport of food aid – recommendations on the selection of type of sack and the liner in relation to the product to be packed
EN 645:1993	Paper and board intended to come into contact with foodstuffs – preparation of a cold water extract
EN 646:2006	Paper and board intended to come into contact with foodstuffs – determination of colour fastness of dyed paper and board
EN 647:1993	Paper and board intended to come into contact with foodstuffs – preparation of a hot water extract

ISO is the International Organization for Standardization which issues ISO Standards. CEN is the European Committee for Standardization which is a body which issues EN Standards.

Table 8.6 Generally accepted multiwall sack manufacturing tolerances

Item	Tolerances Plus	Minus	Notes
Sack kraft grammage	4%	4%	
Sack length	10 mm	10 mm	
Face width	3 mm	3 mm	
Circumference (girth)		10 mm	The upper limit controlled by upper limits on the face width and gussets
Gusset width	3 mm	3 mm	
Bottom widths	5 mm	5 mm	
Valve width	3 mm	0	
Valve sleeve position	5 mm	5 mm	
Seam overlap of 15 mm		3 mm	
Stitch-line position	Not less than 12 mm from sack transverse edge		

Note: This table conforms largely to ISO 8367/1 1993 but the ISO standard has a face width tolerance of +5 mm and −5 mm.

8.7 Environmental position

The paper sack is a product invented and developed in the twentieth century. It continues to be a significant form of packaging. From its humble origin as a sheet of paper, rolled and tied at one end, filled with product and tied again at the other end, it has developed into one of today's leading packaging products. It is ideal for handling and containing a wide range of products in a packed weight ranging from 5 to 50 kg.

The essential element in all paper sacks has to be virgin fibre. Their strength, handling and efficient manufacture depend mainly upon raw material derived from sustainable temperate forests in North America and Europe, including Scandinavia and the Russian Federation. Secondary or recycled fibre at the moment is not an option as such material can result in an inconsistent pack which is weaker and is unsuitable for automated filling and in handling systems. The sacks would be more costly to produce and in many cases would be unacceptable for direct contact with food products.

As a result of continuing research and development, there have been substantial reductions in the average construction weight of a paper sack. In conjunction with paper mills and machinery manufacturers, this work has enabled the industry to maintain an edge against competition from new materials and systems. Subsequent contribution to the overall reduction of packaging and packaging waste has largely not been recognised by governmental interests and also by those involved in the environmental lobby.

Recovered sacks can be recycled. The high quality long fibre they contain should be suitable to paper makers, provided economical recovery conditions prevail. The paper-sack industry is willing and keen to participate in any viable scheme or schemes of recovery.

In Europe, intensive discussions are on-going on the implications of the EU Packaging and Packaging Waste Directive, involving legislators, affected authorities and trade associations.

To summarise:

- Paper sacks are made from renewable raw materials.
- Paper sacks are lightweight materials.

- The inherent energy of paper sacks can be recycled in energy from waste systems.
- Paper sacks are compostable, biodegradable or both.

And, finally, the paper-sack manufacturing industry accepts its responsibility to contribute to the reduction of the overall volume of packaging.

References

Kirwan, M. (ed.), 2005, *Paper and Paperboard Packaging Technology*, Blackwell Publishing Ltd., Oxford, UK.
Mondi, 2012, *Mondi Industrial Bags Full Range*, visit http://www.mondigroup.com/products/desktopdefault.aspx/tabid-1073/

Useful contacts

European Federation of Multiwall Paper Sack Manufacturers
23/25, rue d'Aumale – 75116 Paris, France
Tel: (33) (0) 1 47 23 75 58
Fax: (33) (0) 47 23 67 53
info@eurosac.org

Paper Shipping Sack Manufacturers' Association, Inc.
505, White Plains Road, Suite 206
Tarrytown, NY 10591
Tel: 914/631-0909
Fax: 914/631-0333

Websites

European Federation of Multiwall Paper Sack Manufacturers, www.eurosac.org
Paper Shipping Sack Manufacturers' Association, Inc., www.pssma.com
Mondi Industrial Bags, http://www.mondigroup.com/products/desktopdefault.aspx/tabid-1073

9 Rigid boxes

Michael Jukes

London Fancy Box Company, Dover, Kent, UK

9.1 Overview

A rigid box is a box set-up ready for use without further fixing when received from the box manufacturer (BS, 1986). The origins of the rigid box and its use go back centuries but its appeal today is largely based upon perceived quality and its ability to handle weighty contents – often of high value – and to afford physical protection.

With the dominant position of cartons and corrugated cases within the paper packaging industry, it may appear that rigid box use is in steep decline. However, for numerous boxmakers, both large and small, the market niches that rigid boxes and their derivatives serve is actually widening. Whilst it is a long time since the rigid box was the packaging container of choice for the myriad of products which are now not even in cartons but in cellophane wrappers, for example, there are probably two major factors which underpin the appeal of the rigid box – quality and perceived luxury.

The range of the rigid box is not simply comprised of lift-off-lid (LOL) boxes. We should consider the wide range of products made by boxmakers and their box-making machinery in order to get a full appreciation of the market appeal of the rigid box. All products have their strengths and weaknesses/limitations but, in a world where packaging waste is an international environmental issue, the 'greenness' of the rigid box, together with its versatility, answers many of our present concerns.

Strengths

- structural strength and ability to provide protection for its contents
- perceived luxury by the ability to combine materials such as board, fabric, metal, plastics, etc.
- wide range of styles, sizes, covering materials and accessories, for example hinges, and magnetic closures, etc.
- design freedom – combination of rigid jackets, boxes, slip cases
- wider range of print and surface finishes compared with carton board

- compatibility with on-counter display of retail products
- ability to make 'small quantity' production runs
- second and alternative use when empty
- typically >80% of material content is recyclable and recycled.

Weaknesses and limitations

- large 'empty size' compared to the fold flat carton type offerings, with concomitant implications for transportation costs within the supply chain
- relative higher cost compared with cartons
- unsuitability for automated processing in volume applications
- volume restrictions – production rates limited to around 2500 per hour.

9.2 Rigid box styles (design freedom)

A defining characteristic of the rigid box is the freedom it allows with respect to shape, materials, use of accessories and overall presentation. Rigid boxes may be large or small, square, rectangular, round or elliptical in shape. The simplest rigid box forms are based upon the basic LOL box as shown in Fig. 9.1.

The shell and slide, book style and flip top, as indicated in Fig. 9.2, show other distinctive closure designs. The shell and slide has been expanded into a case with drawers to hold, for example, assortments of chocolates or surgical instruments.

Combining trays (box bases) with different lid constructions results in hinged and shouldered (jewel) boxes as shown in Fig. 9.3.

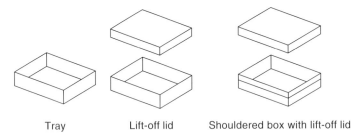

Figure 9.1 Basic lift-off-lid (LOL) boxes.

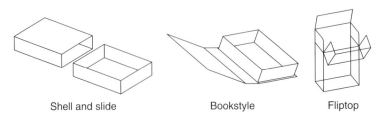

Figure 9.2 Examples of distinctive box styles.

Figure 9.3 Hinged and shouldered boxes.

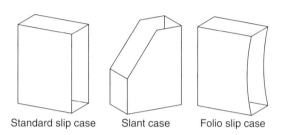

Figure 9.4 Slip cases.

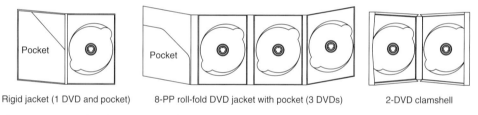

Figure 9.5 Rigid jacket with pocket for DVD.

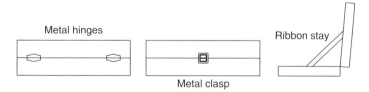

Figure 9.6 Incorporation of accessories in box design.

Turning trays on their sides results in slip cases with a variety of shapes and design features as shown in Fig. 9.4.

Rigid board jacket with pockets and plastic fitments are used to pack digital video discs (DVDs) (Fig. 9.5).

The addition of accessories in the form of ribbons, bows, clasps and hinges can provide 'special' appeal as shown in Fig. 9.6. Plastic and rigid board fitments are also used.

9.3 Markets for rigid boxes

The UK market sectors that are the largest users of rigid box packaging include:

- tableware – ceramics, pottery, glassware and cutlery
- jewellery and watches
- perfumes and cosmetics
- music, video and games
- stationery and office storage
- luxury drinks
- publishing
- chocolate confectionery and gifts
- boxed greeting cards
- photographic trade products
- engineering and DIY hand tools
- medical equipment.

At current price levels, the UK market, including imports, is in the region of £90 million with UK manufacturers producing around £50 million. Exports from the UK mostly into near continental markets is low, probably in the region of <10% of home production. The proportion exported is low because it is relatively expensive to transport empty rigid or 'set-up' boxes over long distances because of the space required. Exported rigid boxes are usually those with more complex design where there is a high added-value component. Imported boxes (largely from the Far East) are generally confined to products where delivery lead-times are long and often they are packed in the country of manufacture with local source contents, for example tableware greeting cards, etc.

Within many of these sectors, the luxury 'gifting' nature of the contents of the box brings a large seasonal element into play. Typically the run up to Christmas, which may last 4–5 months, gives the boxmaker huge challenges as it is necessary to increase production dramatically. The typical box-making factory is both machine and labour intensive. Many of the box-finishing processes utilise hand skills, whereas the setting, operating and maintenance of wrapping, laminating and material preparation machinery demands the highest level of expertise and skill.

In value terms, the use of one-off special boxes probably exceeds that of standard, repeated products. With short delivery lead-times (10 days would be usual in the music sector), the rigid boxmaker generally works directly with the client throughout the design stage, i.e. the use of sales agents is rare. Clearly the use of computer-aided design (CAD), communication through broadband lines, the digitisation of print and other 'modern' processes and systems are key tools in the boxmaker's range of competencies. It is common practice for multiple iterations of new designs/aesthetics to be presented to clients before design finalisation.

9.4 Materials

9.4.1 Board and paper

The main raw materials for paper and board are cellulose fibres, water and energy. The cellulose fibres mostly come from soft and hard woods, but esparto grass and cotton fibres are also used for the finest papers. Boards are principally made from recycled fibres to give maximum humidity-related stability. Board thicknesses are typically in the range of 1000–2500 µm with weights of 400–1600 g/m^2. In 'composite boxes', where folding box board (FBB) is used in conjunction with recycled chipboard, lower weights in the range of 270–400 g/m^2 are used. As well as boards made from recycled fibre, solid bleached boards (SBB) and solid unbleached (brown) boards (SUB) are also used; they may be lined or unlined depending on their role in the box construction.

9.4.2 Adhesives

The rigid box industry usually affixes printed paper to recycled board and generally, depending on the paper, uses glues of both gelatine and polyvinyl acetate (PVA) varieties for finishing. In all cases, both the paper and the board absorb moisture to differing degrees and the 'wet expansion' of each material is a key consideration, not only in material selection but also in design and construction. The absolute level of humidity in each adjacent material is not the issue but rather the differential between each material.

Gelatine glues are almost always preferred, but their inability to adhere to laminated paper surfaces is where the PVA type comes into play. Adhesion may be equal, but gelatine provides a cleaner surface not only to the box but also to the production machinery. Precise glue selection is also dependent on the box construction with open times being an important consideration. Clearly, glue temperature and the dispersion of the solid adhesive are features which the boxmaker must take into account.

9.4.3 Print

Typically rigid boxmakers are not printers; they buy print in, often from local printers, who are familiar with printing on flat sheets. This is quite different from the carton-making industry that generally specialises in printing and then adds extra value by cutting, creasing and gluing to the final carton shape.

Standard sheet-fed printing presses are used with the majority of paper sizes in the B2 (520 mm × 720 mm) range. With the largest production volumes, often associated with the music industry, B1 (720 mm × 520 mm) size sheets are also used. Both 'standard' finishes and UV printing techniques are employed, all driven by the application, with print cost also being a major driving force in the print-finish selection.

Other embellishments are then added in the box-finishing processes, for example foil blocking, embossing, debossing and selective lamination or UV varnishing (where the free areas allow for subsequent gluing). UV varnish is increasingly specified whilst film lamination is declining, perhaps for a combination of environmental and cost-driven factors.

9.5 Design principles

The most obvious need for any form of packaging, including the rigid box, is the need to offer protection for its contents. Depending on the market use, probably a close second is the need to present, in marketing terms, the packaged product to its best advantage. This is sometimes achieved by making the product visible whilst in the box, especially where display at the point of purchase is important, but more usually it is achieved by the use of arresting graphics and print techniques. Other display techniques might include window patching and the use of transparent plastic lids.

Two examples, at opposite ends of the spectrum, might be packaging for the industrial use of acetate films (overhead projection (OHP) slides) and the packaging of luxury cosmetics or chocolates. In both cases, the contents may have significant mass, both need protection prior to use or consumption, but their presentation is critical in the eyes of the customer. A simple LOL box with good print may suffice for the A4-sized box of film, but a luxury appearance as, for example, provided by a round-, oval- or heart-shaped box is required for the beauty preparation or assortment of Belgian chocolates.

Box-making through the use of machinery (as opposed to hand-making techniques) imposes design limitations. The first is that boxes start as flat sheets of board and paper, and subsequent operations to form the box generally require right angles, i.e. 90° bends. They are usually rectilinear in shape although some polygonal shapes can be partially made on machines, for example hexagonal.

From the examples shown earlier in this chapter, more elaboration may be achieved by the combining together of trays and lids, with or without platforms to hold the contents in position. The platforms can have complicated shapes and contours which are achieved by using vacuum formings, or expanded polystyrene, tailored to hold the contents securely. Transit testing, including drop testing, can be onerous on rigid boxes, so each element of the design has to be carefully analysed to ensure structural rigidity and provide protection.

Rigid box construction uses a variety of glues and base materials, varying from recycled board through to covering papers, fabrics, etc., which must be 'stuck' together. The widely differing material/moisture expansion issues have to be borne in mind. Not only can glue/adhesive viscosity have a significant effect on the overall box dimensional sizing but also the climatic conditions within the box-making factory can affect the wet expansion of paper, etc. This in turn can, in its simplest form, affect the clearances and tolerances of the box and lid. The rigid box is not, and cannot be considered, a precision product and tolerances of 0.5 mm are not unusual.

The main characteristic of the rigid box is its appearance. It is the variety of covering materials, including foils, papers, flock or fabric linings, leather – in fact any material that can be stuck to cardboard and take and hold its shape after the adhesive is dry – and a wide array of printed effects that give the rigid box its unique appeal. Lids may be flanged, domed or padded. Box corners can be metal edged. Component parts can be hinged together using paper, fabric, metal or plastic; fastened using many techniques and decorated with bows, ribbons, handles, magnets, etc. All these attributes can combine to make the finished pack suitable for gracing the most prestigious point-of-sale display and frequently provide a second use (often in the home) after the contents are consumed or used.

Boxmakers, like carton producers, generally provide their customers and their design teams templates that show where the graphics are to appear on the sheet of print. These are essential as they show the relationship between those visible faces, flaps, turn-ins, etc. and the parts of the box that are not visible after construction. The addition of printed platforms, sleeves, drawers, etc. all add to the print complexity but all must be considered during the design stage.

9.6 Material preparation

Rigid boxes typically start as flat sheets that are cut to a predetermined shape before the box assembly process. 'Preparation' machinery and equipment can be as simple as scissors, continuing through a range of machinery up to computer-controlled guillotines and continuous cutting machines utilising a stamping-type process.

In the latter case, cutting dies are typically made from steel rule and plywood. They are often very complicated, including the use of make-readies, and not only prepare the board for box assembly but also make the scrap/unwanted board easy to handle and recycle. There are also simple platen dies with different heights of rule depending on whether a cutting or scoring action is required. (Scoring in this context means a knife cut in the surface, which penetrates about 50% of the way through the board.)

Additional equipment is used to cut large boards into manageable sizes for subsequent cutting operations and the forming of grooves and creases on rotary equipment. The boxmaker is always trying to optimise material usage, i.e. minimise the scrap cut to waste. This may be by the simple expedient of cutting boards in half or, more usually, by specifying specific board sheet sizes which the mill then produces to order.

Covering papers use similar processes, but as they are generally printed, the boxmaker often performs embossing, debossing and foil-blocking operations, all of which may be vital ingredients of the finished box's design appeal.

The most simple, covered LOL box has four main components, each with its own profile, comprising box card, lid card, box cover and lid cover. Typically the box and the lid cards will be cut together using a 'one set-on cutting die'. Simple cutters for such a construction are relatively inexpensive at around £100 each. This makes short runs affordable for an extremely wide range of applications.

Modern boxmakers make extensive use of CAD systems for samples and limited production runs which are often cut on computer-driven plotter tables. The component parts are then hand-assembled. This obviously eliminates the need to buy unique cutters for each box type, size, etc.

Whilst small tooling, jigs and fixtures are generally produced locally, box wrapping, gluing, spotting and material preparation machinery is available internationally. Special purpose machinery, i.e. custom designed and built, also provides equipment manufacturers with unique challenges!

9.7 Construction

At its most simple level, prepared paper is coated with a thin film of adhesive and applied to the board, with the paper itself wrapped around the corner flaps to secure them. The next level of complexity is to first 'stay-up' the corners of the board (the box or lid sides) using a heat-activated tape, appropriately called stay tape, before the covering paper is applied.

Adhesive application may be by hand, simply by passing the paper over a glue-covered roller and the paper then being wrapped around the box: this is suitable for samples, very limited production runs or complex packs/components which cannot be made by machine. For volume runs, the staying-up of the box is carried out automatically on a machine. This is then conveyed under a wrapping head to which the glued-out paper is presented, and through a complex series of rollers and brushes, the paper wrap is pressed to the back and sides, both on the outer surface and a turn-in on the inside.

Ideally, two wrapping lines are placed side by side, one producing the box and the other the lid, with the two parts being put together (lidded-up) by hand or by machine. Where a book-style box is required, these box-wrapping lines can be linked to laminating lines (adhering paper sheets to flat boards, grooved or multi-pieced) to produce a jacket which

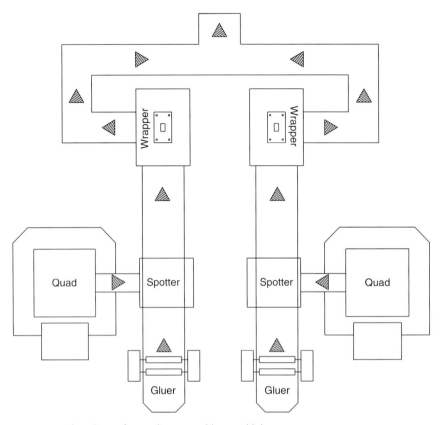

Figure 9.7 Machine layout for producing rigid box and lid.

can then be fixed to the box base. Wrapping lines are complex and one modern line combined with a staying-up machine costs in the region of £250 000.

A typical high-volume production line is synchronised to perform box staying, gluing, spotting and then cover-wrapping. A typical machine layout producing both box and lid is shown in plan elevation in Fig. 9.7.

The addition of platforms, both cardboard and vacuum formed, often requires long conveyorised production lines and throughput levels of 1500–2500 per hour. The boxmaker is often also called upon to pack the customer's contents/product. Box-making factories are capital-intensive with the best being air-conditioned or, at the very least, humidity controlled. Cleanliness and overall pest control are as necessary for rigid boxes as for any other paper-based packaging.

Also the wide variety of box styles inevitably also means that boxmakers are employers of proportionately more semi-skilled employees than many other manufacturing industries. Multi-skilling and a flexible working schedule in manufacture is essential because of the seasonal variability of the sales of many luxury, gifted, products.

Surrounding the main wrapping lines are special purpose machines for miscellaneous operations, such as hinging, stitching, labelling, sealing, hotmelt fixing, corner cutting and many more.

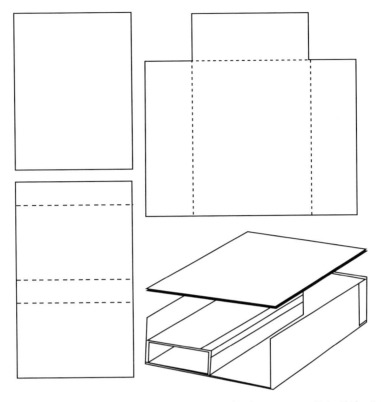

Figure 9.8 Preparation of outer and inner sleeves. (Reprinted with permission of John Wiley & Sons, Inc.)

An alternative method of box covering is known as envelope or loose wrapping. This uses a cover paper which is not pre-shaped but is folded and pleated around the box with only the edge of the paper being secured to the board by adhesive.

9.7.1 Four-drawer box

The construction of a four-drawer rigid box demonstrates the range of operations which may be necessary in order to meet a specific design for a luxury or novelty pack. The example is taken, with permission, from *The Packaging Designer's Book of Patterns* (Roth and Wybenga, 1991).

Step 1: The first set of operations, Fig. 9.8, result in the preparation of an outer sleeve and four inner sleeves, which form the drawer guides. Adhesive tape is used to make the various joins.

Step 2: Then the face of the drawer guides and the outer sleeve are covered with decorative materials (Fig. 9.9).

Step 3: The drawers are cut, scored and corner-stayed. Finally, the drawer pulls are applied, and the assembly checked for a snug fit. The complete assembly is shown in Fig. 9.10.

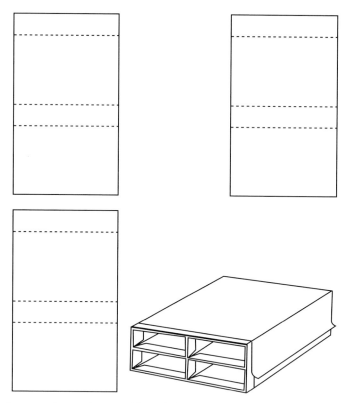

Figure 9.9 Covering of outer sleeve and face of drawer guides. (Reprinted with permission of John Wiley & Sons, Inc.)

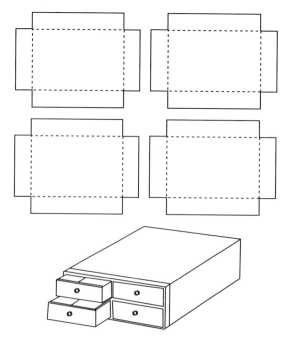

Figure 9.10 Production of drawers and final assembly. (Reprinted with permission of John Wiley & Sons, Inc.)

9.8 Conclusion

Use of the rigid box continues to compete and find a multitude of uses in the packaging world of today. Its raison d'être remains the need to engender the feeling of quality whilst providing physical protection for its contents. Design versatility, the use of accessories and widely varying print finishes provide a unique packaging proposition that continues to delight its users and consumers.

References

BS, 1986, Standard BS 1133-7.3, London.
Roth, L. and Wybenga, G.L. 1991, *The Packaging Designer's Book of Patterns*, Van Nostrand Reinhold, New York, pp. 342–344.

Websites

www.nationalpaperbox.org
www.boxpackaging.org.uk
www.londonfancybox.co.uk

10 Folding cartons

Mark J. Kirwan

Paper and Paperboard Specialist, Fellow of the Packaging Society, London, UK

10.1 Introduction

Cartons are small- to medium-sized boxes made from paperboard. They comprise a significant proportion of the packaging found in the retail sector, i.e. in supermarkets and shops of all kinds, in internet sales/mail order, in vending machines, in the service sector (hospitals, schools, catering, etc.) and in the dispensing of medicines.

Cartons meet packaging needs cost effectively by providing product protection and information, visual impact and convenience appropriate for the product concerned and its method of distribution and consumer use.

Folding cartons are delivered to the packer in a flat form. This may be either as a flat printed blank or as a side seam-glued carton which is folded flat. The carton packer erects, fills and closes the carton either manually, mechanically or by a combination of manual and mechanical means. The flatness of the folding carton prior to the packaging operation is a major space-saving benefit in distribution and storage before use and distinguishes *folding* cartons from *rigid* cartons, which are 'set up', ready for use at the point of manufacture.

Cartons increased in popularity with the changes in retailing, whereby the shopkeeper trading locally, apportioning the product on demand from bulk, was replaced by manufacturers marketing brands and trading nationally and internationally. This was facilitated by the development of machinery to apportion the product and to erect, fill and close cartons.

Folding cartons are made from paperboard, often referred to as 'cardboard' or 'cartonboard'. There are several different types of paperboard differing in the types of pulp used in their construction. The main types of pulp, as has already been discussed in Chapter 1, are chemical bleached, chemical unbleached, mechanical and recycled. Paperboard is usually multilayered and may comprise layers of more than one type of pulp. Surfaces are usually, but not always, coated with a mineral-pigmented coating to enhance appearance and printability.

The properties of paperboard can be extended by the use of other materials, plastic extrusion coatings and other functional coatings and treatments.

Handbook of Paper and Paperboard Packaging Technology, Second Edition. Edited by Mark J. Kirwan.
© 2013 John Wiley & Sons, Ltd. Published 2013 by John Wiley & Sons, Ltd.

Paperboard is characterised by its thickness and weight per unit area. Packaging trade and national and international standards organisations give different guidance as regards the thickness range for defining paperboard. The upper limit for practical purposes would, however, be 1000 μm. The lower limit has been defined as 250 and 225 μm, but there are cartons in use at 200 μm and even lower. The units of measurement are microns (0.001 mm), or thousandths of an inch (0.001 in.). Weight per unit area is measured in either grams per square metre (g/m^2) or pounds per 1000 ft^2.

The carton manufacturer sells cartons by number of cartons, 1 million or whatever, and hence is interested in buying by the *area*, rather than the weight required to make a given number of cartons, taking account of set-up and in-process waste.

In practice, this is facilitated in modern paperboard manufacturing:

- on the paperboard machine by on-line computer-control of the weight per unit area, i.e. yield
- in the finishing department by accurate sheet counting, in the case of sheeted orders, and length, in the case of reels of a given width.

As has also been discussed in Chapter 1, the compression strength of paperboard is highly dependent on stiffness that correlates with thickness to a much higher degree than weight per unit area.

Packaging technologists, packaging buyers and brand managers in end-using companies often ask what type of paperboard they should use for a particular type of product.

The answer, however, depends on a number of factors and the starting point is to examine all the packaging needs. The use of a checklist is advised to ensure that all relevant factors are considered, see Appendix, Checklist for a packaging brief. This review will provide guidance on the required appearance, for example surface, and performance, for example strength and product protection characteristics, required to achieve the desired appearance, printability, conversion and usage. These needs in turn depend on:

- *Product* – type, weight, volume, shape or consistency, i.e. powder, granule, etc., and whether it is pre-packed in a pouch, jar, etc. or whether it is packed in direct contact with, or in close proximity to, the paperboard. Any special functional or legal requirements relating to the protection of the product in packing, storage, transit, merchandising and consumer use must be identified.
- *Presentation* – surface and structural carton design, standard of printing and overall visual impact. It may be that the reverse side of the paperboard is seen or printed. Presentation is closely related to the positioning of the product in its particular market. Each product market will have specific characteristics. All products, for example breakfast cereals, chocolate confectionery, ethical prescription-only medicines, proprietary over-the-counter medicines and tobacco, are presented in their own characteristic ways.

There are also international differences in the ways products are presented. The ways products are presented also depend on the point of sale or use, i.e. whether in self-service display, vending, websites, mail order catalogues, catering, etc., and whether the product is bought by one person as a gift for another. Additionally, within specific product markets, there will be luxury or expensively priced top-of-the-range products, average-priced middle-of-the-range or medium-priced products and budget-priced, value-for-money products at the lower-priced end of the price range. Packaging, including the choice of paperboard, will reflect this segmentation and product positioning:

- *Packing, distribution and use* – details of the number of cartons that are required to be packed in a given time and whether production is variable through the year are needed at the outset to indicate the extent of the mechanically assisted cartonning required. Distribution and storage factors include consideration of the transit packaging and whether any special environments are involved, such as deep freeze or chilled distribution. Consumer use could include how the pack is used and whether this imposes specific needs such as the reheatable requirements of a convenience food product or the way the carton is handled in use, for example the compression needs of a cigarette carton.

In practice, many of the requirements imposed by these needs interact. The carton structural design contributes to the carton compression strength, and the surface strength relates to printability and glueablity, etc.

Folding cartons are usually printed on the outside surface with text, illustrations and decorative designs. Varnishing is used to enhance appearance and to protect print. Graphical design reinforces brand values, particularly in self-service merchandising. Cartons are sometimes printed on the reverse side, for example cartons for chocolate confectionery. Printing provides information on the product and how to use it safely. Hot-foil stamping and embossing are also used to create visual impact.

Folding cartons are produced in a wide range of structural shapes having a choice of closure design, including reclosure features where necessary, together with 'add-on' features, such as handles, windows and internal platforms.

The structural design of most folding cartons is based on a rectangular or square cross section. The type of product to be packed, the method of filling and the way the pack will be distributed, displayed (merchandised) and used will influence the dimensions and design in general. Rectangular shapes are easy to handle mechanically, especially when packed in large volumes at high speeds, easily merchandised, and conveniently handled and stored by the consumer. Other shapes are, however, possible and are used to meet specific needs, for example round, elliptical, triangular, pyramidal, domed and wallet shaped.

All this is possible in manufacture because of the fundamental paperboard properties whereby it is an inherently strong material which is printable, creasable and glueable.

The overall result is that folding cartons are used to package a wide range of products from foodstuffs – such as cereals, frozen, chilled foods and ice cream, confectionery, bakery products, tea, coffee and other dry foods – to non-food products, such as pharmaceuticals, medical and healthcare products, perfumes, cosmetics, toiletries, photographic products, clothing, cigarettes, toys, games, laundry, paper goods, household, electrical, engineering, sport and leisure, gardening and DIY products, etc. Around 55% of all cartons are used in the packaging of food products (Pro Carton, 1999).

10.2 Paperboard used to make folding cartons

The main types of paperboard, i.e. solid bleached board (SBB or SBS), solid unbleached board (SUB or SUS), folding box board (FBB or GC1 or GC2) and white-lined chipboard (WLC, GT or Newsboard), have been described in Chapter 1.

The outer surface is usually coated with a mineral-pigmented coating based on china clay and/or calcium carbonate to enhance the appearance (colour, smoothness, surface structure and gloss level) and printability. The reverse side may also have a mineral-pigmented coating for appearance and printability.

The performance of paperboard and cartons can be extended by the use of additional materials and by using processes such as lamination, plastic extrusion coating, impregnation and functional-surface coating. Such treatments are used to ensure that cartons meet the needs of product protection throughout the distribution and storage chain up to, and including, the ultimate consumer.

Performance laminations of paperboard are possible with greaseproof and glassine paper, polyvinylidene dichloride (PVdC or Saran®)-coated paper and aluminium foil. These surfaces provide grease/fat resistance. Wax used as adhesive for a greaseproof or glassine lamination enhances the barrier of these laminates to moisture and moisture vapour. For product release, greaseproof paper or glassine coated with a silicone-release coating is used. Aluminium foil provides a number of functional benefits, such as barrier to moisture and moisture vapour, grease/fat barrier, odour barrier, product release and heat resistance. PVdC provides heat sealability and a barrier to moisture and moisture vapour, grease/fat, flavours and odours; it is not, however, used widely in carton constructions. It is used as a coating on oriented polypropylene (OPP) film which is used as a barrier overwrap for chocolate confectionery, tea and cigarette cartons.

Coated paper with good printability can be laminated with paperboard using microcrystalline wax blends as the adhesive to provide a paperboard with moisture vapour resistance.

Paperboard can be plastic extrusion coated to achieve a range of barrier, heat sealing and heat-resistant properties with polyethylene (PE), polypropylene (PP), polyethylene terephthalate (PET or PETE), polymethylpentene (PMP) and ionomer (Surlyn™). Bio-polymer PE is now available, which is made from a renewable resource, and it is biodegradable meeting the EN13432 standard for compostability (Packaging News, 2008).

Paperboard may be coated with silicone, wax or a heat-seal varnish. It can be impregnated with wax during paperboard production. Water-based coatings can provide heat sealing and grease resistance. Wax is a good barrier to volatile contamination from outside and for the retention of flavour and aroma within packaging.

10.3 Carton design

10.3.1 Surface design

This involves all the aspects of a carton which relate to its overall visual impact. The basis is the colour of the surface of the paperboard, its smoothness and surface finish. The colour is usually white, though other colours are possible by the use of coloured mineral-pigmented coatings. High whiteness enhances print contrast. Surface finish can be either high gloss, satin or matt. There are variations in appearance between paperboards, and the existence of the ultra-high gloss cast-coated products, also available in different colours, extends the gloss range further. A matt finish may increase print contrast and make text easier to read whereas gloss and sparkle can attract attention. It is also possible to apply an overall embossed design to the paperboard surface, for example linen effect. All other variations in surface appearance are achieved by conversion, printing and varnishing.

There are many possible types of lamination which are used to create a visual effect. Plastic film, for example cellulose acetate and polypropylene, may be applied to enhance gloss and provide a luxury appearance. Aluminium foil or metallised polyester plastic film is used to produce an overall metallic effect. Papers, such as a high specification coated paper or coloured glassine, for example chocolate coloured, may also be laminated to a paperboard base. In the latter example, there is also a performance element relating to product protection.

Whilst plastic extrusion coating is primarily used as a functional addition to paperboard, it can also have a visual effect. This is usually to impart gloss.

Printing offers a very extensive range of possibilities for creating visual impact. It is used for brand names, text, illustrations and diagrams. This range can include tactile text and decoration. Printing can provide solid colour or patterned decoration.

Varnishing is applied over print to improve rub resistance, and it can also be used to create a visual effect. This effect can result in a gloss, satin or matt surface, and may either be all over the surface or limited to specific parts of the design. Pearlescent varnishes can also be applied to create a special effect.

Localised effects can be produced by hot-foil stamping or embossing or by a combination of both. The hot-foil effect is achieved with an aluminium (coloured) foil applied to a heat-resistant film and a heated die profiled to the required design. Embossing can also be used to create an overall decorative effect.

Braille is a form of surface design consisting of embossed letters, numbers, punctuation, symbols and special characters which allow the text to be easily identified by visually impaired people. Currently, it is used to identify pharmaceutical products since according to Article 56(a) of European Union Council Directive 2001/83/EC, it has been mandatory to include Braille on all EU pharmaceutical packaging since October 2010. This includes pharmaceutical products which were launched prior to October 2005.

Braille must show the name of the medicine and the appropriate strength, where more than one strength is available.

It has been mandatory to include Braille on all new pharmaceutical EU Marketing Authorisations since 30 October 2005.

A standard for Pharmaceutical Packaging Braille was adopted by the European Standards Organisation (CEN) as EN 15823:2010 entitled 'Packaging – Braille on Packaging for Medicinal Products' (CEN, 2010).

This European Standard is the result of a number of years of work by a CEN Working Group encompassing organisations for the visually impaired, health agencies, the pharmaceutical industry, and packaging and materials suppliers. It specifies requirements and provides guidance for the application of Braille to the labelling of medicinal products.

In North America, the International Association of Diecutting and Diemaking (IADD) has created 'Can-Am Braille', a set of guidelines and recommendations for the use of Braille on pharmaceutical packaging (PharmaBraille, 2009).

'Scratch and sniff' can be used in a surface-printed design. The inks are available for printing in offset litho, flexo and screen printing which are ultraviolet (UV) cure, solvent and aqueous compatible. These inks contain encapsulated flavours, for example orange, lemon and chocolate, the flavour and aroma of which can be released and experienced by the customer by scratching the surface of the print.

10.3.2 Structural design

Structural design is based on the creative and functional needs of the pack. A carton has to perform efficiently in the packaging operation and thereafter to protect the product

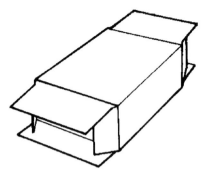

Figure 10.1 End load carton. (Reproduced with permission from Alexir Packaging Ltd.)

throughout its life during distribution to the point of sale or dispensing and to its use by the ultimate consumer. This represents the functional requirement. The function may need to include convenience in-use features, such as opening and reclosure features, carrying handles, windows, internal platforms to locate and display the product, etc.

Structural design must also take into account the dimensions, volume and, if a solid, the shape of the product. The design is usually in three dimensions though there are examples of two-dimensional paperboard packaging, for example blister cards and wallet-style packs for retort pouches.

In addition to the development of a functional design, paperboard gives the designer the scope for creating more imaginative designs whilst not compromising the functional needs of the pack. Such scope arises in, for example, confectionery packaging for chocolate assortments, Easter Eggs or chocolate bar cartons with after-use play value. Examples exist in other end-use areas, for example cartons for luxury cosmetics and gift packs of all kinds.

The structural design is facilitated by the strength and toughness of paperboard and by the ease with which paperboard can be accurately and cleanly cut, creased, folded and glued.

Several comprehensive guides on the mainstream types of carton construction are available in, for example, the European Carton Makers Association Code (ECMA, 2009) and *The Packaging Designer's Book of Patterns* (Roth, 1991a), published by John Wiley & Sons, Inc., from which several examples are quoted later. Computer-aided-design (CAD) packages are also available.

As already noted, the most common shape of folding carton is based on a rectangular or square cross section. It has four rectangular or square panels. There is a fifth, narrow panel which when fixed, using an adhesive, to the underside of the first panel creates a tube. This structure is said to be side-seam glued and it is folded flat by the carton maker for shipment to the packer.

The packer usually erects this style mechanically. The product is loaded, either horizontally, or from the vertical direction, and the end flaps are closed either by tucking, locking slits or with the help of an adhesive. It is referred to as an 'end load' or tube-style carton. An example of this design is shown in Fig. 10.1.

Another popular rectangular shape is based on a tray style. In this instance, the carton maker supplies a flat blank. The base of the tray is a flat area of paperboard to which four panels are attached by creases. The tray is erected mechanically by folding these four panels through 90°. This structure is secured either by the use of a hot-melt adhesive or with locking, hook-shaped tabs which fit into slots in the adjacent panels. There are many variations to this style. Lids can be incorporated as extra panels and both lids and side panels can have

Folding cartons **271**

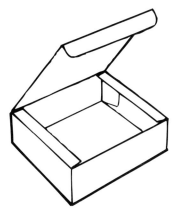

Figure 10.2 Top load carton. (Reproduced with permission from Alexir Packaging Ltd.)

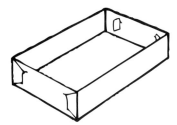

Figure 10.3 Shallow-depth tray with locked corners. (Reproduced with permission from Alexir Packaging Ltd.)

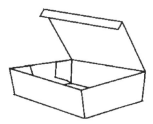

Figure 10.4 Tapered carton with lid. (Reproduced with permission from Alexir Packaging Ltd.)

extensions which are folded into position after the product has been loaded, as in Fig. 10.2. In the case of the lids, they can be closed by tucking in a folded extension to the lid or by hot-melt gluing one or three closure flaps. This style is often referred to as a 'top load' carton. A point worth considering is that, depending on the product, it makes sense to load the product through the largest opening.

A simple tray style is the shallow tray with side walls 25–38 mm deep with glued or locked corners, which are used to collate groups of cartons for stretch or shrink-wrapping, as in Fig. 10.3.

The side walls do not necessarily have to be vertical – they may be tapered as shown in Fig. 10.4.

Figure 10.5 Tray with tapered sides and heat-sealed corners. (Reprinted with the permission of John Wiley & Sons, Inc., from *The Packaging Designer's Book of Patterns*, Van Nostrand Reinhold, New York, 1991.)

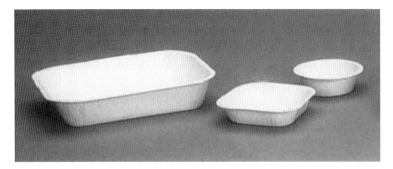

Figure 10.6 Deep drawn, pressed trays. (Courtesy of Iggesund Paperboard.)

Where these trays are plastic extrusion coated with PE, PP or PET (PETE), they can be erected by heat sealing. The method of corner-folding and heat sealing makes this design leakproof as shown in Fig. 10.5.

Trays can also be made by pressing and deep drawing, as shown in Fig. 10.6. This process places high demands on the paperboard. The normal depth of drawing is around 25 mm, and with additional moisture application and two-stage drawing, it is possible to extend this to 45–50 mm. Deep drawing can be used with paperboard extrusion coated with PET (PETE), which can be used to pack food products that are cooked and/or reheated in the pack. Reheating is possible using microwave or radiant heat. The trays can be lidded with plastic snap-on lids, heat-sealed plastic film or heat-sealed plastic-coated paperboard.

By changing the cutting and creasing of the paperboard carton blank and by corner-gluing, tray designs can be made by the carton maker, which are folded for packing and shipment to the carton user who then erects these cartons by hand. These can be four-corner glued or six-corner glued, the latter, which incorporates a lid, being traditionally used by cake shops (Fig. 10.7).

A modification of the tray design with hollow walls and a double-thickness lid, as shown in Fig. 10.8, is a popular choice for chocolate assortments. This carton can be erected mechanically on the packing line. The hollow walls provide excellent rigidity, enabling the carton to be held by one corner when offering the contents around. The double-thickness lid results in the same high-quality printing which is expected with this type of application to be achieved on the inside surface of the lid.

A popular way of providing product protection is to overwrap a folding carton with a barrier-coated PP film with envelope-folded and heat-sealed end folds. Typical applications include cartons for chocolates, tea bags and cigarettes. Another material used in this way is high-gloss wax-coated paper used to overwrap unprinted cartons for frozen food.

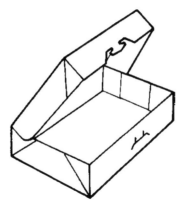

Figure 10.7 Six corner-glued cake box. (Reproduced with permission from Alexir Packaging Ltd.)

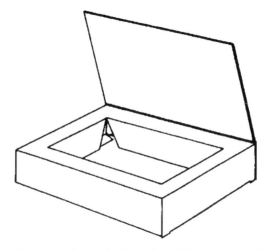

Figure 10.8 Carton with cavity walls and double-thickness lid. (Reprinted with the permission of John Wiley & Sons, Inc., from *The Packaging Designer's Book of Patterns*, Van Nostrand Reinhold, New York, 1991.)

A popular carton style is the hinged-lid cigarette carton as shown in Fig. 10.9. This carton is erected at high speed from a rectangular blank. It incorporates a small additional piece of paperboard, an inner frame, which provides strength and facilitates the design of the closure.

Other shapes are, however, possible and are used to meet specific needs, for example round, elliptical, triangular, pyramidal, hexagonal, domed and wallet shaped (for retort pouches) (Figs 10.10 and 10.11).

Where a product requires protection from moisture, oxygen, contaminating odours and taints, etc. cartons may be lined by the carton maker with a flat tube of a flexible barrier material which is inserted during carton manufacture. The flexible material is usually heat sealable – examples include paper/aluminium foil/PE and laminations involving plastic films. The lined cartons are supplied folded flat to the packing/filling machine. One end of the liner is heat sealed, the associated carton flaps closed and then, after filling, the other end is sealed and the carton flaps closed. This type of carton is used for ground coffee, dry foods

Figure 10.9 Hinged-lid cigarette carton. (Reprinted with the permission of John Wiley & Sons, Inc., from *The Packaging Designer's Book of Patterns*, Van Nostrand Reinhold, New York, 1991.)

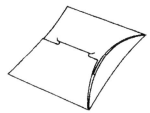

Figure 10.10 Wallet or pillow design. (Reprinted with the permission of John Wiley & Sons, Inc., from *The Packaging Designer's Book of Patterns*, Van Nostrand Reinhold, New York, 1991.)

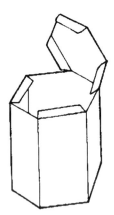

Figure 10.11 Hexagonal design. (Reproduced with permission from Alexir Packaging Ltd.)

and liquids. Packing machines can vacuum pack or gas flush a product such as ground coffee, (Fig. 10.12). A lined carton of this nature may be fitted with a plastic hinged lid incorporating a moisture vapour barrier and tamper-evident diaphragm.

Another type of lined carton, which can be formed by the packer on the packing machine, takes flat carton blanks and a roll of the material to be used as the liner, frequently bleached kraft paper. First, the liner is formed around a solid mandrel. The side seam and base are either heat sealed or glued with adhesive, depending on the specification. The carton is then wrapped

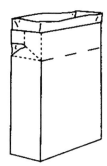

Figure 10.12 Lined carton. (Reproduced with permission from Alexir Packaging Ltd.)

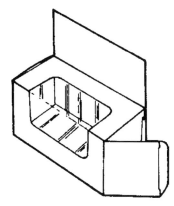

Figure 10.13 Windowed carton. (Reproduced with permission from Alexir Packaging Ltd.)

around the liner with the side seam and base sealed with adhesive. The lined carton is removed from the mandrel, the product is filled and both liner and carton are sealed/closed. This type of pack is suitable for the vertical filling of powders, granules and products such as loose filled tea.

Folding cartons can have windows or plastic panels for product display, for example spirits and toys (Fig. 10.13).

A *display outer* is a carton which performs two functions. At the packing stage, it is used as a transit pack or outer. When it arrives at the point of sale, the specially designed lid flap is opened and folded down inside the carton at the back of the product to become a display pack with a printed header display. This design of carton, as shown in Fig. 10.14, is used to pack a number of smaller items which are sold separately, for example confectionery products, such as chocolate bars, also known as countlines.

In the case of the end load or tube-style carton, the carton flaps on the base are closed by either tucking, locking, gluing or heat sealing. A heat-sealed closure requires the presence of a thermoplastic layer, for example PE, PP or PET (PETE).

The crash-lock base is made by the carton maker. The bottom flaps are creased and glued in a special way which enables the carton maker to fold the carton flat for shipment to the packer who erects the carton by hand in such a way that the bottom flaps snap, or crash, into position irreversibly. This type of base can support a considerable weight, for example, a bottle of alcoholic liqueur.

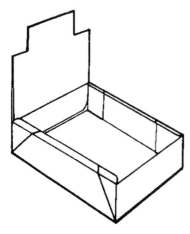

Figure 10.14 Display outer. (Reproduced with permission from Alexir Packaging Ltd.)

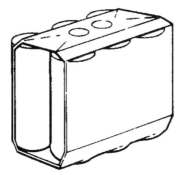

Figure 10.15 Multipack for six beverage cans. (Reproduced with permission from Alexir Packaging Ltd.)

The lid flaps may be closed with a tuck-in-flap or flip-top, or may be locked, glued or heat sealed. Closures may be made tamper-evident. The top may have an easy-opening device and, where necessary, a reclosure feature. Closures which are repeatedly opened and closed during the life of the contents, for example perfume, require their creasing to have sufficient folding endurance. This requires careful consideration of the type of paperboard used.

Additionally, cartons can have integral, internal display fitments or platforms. Sleeves can be used for trays of chilled ready meals and multipacks are used for bottles, yoghurt pots (Fig.10.15) and drinks cans (Fig. 10.16).

Cartons can incorporate dispensing devices, carrying handles and easy-opening tear strip features for convenience in handling and use. Cartons can be made into non-rectangular, innovative shapes, such as packaging for Easter Eggs.

This chapter is not comprehensive. It is intended to indicate that folding cartons can take many shapes and meet many packaging needs. A carton designer working on a project brief has many options and considerable freedom to express imaginative and creative ideas. The paperboard features that make this possible depend on the strength and toughness of paperboard and its creasing, folding and glueability properties.

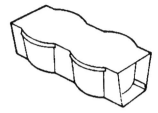

Figure 10.16 Sleeve for two plastic tubs. (Reproduced with permission from Alexir Packaging Ltd.)

Reference has been made to the use of CAD packages – an added advantage of CAD is that it can be linked to plotters which can quickly reproduce drawings and even make actual samples for evaluation.

Together with an appropriate coating, lamination or separate carton lining, paperboard can be used in conjunction with the design of the pack to produce cartons, sleeves, trays and blister packs which meet demanding performance requirements, such as liquid tightness, sift proofness, heat sealability, grease resistance, moisture and moisture vapour resistance, product release and reheating by microwave, convection and radiant heat. Folding cartons can meet the needs of a wide range of distribution, storage, packaging and use conditions, for example frozen, chilled, ambient, tropical and wet conditions.

Once a specific type of paperboard has been selected, it is necessary to choose the grammage (basis weight) and thickness to provide adequate carton strength at each stage of the packing chain from packing through to use by the consumer.

The outcome of the design phase for both surface design and structure is to provide one or more design proposals for evaluation by marketing, production, R&D and for costing and functionality at each stage of the packaging chain.

10.4 Manufacture of folding cartons

10.4.1 Printing

The main processes for carton printing today are offset lithography, flexography and gravure. Letterpress and silk screen are used to a limited extent. The most recently introduced process, digital printing, can be used for short print runs and for customising packaging already pre-made in bulk.

All these processes are discussed in Sections from 4.6.1 to 4.6.8.

The answer to the question as to which print process to use is complicated. The considerations are:

- product, distribution and usage
- quality of reproduction
- run length
- lead time
- cost.

Printing involves solid print, text, illustrations and diagrammatic representations appropriate for the type of product concerned and its market positioning. As already noted, product positioning and specific brand values will have a major influence on the print design and the quality of reproduction required.

Functional needs of the packaging, depending on the product or method of distribution, may impose constraints. Products such as detergents can be aggressive to print, particularly in conditions of high humidity. Powdered products packed hot may impose the need for highly scuff- and rub-resistant print surfaces. Products such as chocolate and tobacco are sensitive to retained ink odours, and this will influence the choice of inks and print process. In every case, the overall needs should be discussed with printing experts.

With respect to quality of reproduction, there is overlap in what can be achieved today with the different printing processes.

It used to be said that flexo was poor for solids, half tones, as required for illustrations, and varnish gloss. This is not true today. Excellent results have also been demonstrated using UV-cured varnish. Flexo printing is available on reel-fed machines with in-line cutting and creasing.

Gravure was considered the best for solids, and offset litho the best for half tones. Offset gravure and conventional gravure with electrostatic assist may improve gravure half-tone reproduction, but the benefits are academic if the capacity to print large volumes is not available or if run lengths are too low to justify the high cost of gravure cylinders. Gravure printing is also available on reel-fed machines with in-line cutting and creasing.

There was a time when, for example, a rose would be printed by offset litho on a carton for chocolate assortments and the rest of the carton, with an overall solid print, would be printed by sheet-fed gravure. There were specific reasons for this. Gravure, provided the retained solvent level met the required standards, was the better choice since in those days the residual odours from oxidation–polymerisation drying oil-based litho inks were a potential hazard for chocolate packaging. Moreover there was a risk of set-off and poor rub resistance with litho on such large solid areas of print. Today there is a wide choice of offset litho ink and drying systems, particularly with UV and EB (electron beam)-assisted drying.

It used to be the case that varnishing in-line by offset litho was poor in terms of colour and gloss. Today, this no longer applies as presses are fitted with coating units and can apply UV-cured varnishes which are water clear and have high gloss and rub resistance.

It is still the case that the cylinders are relatively dearer for gravure than plates are for litho and flexo, but they are longer lasting and hence are competitive when long runs are required.

Silk screen printing has always been known for its ability to print thick films of ink – it gives the best fluorescent effect, for example. It is possible to use UV systems in silk screen printing. This has the advantage of rapid ink drying. A recent review lists the range of effects which are possible – raised images, including Braille text and warning symbols, highly opaque prints, high-lustre varnish finishes and textured finishes. Examples of the latter include a colourless varnish containing large particles to create a coarse feel and give an 'ice look' to a pack. The examples demonstrate the ability to mimic pastry and a luxury, metallic embossed leather look (Packaging, 2004). Silk screen is not going to challenge the high-volume package printing market, but it has the ability to print special effects.

Printing continues to innovate and surprise.

Overall pre-press, plate/cylinder and make-ready costs are relatively lower than 10+ years ago due to technical changes in scanning, digitalisation, processing, proofing and on-machine improvements in make-ready. If one wants to check out the printing industry for its application of new technology and higher productivity, one has only to visit one of the leading exhibitions, such as DRUPA (Druck und Papier) held in Düsseldorf.

Pre-press processing times including off-press proofing have greatly reduced lead times to meet end-users' needs for a quick response. Not only is there a need for rapid response, but run lengths have reduced with the result that a higher proportion of medium size presses are used today in sheet fed litho. There has also been an increased interest in narrow web gravure/flexo with cutting and creasing in-line. Speeds continue to increase for all types of printing press.

Despite all the high technology and investment to be seen in the press room today, the quality needs for paperboard to be flat, accurately and squarely cut, with dust and debris-free surfaces remain just as important as ever. In this respect, it is worth mentioning two areas where attention is still very important.

1. The moisture-resistant wrapping, which protected the paperboard in transit and storage after manufacture, must not be removed until the mass of the board has assumed temperature equilibrium with the atmosphere of the press room. When cold board is unwrapped, moisture can condense on the edges in the same way as an inside window mists up when it is cold outside. Moisture which condenses on the surface of the paperboard is not visible but its effect may be, and can result in, a curled or wavy sheet.

    A problem experienced by the author occurred when paperboard having been delivered and immediately printed for an urgent (JIT) order during very cold weather had the liner *completely* ripped away by the litho inks. The temperature in the middle of the pallet of printed paperboard was still only 11°C several hours after printing compared with a temperature in the pressroom of 20°C! The litho inks which are already tacky became even tackier when they contacted the cold paperboard surface.

    This paperboard subsequently, nearly 48 h later, was printed satisfactorily. Warming-up times for various temperature differences and weights of paperboard are published by paperboard mills.

2. Spots, also known as hickeys and bullseyes, result in poor print appearance in solid areas of colour when printed by offset litho. They are variously referred to as 'dust and debris' problems. There tends to be a general belief that these only arise from the paperboard and it is a fact that particles which are distinctive in appearance from this source do occasionally occur, and examples are as shown in Fig. 10.17.

Particles may also originate from ink, the press and the pressroom environment. Examples in these categories are shown in Fig. 10.18.

In theory, none of these particles should be present. A particularly memorable 'loose fibres from clothing' problem occurred when a man loading sheets into the in-feed of the press and wearing a thick red woollen sweater was surprised when the particles causing lines in the print were found to be red and woollen. Today, the press has a direct feed, so there is no need for such close contact and the operator will be wearing suitably protective clothing. When a problem of this sort arises, it is always essential that the actual spots/particles be retrieved from the press and examined and identified microscopically so that the correct corrective action can be taken.

A major change in attitude within the press room has occurred in recent years, which has resulted in the certification of carton manufacture by appropriate authorities to food and pharmaceutical packaging standards.

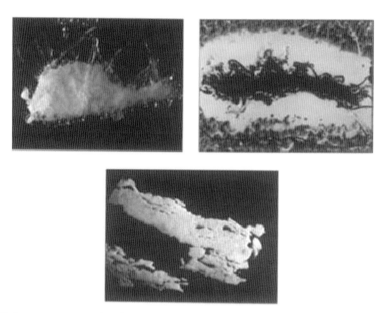

Figure 10.17 Coating particles, slit and chop edge particles. (Courtesy of Iggesund Paperboard.)

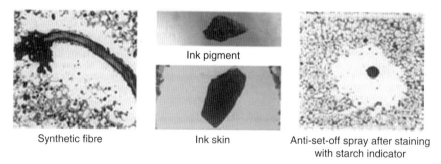

Figure 10.18 Anti-set-off spray, undispersed pigment dried ink skin and synthetic fibre. (Courtesy of Iggesund Paperboard.)

10.4.2 Cutting and creasing

The process of cutting and creasing converts the printed paperboard into flat individual profiles, or blanks, of the intended cartons with, in the case of printed cartons, all the cut edges, creasing grooves, panels, flaps, interlocking features, localised embossing, etc. in register with the print.

Cutting must ensure that the edges of the printed paperboard blank are clean and free from fibrous debris, such as loose fibres, fragments of fibres, clumps of fibres or thin wispy slivers of paperboard. This is important as otherwise loose material may be shed during gluing, if that is the next process, or on the packing line where it can interfere with efficient machine operation and contaminate the product.

A crease (score) is a groove in paperboard which facilitates bending or folding along a clearly defined line. In a carton blank, creases (scores) define the edges of the panels and flaps which are subsequently folded during gluing and carton erection, filling and closure. The action of folding or bending the board along the crease lines causes the carton to assume its three-dimensional shape and contribute to the compression strength of the carton in storage, distribution and consumer use.

'Creases' are also referred to as 'scores', and this interpretation is used in this text. This has been noted because in some parts of the world, a 'score' has a different meaning, i.e. a score is an actual cut, part way through the paperboard, which has been made in the surface of the paperboard. A cut score also facilitates bending and folding, but the surface is weakened and the visual effect may be unsightly on printed board. An important use of the cut score is where it is used as part of an easy-opening tear strip. Perforations are used to facilitate bending, for example on 45° glue flap creases for folded, glued trays and crash-lock bottom closures.

A crease (score) should operate as a hinge. It is possible to measure the force necessary to bend a creased panel through any given angle up to 180° and the spring-back force on a panel as it tries to resist being held after folding. Both aspects are relevant to the performance of a carton during and after the packaging operation.

Cutting and creasing are very different operations. They are clearly interrelated with respect to the carton profile and they are carried out simultaneously by a tool known as a die.

There are two types of cutting and creasing equipment, namely flat bed and rotary. The main difference between them is that whilst cutting takes place, with flatbed, the paperboard is stationary, and with rotary it is moving. The rotary method is usually operated in-line following printing from the reel. Flatbed cutting and creasing can either be sheet-fed or take place, in-line, after printing on a reel-fed machine.

Despite the fact that the criteria for cutting and creasing are very specific, there are many different ways of achieving these criteria in practice. This has significant commercial implications relating to order size (number of cartons).

10.4.2.1 Flatbed die

The die is made by setting cutting knives and creasing rules, in a stable wooden forme, in pre-cut channels which have been accurately cut using a laser working to a carton design in a CAD system. Before lasers were introduced, other methods such as cutting with a jigsaw or assembling with the help of accurately cut wooden blocks were used. Cutting is carried out with knives which have sharp edges, and these knives cut vertically through the paperboard. Creasing (scoring) is carried out with creasing rules which have rounded ends. They form a groove by indenting the paperboard surface by pushing it into a groove in a material known as the 'make-ready'. Knives are longer than creasing rules because they have to cut right through the paperboard.

The flatbed die is mounted in the upper platen of a cutting and creasing machine. Figure 10.19 shows a schematic layout. Sheets are gripped at the leading edge and the gripper bars are connected to chains which move them through the machine. The motion is intermittent to allow the sheet to be processed at each stage.

The die also contains a compressible material, usually rubber of specified hardness, mounted on the die in close proximity to each side of every knife. The purpose of this material is to press the paperboard against the bed plate of the lower platen and hold it securely during cutting. It continues to push against the paperboard as the knife is retracted and ensures that pieces of paperboard do not adhere to the dieboard. The type of die described here is secured between parallel steel plates, one of which moves vertically and intermittently

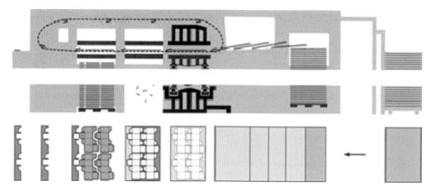

Figure 10.19 Automatic platen machine showing stages of sheet feed, cutting and creasing, stripping and carton blank separation. (Reproduced with permission from Bobst S.A.)

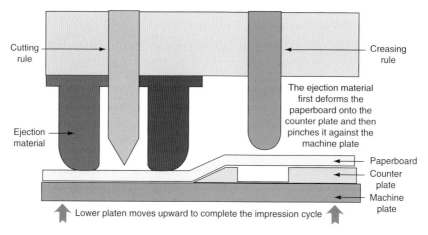

Figure 10.20 Stage 1 – cutting and creasing – lower platen moves upwards. (Reproduced with permission from DieInfo, Inc.)

in a platen press. The sheet is inserted when the platen is open. In Figs 10.20 and 10.21, stages 1 and 2 respectively, the lower platen moves upwards to cut and crease the paperboard. In Fig. 10.22, stage 3, the lower platen moves downwards with the cut and creased sheet which is then pulled clear by grippers thereby allowing another sheet to move onto the lower platen. The cycle then repeats.

When the platen closes, the die cuts through the paperboard with the knives making 'kiss' contact with a backing steel plate. The backing plate is important in that it is possible to adjust the kiss contact of the knives by what is known as 'patching up' behind this plate. This is done with thin tissue, along the line of the cutting, as required, in localised areas of the die.

As the printed sheets, shown in Fig. 10.23, may be passing through the platen at speeds up to around 7000 sheets per hour and as the individual cartons on each printed sheet are cut with every vertical cycle of the platen, it is important that the carton profiles remain attached sufficiently strongly together and to the gripper or front leading edge of the sheet, which is pulled through the machine, until the point when they are removed from the sheet. This is achieved by leaving the minimum number of very short uncut notches in the cutting profile. Keeping the number of notches as low as possible is important in edges which remain visible, and therefore possibly unsightly, when the carton is erected in its final shape.

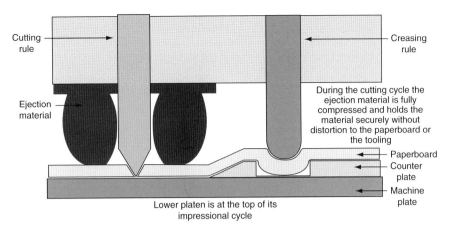

Figure 10.21 Stage 2 – cutting and creasing – sheet is cut and creased. (Reproduced with permission from DieInfo, Inc.)

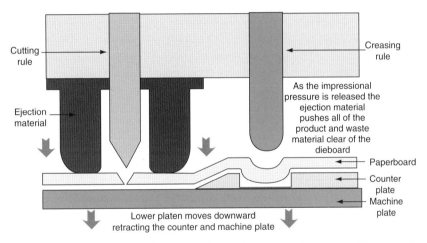

Figure 10.22 Stage 3 – cutting and creasing – lower platen moves downwards. (Reproduced with permission from DieInfo, Inc.)

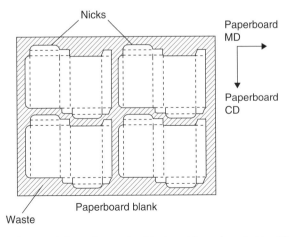

Figure 10.23 Cut and creased sheet, 4-up carton blanks, nicks and waste trim. (Courtesy of Iggesund Paperboard.)

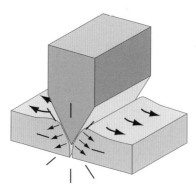

Figure 10.24 Crush cutting – in platen die-cutting paperboard is burst apart by using a steel wedge (knife) to convert downward force into lateral separation force. (Reproduced with permission from DieInfo, Inc.)

The overall process is known as 'platen die cutting' and the method of cutting is 'crush' cutting (Fig. 10.24) because the knife is forced through the paperboard. This type of platen requires high pressure in order to make all the cuts and creases simultaneously. It is important that the platen is evenly levelled. The sheet being cut is usually smaller than the area of the platen and it is important that the die is 'balanced', the latter being achieved by inserting knives in those areas of the platen area outside the area of the sheet being cut and creased. As noted, the pressure required is high, and it is important to maintain an even pressure across the platen as a whole.

The top edge of the cutting knife, whilst sharp, will be slightly rounded, and horizontal stresses will occur in the paperboard as the knife is forced through the paperboard (Fig. 10.24). These stresses are greatest where there are creasing grooves situated in the vicinity of a knife, as is the case with the relatively narrow glue flap panel. Where the cut in such a situation is in the cross direction (CD) of the board, as it usually is for a glue flap, there is a greater tendency for 'shattering', or ragged tearing, along the line of the cut on the reverse side (back) liner, on the underside of the paperboard. This is because the simultaneous forming of the adjacent crease, also in the CD of the paperboard, stretches the board in the machine direction (MD). Dennis Hine (1999, p. 233) has shown that the increase in width which occurs under the creasing rule as the crease is formed is 57%, i.e. the difference between the length (semicircle) of the rounded end of the creasing rule compared with the width. This amount of stretch relaxes after the creasing rule is withdrawn, but because the elongation properties of the paperboard are lower in the MD, tension is applied to the adjacent narrow panel, which may result in the reverse side liner rupturing just *before* the crush cut is completed.

There are techniques for avoiding this effect. These include ensuring that the knives are sharp and the board is not allowed to dry out prior to cutting – as this reduces the 'stretch' or elasticity of the paperboard. It is also important that attention is paid to the setting and choice of the ejection rubber alongside the knife which holds the sheet in a fixed position during cutting and facilitates the removal of the knife from the paperboard.

10.4.2.2 Rotary die

The traditional rotary die comprises two sets of steel cylinders: one set for creasing and another for cutting, shown schematically in Fig. 10.25. Metal is removed from a solid metal

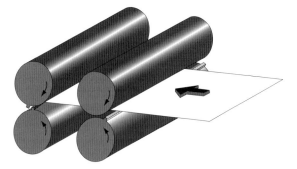

Figure 10.25 Rotary die cutting showing concept of separating the cutting and creasing operations whereby each unit can be adjusted independently. (Reproduced with permission from DieInfo, Inc.)

cylinder to create slots for the insertion of knives and creasing rules. The cutting die knives are set to have a 'kiss' contact with the backing roll. Grooves are cut into the creasing roll backing cylinder in line with the creasing rules. A modern method for making these dies is by electronic discharge machining (EDM) (Bernal, 2012).

The pressure for cutting is much less, compared with that needed on a platen press because the carton profile is cut incrementally as the die rotates at the same linear speed as the paperboard web. The method of cutting is, however, the same as with the platen die cutting already described, i.e. 'crush' cutting.

This type of rotary die is expensive though they are cost-effective for large orders of cartons in respect of the number of impressions they can make, for example 1.25 million. At this point, the die is resharpened and a further 1.25 million impressions are possible. The solid rotary die can be resharpened up to four times, i.e. over 6 million impressions overall.

Bernal has introduced a simpler, less costly version of this type of solid die which is journal-less. This means that less metal and less machining are required with respect to the bearing journals (Bernal, 2012).

There have been several important innovations to adapt the rotary die for shorter numbers of impressions. Wrap-around plate rotary dies are thin steel sheets which have been chemically etched. These plates can give up to 800 000 impressions after which they have to be replaced. These plates are fixed to mandrels on the machine, either mechanically or magnetically.

There is another type of cutting which is used, known as 'pressure' or shear cutting, as shown in Fig. 10.26. This can also be carried out using rotary dies. To make a cut, the knife, as used in 'crush' cutting, is replaced by two flat metal strips, known as 'lands'. One is located on the upper cylinder, the other on the lower. They are offset from each other, as shown in the diagram. The paperboard is squeezed between the plates causing a clean and dust-free cut. The cutting action is similar to the way scissors cut paper and paperboard. The method was developed by Marathon Corporation in the USA. The original dies were in the form of plates wrapped around cylinders. They were made using photographic and etching techniques and went under the name 'B11'. The original method was subject to problems of inaccuracy and life expectancy. After the patents ran out, Bernal redeveloped the concept using solid hardened cylinders and trademarked their system as the 'RP system', where RP stands for 'rotary pressure'. Today, this tooling is accurate, produces consistent quality over very long production runs and very little fibre debris is produced.

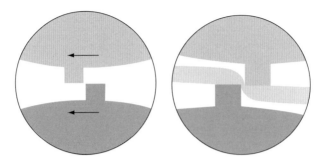

Figure 10.26 Pressure cutting lands – paperboard is 'cut' by raised 'lands' on the rotating cylinders. (Reproduced with permission from DieInfo, Inc.)

The pressure, and hence the knife wear, on these dies is much less than with crush cutting and the dies have a much longer life before they need resharpening. Runs of 10 million impressions would be typical before resharpening and up to four resharpenings thereafter would be possible (Pfaff, 1999).

Pressure-cut wrap-around dies based on thin sheets of steel are cheaper and are made by chemical etching. Such dies are cheaper than solid rotary dies and can give over 6 million impressions after which they are replaced.

Michael Pfaff (2000) also emphasises the fact that the important cost figure, which should be calculated for the various rotary die specifications, is the die cost per 1000 cartons per minute (cpm). This reference explains the methodology for calculating die cpm. Both crush and pressure cutting rotary die sets can incorporate creasing within one set of cylinders with a saving in die costs and machine set-up time (Bernal, 2012).

The most effective tear strip is achieved with cut scores. This is achieved by cut-scoring the surface from above the sheet and, in a slightly offset line, cut-scoring the reverse side using a shallow, chemically etched blade set onto the bed plate. The reverse cut score cuts against a flat anvil placed in the die. If this cut-scoring is carried out on either side of a narrow strip of paperboard, the strip can be easily and cleanly removed from the carton (Atlas Die, 2012).

Perforations can be cut into the board with a serrated knife. They are used in place of creases on the short 45° folds which are used to form trays and crash-lock bottoms. After gluing these creases, the adjoining panels are folded back on themselves, i.e. towards the print, so that the tray or carton is folded flat.

The carton profiles, or blanks, are removed from the sheet by 'stripping'. This is carried out automatically on rotary and platen presses. The stripping unit needs to be carefully designed and set up to ensure that an upcurl (towards the print) is not induced as the blank is forced downwards away from the plane of the sheet. Where it is not carried out automatically on platen presses, stripping is, subsequently, carried out manually, using rubber-headed hammers. The automatic approach requires a system by which the waste, which surrounds the carton profile on the sheet, is efficiently separated and removed.

A key feature for ensuring a high cartonning machine efficiency is that the carton dimensions conform with the agreed specification drawing. Modern die and make-ready technology provides for this need. Many years ago, Pira introduced a measuring table with a travelling microscope and it was used by the author in the 1960s. Indocomp Systems introduced a computer-based system (ACT II) in the late 1980s, which automatically checks the profile of creases and the dimensions of panels. On a gable top milk carton, for instance, it would

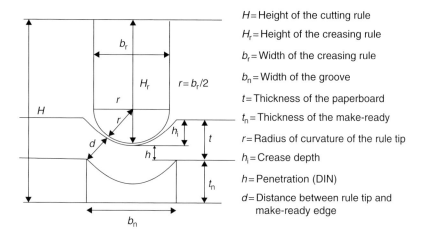

Figure 10.27 Critical dimensions for the creasing operation. (Courtesy of Iggesund Paperboard.)

H = Height of the cutting rule
H_r = Height of the creasing rule
b_r = Width of the creasing rule
b_n = Width of the groove
t = Thickness of the paperboard
t_n = Thickness of the make-ready
r = Radius of curvature of the rule tip
h_i = Crease depth
h = Penetration (DIN)
d = Distance between rule tip and make-ready edge

precisely locate the position of 74 creases (scores) and 32 edges (FCI, 1988, 1996). The system can provide a variety of management reports and has introduced ACT III with a Microsoft Windows-based system. This is faster and has several enhancements, such as a three-dimensional crease (score) and edge profiling (Indocomp Testing Systems, 2012).

10.4.3 Creasing and folding

Creases are made using creasing rules. These rules are thin strips of metal with smooth rounded edges which indent the board surface and push it into an accurately cut groove on the underside of the paperboard. The groove is formed in a thin hard material called the 'make-ready' matrix or counter die. The critical, or important, dimensions which are relevant to the creasing operation are shown in Fig. 10.27. The depth and width dimensions of the creasing grooves depend on the paperboard product being used together with the width of the creasing rules and the difference in height between the creasing rules and the knives used in the die.

Most paperboard manufacturers provide guidance on groove width and matrix thickness for given heights of cutting knife and the associated heights/thicknesses of creasing rule. The groove width is usually 1.5 times the thickness of the paperboard plus the width of the creasing rule and is slightly narrower for creases parallel with the MD of the board.

A note of caution should be made concerning the convention for describing a crease in terms of MD and CD. Some publications define an MD crease as a crease *parallel* with the MD of the paperboard, whereas this publication and others, including Dennis Hine's *Cartons and Cartonning*, define an MD crease as one where the *MD stiffness* is involved when the crease is *folded*. In this case, an MD crease would be a crease at *a right angle* or 90° to the MD of the board. Some writers refer to this as an 'across the grain' crease. Therefore to avoid confusion, it is always advisable to define the terminology used (Fig. 10.28).

There are several types of matrix material available in a range of thicknesses in common use. Platen counter dies can be made from hard phenolic plastic sheet, vulcanised fibre sheet (Presspahn) or steel depending on the length of run required, i.e. number of impressions. An alternative approach is to use polyester channel (pre-made) of fixed width and depth. Creasing grooves in rotary metal cylinders have the longest life.

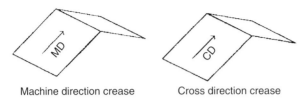

Figure 10.28 Crease definitions – the crease line is at right angles to the indicated board grain direction. (Reproduced with permission from PIRA)

There are many possibilities which affect the commercial considerations for any particular carton estimate/enquiry requiring the same paperboard and printing specification. The choice of die, platen or rotary, the various make-ready options and, in the case of rotary, the various die options discussed have different commercial implications depending on order and run length. Add to this, the different machine make-ready times for the various options, stripping waste and the number of cartons possible with the various formats, which are set by the maximum platen area, and the position looks very complicated.

In practice, any given carton maker will have a limited range of machine sizes. This is relevant as the machine size will determine the number of cartons per impression with the die area or format. Format in this sense is the arrangement of the cartons on the area available. Sheet-fed machines have a maximum area which they can print and cut and crease. Reel or web-fed machines have a fixed maximum width and the repeat length, or cut-off limit for the dimensions in the MD, is controlled by the circumference of the cylinders. In any factory, machine sizes are matched for sheet area and production output as between printing and cutting and creasing (scoring). In the case of rotary, it is likely that both processes will take place on the same machine, in-line. Just to make it a little more complicated, depending on the rotary machine concerned and how it is equipped, one can have either a flat bed die or a rotary die. Hence the price for the same enquiry from carton makers with different types and sizes of machinery can be significantly different.

The format area is not only controlled by the maximum and minimum sheet size that will fit on the machine. The way the cartons can be laid out in the available area also has to take account of the carton grain direction. An interesting case study where this went wrong for a carton maker was as follows.

Two companies were supplying a carton design to an end-user. Company A printed and cut and creased with a format of nine cartons on a sheet; the price and the quality were satisfactory. Company B only had a larger format machine available and produced 15 cartons per sheet in three rows of five cartons. Quality was satisfactory but the cost of production was too high. Company B re-costed at 16 per sheet, which was theoretically possible with 4 rows of 4 cartons. But to keep the grain direction the same on the cartons they had to change the grain direction on the sheet of paperboard. Printers normally use a sheet with the grain direction parallel to the axis of the cylinders of the printing press – this is known as a long grain sheet. In this case, not only was the grain direction of the sheet changed but the front-to-back dimension was now right on the limit for the printing press. This might have been all right except for the fact that the carton had a very heavy, i.e. overall solid, ink coverage, which at the back edge of the sheet was right on the edge of the sheet – so near in fact that the vertical side of the pallet of printed board at the back was the same colour as the print. This resulted in a severe tail-edge hook, or hump, in the accumulated sheets on the pallet, along the back row of cartons. When these cartons were cut and creased, they had a severe down curl (away from the print) which rendered them useless for use on the packing

Folding cartons **289**

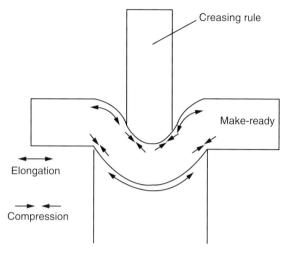

Figure 10.29 Strains (forces) induced in paperboard during creasing. (Courtesy of Iggesund Paperboard.)

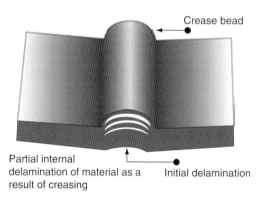

Figure 10.30 Creasing causes internal delamination. (Reproduced with permission from DieInfo, Inc.)

machine. Hence the conclusion is that those concerned must take care to recognise the limitations of what can be achieved – they may not simply be dimensional.

The effect of the creasing process on both the surfaces and the internal structure of the paperboard is complex (Fig. 10.29). Across the crease there are:

- tensile strains, which are greatest in the surface and reverse side liner plies
- compression in the direction perpendicular to the surface
- shearing strains within the paperboard, parallel with the paperboard surfaces.

Reference has been made to the fact that depth of the creasing groove in the surface causes a certain amount of stretching in the surface. Moreover, the initial depth of the groove formed in the surface reduces after the creasing rule is withdrawn from the paperboard. These forces have been studied by a number of researchers (Hine, 1999, pp. 232–241).

The internal shearing forces which occur during the formation of a crease cause some internal delamination of the interply adhesion. This results in a bulge on the reverse side of the board, as shown schematically in Fig. 10.30.

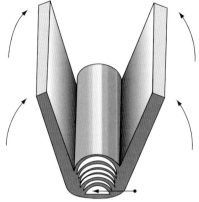

Delamination is completed
as the carton folds and the hinge is formed

Figure 10.31 Further internal delamination occurs on folding. (Reproduced with permission from DieInfo, Inc.)

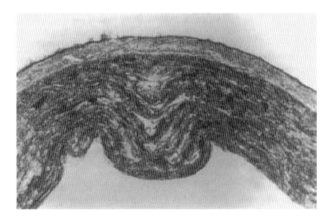

Figure 10.32 Photomicrograph showing delamination. (Reproduced with permission from PIRA)

When the crease is folded, further internal delamination occurs as shown schematically in Fig. 10.31. A good crease should not show any liner cracking on the printed side and an even symmetrical rib or bulge, with no signs of crumpling, on the reverse side. These conditions should be maintained when the crease is folded through 180°. After such folding, it will be noticed that the bulge on the reverse side has expanded and the thickness of the paperboard in the middle of the crease is much thicker than the nominal thickness of the paperboard. The delamination which occurs inside the bulge is confirmed by photomicrographs of the folded crease (Hine, 1999, p. 226) as shown in Fig. 10.32.

A number of ways of examining the creasing groove have been described, such as by microscope with calibrated graticule, use of lamp and lens assembly to project a shadow profile of the groove and electromechanical devices used in engineering surface examinations where they traverse the groove (Hine, 1999, p. 226). In the author's experience, the latter quickly demonstrates the differences in creases resulting from the misalignment of a rule with the make-ready groove and early warning of the make-ready deteriorating as a result of wear. The Indocomp ACT testing equipment also profiles the crease outline, using a 'high precision LVDT (Linear Variable Differential Transformer) probe' (FCI, 1996).

Good creasing is necessary for the following reasons:

- visual appearance of the carton
- efficient performance on the packing line
- maintaining the compression strength of the carton in storage, distribution and use.

Poor creasing is apparent when the folded crease shows liner splitting. This is particularly obvious if there is solid print colour covering the crease because the internal layers of the paperboard are exposed. A possible cause of a crease bursting along its length can be that the paperboard has dried out as a result of excessive heat being applied during radiation-assisted print drying. A burst over a very short distance in a folded crease can occur if some foreign material jams in the make-ready groove. Other visual defects could be apparent in the bulging of panels and the shattering of the back ply adjacent to a narrow panel such as the glue seam.

The creasability of paperboard can be assessed in the laboratory using a small platen or an instrument such as the Pira Cartonboard Creaser using BS 4818:1993. This is a press which simulates creasing whilst the sample of paperboard is clamped with adjacent creases formed at the same time as the test crease. The instrument can make creases at a range of crease depths and widths. The results are evaluated visually. For a paperboard to be considered to have good creasability, it is important that it should give good creasing over a range of crease settings.

The performance of creases on a packing line is very important. Creases behave like hinges which enable adjacent panels to move through specific angles, usually 90°, and remain there. The bending force is an important parameter. It is especially important for creases which have not been broken previously. Folding is carried out as the carton is moving in relation to fixed rails and ploughs, and therefore undue pressure from crease spring back can at least cause rub and, at worst, delays and jams. Flaps may be glued, and during the setting time of the adhesive, the flap may attempt to spring back and hence must be restrained for sufficient time. In these examples, if the force required to make the fold or the subsequent spring back is too high, the efficiency of the operation will be poor.

The question therefore arises as to how creasing at the point of carton manufacture can be measured and controlled. This is even more important as it has been shown that a rise in the resistance to folding can be measured well before any visual change in the appearance of the crease can be detected (Hine, 1999, p. 250). This is because of the wear in the groove width which takes place, depending on the make-ready material, over time.

Consideration of how and what to measure starts with identifying the parameters which are involved in the bending of a crease. These parameters are the angle through which the flap is turned, the force required, the distance between the application of the force and the crease, and the time to complete the folding. There is a further consideration. If the crease resistance is greater than the paperboard stiffness, the panel will bow as the panel rotates around the crease. On the other hand, a certain minimum spring-back force is necessary for the correct operation of certain design features such as the retention of a tuck-in-flap locking slit.

The implication of this is that the ratio of crease stiffness to paperboard stiffness is an important parameter and that it should be maintained within upper and lower limits. The limits suggested are 1.5–3.0 for MD creases and 3.0–7.0 for CD creases, and these limits have been accepted for many years (Hine, 1999, pp. 111–139). (The MD crease convention used here is that MD creases are those where the creases are at right angles to the MD of the paperboard.)

The parameters listed have to be considered when designing a method for testing crease resistance at the point of manufacture. Several methods have been used successfully, such as the Pira Crease Tester and the Marbach Crease Bend Tester. The latter has the advantage of recording and displaying the bending force (torque) dynamically from 0° to 90° and from 0° to 180°. There is a choice of folding time, either 1.0 or 0.1 s, the latter simulating folding in gluing and cartonning machines.

Faulty creasing results in a poor carton appearance and low packaging-machine efficiency, both of which are easily observed. A further consequence is that cartons affected by faulty creasing will not achieve their optimum compression strength as panels and flaps will not be correctly positioned with respect to each other, leading to bowing and twisting. Damaged creases will not provide the required strength.

10.4.4 Embossing

Embossing is a process which imparts a relief or raised design in the paperboard surface. It can be applied all over the surface, for example a sand or linen pattern. Embossing can also be applied after printing in register with the print, in which case it would be applied as a separate process or during cutting and creasing. Embossing is a design option which enhances visual impact; it is tactile and can impart a luxury feeling. The design could be text or any graphical representation, for example coat of arms (crest), a flower, automobile, fruit, food and bottle. The relief can be raised (positive emboss) or impressed (negative emboss). It is carried out with a shaped metal surface above the paperboard and an inverted pattern underneath using heat as well as pressure (Fig. 10.33).

Whilst all types of paperboard can be embossed, the suitability of any given paperboard for a specific emboss should always be investigated (proofed). The finer the detail and the deeper the emboss, the greater are the demands placed on the paperboard in terms of the strength, toughness, rigidity and elasticity required to achieve the required result. As with creasing, embossing creates forces in the surfaces and the internal structure of the paperboard. The relevant paperboard properties for embossing are tensile strength, percentage of stretch (elongation), toughness, moisture content, stiffness, short-span compression strength and density.

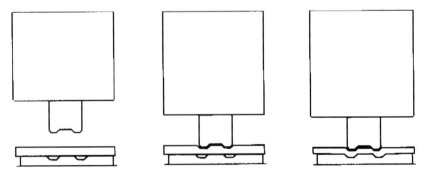

Figure 10.33 Embossing operation. (Courtesy of Iggesund Paperboard.)

As already noted, cartons for medicinal products require details about the product to be embossed on the packaging. The European Carton Makers' Association (ECMA) launched a European code of practice in 2005 for the production of Braille on folding cartons (ECMA, 2005).

The Braille fonts comply with the Marburg Medium range of Braille fonts specification as recommended by the European Commission, the *Can-Am Braille* standard and the ECMA EuroBraille standard for use on pharmaceutical packaging and labels (PharmaBraille, 2011).

Paperboard is embossed using male and female tools, and it is applied to folding cartons in one of two ways. Embossing tools can be set in each carton station on the platen die cutter or it can be applied by a rotary unit mounted on a carton gluing machine (Bobst, 2010).

The rotary unit approach is considered to be more advantageous because embossing in this way results in the following:

- Significant dot height improvement which provides easier reading for end-users.
- Make-ready time of the die cutter is faster (no set-up of embossing tools).
- Die cutting will have fewer stops (less cleaning of embossing tools).
- There are fewer challenges on the folder-gluer because the boxes are embossed in the section following their feeder, which therefore provides improved run-ability of the gluer as the embossing module is mounted after the feeder and it ensures that no blanks are stuck together.
- The set-up of tools in the rotary unit takes less than 5 min and no additional staffing is required (PharmaBraille, 2010).

10.4.5 Hot-foil stamping

Hot-foil stamping is a form of surface printing or decoration. It is applied to paperboard using a heated die, containing the design, from a special film. The colour may be either a pigment or a plain (silver), or coloured, aluminium foil (Fig. 10.34). It is applied either in

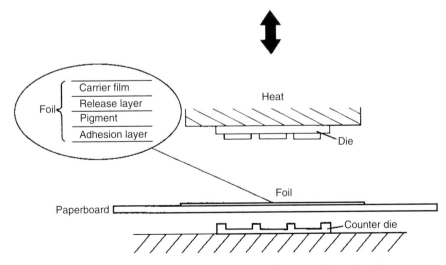

Figure 10.34 Hot-foil stamping and embossing. (Courtesy of Iggesund Paperboard.)

register with the print or to an embossed design feature. It can be applied using a special machine or incorporated with the embossing tool at cutting and creasing.

10.4.6 Gluing

Gluing is the technique used to erect and close cartons using adhesives which are also referred to as 'glues'. Several different types of adhesives are used with folding cartons depending on the surfaces being joined and the pressure–time parameters of the gluing system. The principles are discussed in Section 1.5.3.17. In its broadest sense, gluing includes heat-sealing plastic-coated paperboard, where the molten plastic in the sealing area acts as the adhesive.

In carton manufacturing, the gluing operation is used to seal carton side seams, in corner-gluing and in sealing the base flaps of a crash-lock bottom.

The most common type of folding carton is the straight line, side seam-glued, tubular style with open ends. Flat carton blanks are placed in the feeder of a high-speed folder-gluer, print face down. This operation can be made more efficient by the use of a pre-feeder which has high capacity storage, is easy to load and presents the cartons in a smooth high-speed rate into the machine. Two distinct operations are carried out. First, the glue flap crease 1 and crease 3, i.e. the crease opposite the glue flap crease in the finished carton, are pre-folded or 'broken' by folding them over as far as possible, as near to 180° as possible, and back to the horizontal. Then the adhesive is applied to the glue flap (Fig. 10.35).

The choice of adhesive depends on the nature of the surfaces being sealed and the parameters of open time, setting time and compression time inherent in the system. The choice must also take into account any special environments and product needs, for example frozen food storage, moist humid conditions and soap/detergent resistance.

For most types of paperboard, the adhesive of choice is a polyvinyl acetate (PVA) emulsion applied by wheel to the glue flap. Creases 2 and 4 are then folded over and a bond created between the glue flap and the edge of the overlapping panel. The carton is then compressed to allow the adhesive to set. On exiting the compression section, the cartons are counted and packed, usually, in non-returnable corrugated cases, where the drying process takes place. This operation can be automated and run at up to 200 000 cartons per hour.

Modified machines can form and apply tubes of flexible packaging materials to carton blanks, prior to side-seam sealing for bag-in-box cartons.

The gluing of the crash-lock bottom style is similar to the side seam-glued tube style with the additional gluing of two diagonal flaps attached to the base of either panels 1 and 3, or panels 2 and 4. As with ordinary tube-style glued cartons, they are then folded flat. As

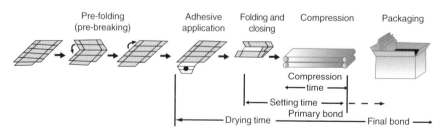

Figure 10.35 Straight line folding and gluing operation. (Reproduced, with permission, from M-Real.)

mentioned earlier, this style can be erected manually by the packer/filler so that the base panels lock into place, after which the carton is filled and closed.

Double-thickness side walls which are derived by the folding of additional panels can be glued on the straight-line gluer. This style is folded flat and erected by the packer/filler to produce a rigid-tray construction. It is also possible to plough, fold and glue additional panels in such a way that integral platforms to support, display and locate products inside a normal end-loading tube-style carton can be incorporated – such a carton may have been fitted with a window.

The 4- and 6-point glued trays are glued by applying the adhesive from overhead glue pots to diagonally folded back flaps. The 6-point glued style (Fig. 10.7) has an integral lid. Where diagonal flaps are folded back on themselves, and folded flat, it is usual to make the creases with perforating rule at the cutting and creasing stage.

Special adhesives are used where the joining surfaces are not suitable for PVA emulsion sealing. Hot-melt adhesive can be applied on the gluer as a coating which solidifies. This coating is reactivated by heat by the packer/filler in a way which allows the hot-melt coating to flow and create sift-proof and pinhole-free seals in the packed carton.

PE-coated cartons can be sealed with hot air on a straight-line gluer. Straight-line gluers can be fitted with detectors which check for the presence of a glue line and the measurement of glue line film weight. They can be fitted with code readers to ensure that multiprint orders of the same size cartons do not get intermixed during conversion.

Other important quality issues concern the avoidance of any skewing of the glue flap, vertical or horizontal displacement of the glue flap, glue splashes and glue squeeze-out which can prevent automatic opening on the packing line. The glue flap panel must be free from ink and varnish in the glued area, and this also implies tight control of print 'bleed' from adjoining panels.

The effect of the pre-folding, the pressure applied to the outside creases in the compression section and storage aspects are all relevant to packing-line efficiency as they relate to the carton opening force.

Glued cartons are counted and batched automatically prior to packing in corrugated fibreboard cases and palletised. Automatic case packers can be fitted at the end of the gluing machine. Pallets may be stretch- or shrink-wrapped in PE film for reasons of hygiene and moisture protection.

10.4.7 Specialist conversion operations

10.4.7.1 Windowing

Cartons can be windowed to enable the contents to be displayed. The windows are made from plastic films, such as cellulose acetate, PVC, PET/PETE and PP. The window is in either one panel or two, in which case it bends around one of the corners when the carton is erected. The paperboard aperture is cut at the cutting and creasing stage. The window patch is applied on a window-patching machine which applies adhesive to the reverse side of the carton blank in line with the perimeter of the window. The film is cut automatically from a reel and applied over the adhesive. The carton is then sent to a straight-line gluer for folding and gluing.

The windowing machine can also be used with attachments to apply flexible packaging tubular material for bag-in-box cartons. Specially designed windowing machines, which can also make creases in plastic film, are available to make paperboard cartons with windows on three or four panels.

10.4.7.2 *Waxing*

In addition to making cartons from paperboard impregnated with wax during paperboard manufacture, it is also possible to apply wax to one or both sides of a cut and creased flat carton blank. Wax can be applied in patterns and can be kept off glue flaps.

Waxing in this way is either 'dry' waxing, where the wax solidifies on the surface giving a matt appearance, or 'wet' waxing, where the carton passes under heaters which remelt the wax before the blank is carried on belts through refrigerated water. The shock cooling produces a high-gloss finish on the surface of the wax. With appropriate wax blends, cartons which are high gloss waxed can be heat sealed. Waxed cartons are used for frozen foods, ready meals and ice cream.

Cartons with tapered sides in tub and conical shapes and with a round or square cross section can be waxed after forming. The first liquid-packaging cartons were made in this way.

10.5 Packaging operation

10.5.1 Speed and efficiency

Cartons are erected, filled and packed by product manufacturers, also known as end-users or packer-fillers. There are, additionally, contract packers who provide a packaging service, which includes cartonning, to manufacturers, particularly in order to meet promotional and test marketing needs.

Carton packing may be either manual, partly manual and mechanically assisted, or fully automatic. Speeds vary considerably from, for example, 10–1000 cpm, though not many would be running at speeds in excess of 500 cpm. Mechanical cartonning with manual product-assisted loading is possible up to 40–60 cpm. Fully automatic cartonners start at speeds of around 60 cpm and most cartonning machinery manufacturers offer equipment which can run at higher speeds, such as 120–240 cpm. Higher speeds are possible in the range of 250–400 cpm, thereafter the machinery is designed with specific products in mind, and in this respect, cigarette cartonning is unique. Packets of 20 cigarettes (sticks) in the special style of carton made from a hinged lid blank (HLB), which has a three-sided inner frame and a flip-top, run on machines at 400–700 cpm. The Focke twin track F8 HLB cigarette cartonning machine has achieved a speed of 1000 cpm (*Tobacco Reporter*, 2008).

Whatever be the speed and overall output needs of a particular operation, there will be a choice of cartonning machinery to meet the needs of the business. Some manufacturers may prefer three machines rated at 60 cpm to one machine rated at 180 cpm. Several factors will influence the choice, such as:

- factory layout or features of the production process
- the need to pack several sizes of product concurrently
- the need to pack different products concurrently.

The sequence of operations on a packing line is:

1. feeding and erecting the cartons from a box or magazine
2. filling with the product
3. closing the carton

4. check-weighing and metal detection depending on the product
5. end-of-line operations preparing the product for distribution.

A survey (unpublished) of different types of packing line in several locations by a multinational FMCG manufacturer revealed that problems associated with the in-feed section of packing machines were the most prevalent cause of stoppages.

10.5.2 Side seam-glued cartons

Side seam-glued cartons are placed in a magazine from which they are removed one at a time by vacuumised suckers (pads). There are two basic methods by which cartons are extracted from the magazine and erected. In the first method, the suckers pull on one panel and transfer the cartons into the moving pockets of a flighted conveyor. The length of each pocket, which is controlled by the flights, reduces automatically to the width of the carton and, in so doing, erects the carton by pressing on the two opposite, folded creases. This method is referred to as 'diagonal' loading. The other method is to use suckers on adjacent panels and pull the carton in opposing directions such that the carton assumes a rectangular cross section by the time it is dropped into the pocket of the flighted conveyor. This is known as 'rotational loading'. Another mechanical opening method inserts knives from both sides which are twisted as the carton is eased into the flighted conveyor in a diagonal loading mode.

The opening of side seam-glued cartons has been studied in depth and the carton opening force measured using methods which replicate both diagonal and rotational loading. For a full treatment, see Hine (1999, pp. 111–139). This reference relates carton gluing, crease pre-breaking, and the variation in carton opening force with storage time.

An important conclusion of this work is that with *diagonal* loading, carton opening force increases rapidly after gluing and packing by the carton maker during the first 6 days of storage, levelling out after 2 or 3 months. However, with *rotational* loading, there was no significant rise in carton opening–force torque with storage, suggesting that this is a superior method of carton erection. Another conclusion is that for both methods of carton erection, the main resistance to opening comes from the pre-folded creases, indicating the importance of this aspect of folder–gluer operation.

An additional aspect of folder–gluer operation is the effect of compression on the folded creases. This can be assessed by measuring the height of a given number of cartons at the end of the gluing operation. The higher the compression pressure, the lower the height and the greater likelihood that the carton will be difficult to open on the packing line. If, however, the compression is too low, the side-seam adhesion may be impaired, and in addition, it would be difficult to load the cartons into the magazine of the cartonning machine. In practice, for a given carton, this height should be maintained within a range established by correlation with the acceptable range of heights at the cartonning stage, i.e. after storage. This height feature is also referred to as the 'bounce'.

One of the main causes of carton-feeding problems at the packing stage is the distortion of folded cartons which may occur in storage. In particular, the shape distortion, which is described as a 'banana' or 'armchair' shape, is virtually impossible to open. Another form of distortion produces an 'S' shape. Hence the recommendation is that cartons are stored on edge and isolated from stacking pressure in non-returnable corrugated fibreboard outer cases. The resulting rows of cartons should not be too tight. Hanlon

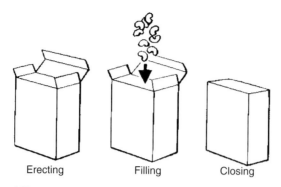

| Erecting | Filling | Closing |

Figure 10.36 Vertical filling and sealing operation. (Courtesy of Iggesund Paperboard.)

suggests that the combined thickness of the row – calculated as three times the paperboard thickness, i.e. the thickness at the glue flap, multiplied by the number of cartons and adding 15% of the result – should be used as the internal length dimension of the case (Hanlon et al., 1998).

Consideration of carton opening force has led in some case to changes in the way the gluing of cartons is organised. Cartons may be printed and cut and creased in large batches, taking account of the cost-benefit of longer production runs. The gluing, however, has been organised in much smaller batches to minimise the risk of a high carton opening force and/or distortion in storage. In some cases, the gluing has virtually been organised on demand in a location and facility remote from the carton manufacturer and adjacent to the cartonning operation. Alternatively, some cartonning machines have included a simply designed side-seam gluer actually attached to the in-feed.

Side seam-glued cartons may be filled horizontally or vertically (see Fig. 10.36), depending on the product. A free-flowing product which is apportioned gravimetrically would have an integral weigh filler with many filling heads. This type of filler can progressively fill the cartons vertically as the filling heads travel around a semicircular (carousel) track. This type of filling can run at high speed, for example hundreds of cartons per minute.

Some cartonning machines are fitted with a pre-feeder. The object is to extend the time given to erecting the carton under controlled conditions. This can be done by designing a circular pre-feeder and fitting it alongside the cartonning machine in-feed whereby *erected* cartons are transferred to the flights (pockets) of the cartonning machine.

Carton machinery, where the product is free-flowing and filled vertically, can incorporate carton tare weighing and product top-up features to achieve very high accuracy in fill weight. This is beneficial when filling expensive products.

The closing of side seam-glued cartons is by hot-melt sealing, as shown in Fig. 10.37, tuck-in-flaps or locking tabs. Sealed closures would usually have an easy-opening feature, the design of which would be dependent on whether a reclosure feature is also required.

10.5.3 Erection of flat carton blanks

Flat carton blanks are erected by the packer/filler using one of the following methods:

- Using a reciprocating tool which is pressed against the base panel thereby forcing the side panels through 90° into the vertical (usually, there are exceptions) position. The side

Folding cartons 299

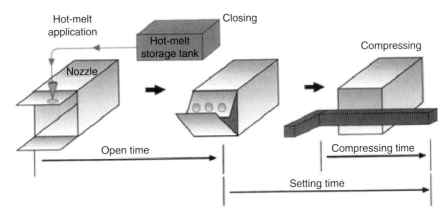

Figure 10.37 Carton closing using hot-melt adhesive. (Reproduced with permission from M-Real.)

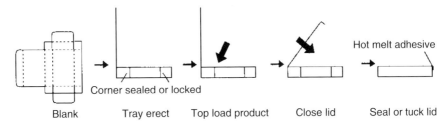

Figure 10.38 Erection of tray-style carton for top loading and closing. (Courtesy of Iggesund Paperboard.)

panels are then secured in this position, forming a tray shape either by means of interlocking tabs or by hot-melt adhesive (Fig. 10.38).
- Applying an adhesive to a side seam and folding the carton blank around a mandrel. This is usually preceded by wrapping paper, paper coated with a barrier coating, such as PE or PVdC, an OPP film with PVdC coating or a film laminate around the mandrel, sealing the side seam and base.
- Forming the HLB together with a reinforcing inner frame – a design mainly confined to the packing of cigarettes.
- Applying adhesive to the side seam as the first operation on the packing line using a simplified side-seam gluing unit.

The efficiency of operation, where the cartons are presented in the form of flat blanks, mainly depends on the flatness of the paperboard blanks being maintained. This is because a blank which has a curl or twist can easily misfeed and cause a stoppage. For example, if the carton blank is pulled out of the magazine, print face downwards, onto a short conveyor, the lugs which are supposed to push the (back) edge of the board, instead, pass under the upturned edge. Curl may also prevent a tuck-in-flap from being pushed in accurately.

The tray-type carton is top loaded, either by hand or mechanically by an automatic 'pick and place' action. The integral lid is closed and sealed on one or three sides with hot-melt adhesive. Where the product is filled hot, water-based adhesives based on PVA, starch or dextrin are required. The efficiency of water-based adhesives depends on the absorbency of the surfaces being sealed and the compression time to allow the adhesive to set, i.e. to become tacky enough so that unrestrained joints do not open.

Plastic-coated trays are usually erected and lidded by heat sealing, but plastic-coated end-loaded cartons are usually sealed, though not always, using hot-melt adhesive.

An important precaution in the use of hot-melt adhesives is that they must not be exposed to air for long periods at the working temperature when the machine is not in production. This will cause heat degradation and subsequent loss of adhesion. A situation was investigated where the packer claimed that the surface of the paperboard was defective because the hot-melt adhesive would not close the cartons permanently. It was found that the heat supplied to the adhesive reservoir had not been switched off when the machine had been left unattended overnight and over the weekend. This, it was claimed, had been standard practice to ensure a quick start-up when production was resumed. There was evidence of severe carbonisation in the adhesive system. The hot-melt adhesive had been degraded. The solution was to use time switches set to remelt the adhesive a short while before packing was required to recommence. Today, pressurised on-demand nozzle applicators are preferred to open-to-air glue pots.

10.5.4 Carton storage

Reference has been made to the fact that paperboard will absorb moisture when exposed to high humidity and lose moisture in low humidity. Moisture content changes are usually accompanied by changes in flatness (shape). Hence reasonable precautions should be taken at all stages where the paperboard may be exposed to changes in RH. The carton manufacturer should provide moisture protection for storage and transit. The packer/filler (end-user) must ensure that cartons are not unwrapped until they have attained temperature equilibrium with the area in which the packaging is carried out.

Problems have been observed where unwrapped pallets of cartons awaiting packing have been left near exits to the outside environment. Also cold cartons have been found to affect the efficiency of hot-melt adhesion due to the fact that the adhesive open time is reduced by being applied to a cold surface and the tackiness is lost before the surfaces being sealed are brought together.

Packer/fillers should also replace the moisture-resistant wrappings of pallets and boxes of cartons left unused at the end of a production run and over a weekend. This is especially important in dry (low RH) packing areas handling dry food products, such as tea and baked products, for example biscuits and cereals. In this dry environment, unprotected flat paperboard carton blanks are likely to develop downcurl, i.e. curl away from the print and this will cause problems on cartonning machines.

10.5.5 Runnability and packaging line efficiency

Good runnability is essential. The requirements of good runnability are many and varied. Good runnability is difficult to define, but everyone knows when it is missing. In a general way, it describes a packaging operation running with minimum disruption, at a specified level of efficiency, which can be measured and monitored.

Packing-line efficiency is dependent on:

- the machine(s) or method of packaging in the case of a manually operated line
- the reliability and maintenance of the machinery

- the product
- the operators, level of training, etc.
- the quality of the cartons.

The packer/filler's aim is to avoid, or minimise, the production of damaged packs, wasted products, wasted packaging and to achieve the rated output of the packing line. The efficiency of a packing line is given by:

$$\text{Line efficiency \%} = \frac{\text{Actual output}}{\text{Expected output}} \times 100$$

In establishing 'expected output', it is important to base this on the real time available for packing. This means that setting-up time and routine maintenance must be eliminated from overall production time.

A packaging line may comprise several, linked, packaging machines, for example form/fill/seal pouch or sachet machine, cartonning machine and case packer. If the efficiency of each machine is 90%, then the efficiency of the line as a whole would be the *product* of these individual machine efficiencies, in this example 72.9%. This must be taken account of when planning a packaging system.

A mistaken poor efficiency complaint arose when a packer reported high carton wastage. The line was fed from a carton erector fitted with a counter which counted every vertical cycle. The carton erector could erect cartons faster than the rest of the packing line could pack and close the cartons – it itself a good feature. When the line was full of partially loaded cartons, an automatic switch stopped the carton feed into the carton erector. However, the carton erector continued to cycle automatically even though no cartons were being erected. The counter ticked away and at the end of the shift the counter figure was taken as the number of cartons erected – clearly this was erroneous. The production of filled cartons was much lower than the figure from the carton erector and the difference was interpreted as *high carton* wastage! Eventually, the correct usage was established by reconciliation with the quantities of cartons in, and issued by, the warehouse, but not before someone had initiated an investigation of the supplier of the cartons and the paperboard!

Some examples of features which affect runnability are less easily defined than others, and one which comes into this category is 'timing'. This relates to the settings on a cartonning machine within which an established carton specification can be run with a satisfactory packing-line efficiency. Settings control mechanical movement whereby the machine interacts with the carton and the paperboard. Some settings can be advanced or retarded in response to, for example, the paperboard stiffness or the resistance to folding or spring back of carton creases.

The importance of timing was highlighted by the following example which occurred when an alternative carton specification was trialled on a well-established packaging line. The alternative paperboard specification was significantly different in that it was based on FBB from a machine fitted with Fourdrinier wires whilst the established carton was based on recycled board, WLC, made on a modified vat machine, and the thickness of the two boards was the same, which meant that the FBB was 23% lower in grammage.

The cartons were medium to large in size. They were erected for horizontal end-loading of the product which was already packed in a PE-coextruded film bag. The cartons were therefore moving in the same direction as the MD of the paperboard. There was a large difference between the MD stiffness of the WLC and that of the FBB, with the MD stiffness of the WLC being about 25% higher. This mainly resulted from the difference in forming on the paperboard machine. The MD/CD stiffness ratio for the WLC was 2.8 and for the FBB 2.1. When the settings which suited the carton made from WLC were used for the FBB cartons, the machine quickly jammed. This could have been the end of the trial and the result recorded as a failure. However, with the cooperation of engineering personnel new *settings and timings* were found for chains, flights and conveyors, which enabled the FBB cartons to run satisfactorily.

Today all settings and adjustments can be logged and retained electronically so that cartonning machines can be quickly reset after size changes. Machines are also fitted with technical support visual displays for troubleshooting to minimise the effect of any stoppages which may occur.

Coefficient of friction, measured in the dynamic mode as opposed to the static mode, is frequently found to be involved in runnability investigations. A note of caution should be made when deciding which surfaces to use in the test method. As the guide rails and ploughs on the machine are likely to be made of steel or aluminium, it is likely that the particular metal surface involved is used in checking the coefficient of friction against the carton surface. This, however, has been found to yield confusing results. It should be recognised that the metal surfaces on the machine can become coated with material which transfers from the cartons, and account of this should be taken.

Surface friction resulting from inks and varnish can be modified with the help of the respective suppliers. The inclusion of silicones, or wax, to improve rub resistance can lower coefficient of friction or angle of slide. It can also reduce gloss levels and hence care is necessary when any changes are contemplated.

Whilst high surface friction is sometimes the cause of poor runnability, it is unlikely that the carton surface will be too slippy as this would give other problems, such as making it difficult to handle a bundle of cartons.

Another property of paperboard where problems have arisen in the past is air permeability (porosity). In a particular case study, the carton blanks were supposed to be picked out of a pile, one at a time, by rubber-vacuumised suckers which contacted the *reverse unprinted side* of the paperboard. A problem arose resulting in misfeeds or partial pick-up which led to misalignment and jamming in the in-feed section of the cartonning machine.

The carton was made from mineral pigment-coated paperboard. This is virtually impermeable to the *rapid* passage of air. In this case, the vacuumised suckers were set quite close to the cut edge of the carton. As a result, air was being sucked in through this edge into the middle plies of the paperboard and from there to the point where the suction was applied. The problem was solved by adjusting the position of the suckers away from the edge of the carton blank.

Where *uncoated* and *unlined* thin paperboard is used, less frequently today compared to years ago, it has been known that the suckers can pull air through two sheets causing a double feed and a machine jam.

The study of cartons and cartonning machine interactions is important and cooperation between the manufacturers of cartons, paperboard, packers and machinery companies should be encouraged. It must always be appreciated that there is an explanation to every problem, and the real explanation must be found if the problem is to be understood and solved. This is often difficult when problems interfere with production.

Sometimes a repetitive sequence is detected in the occurrence of a specific fault. For example, the problem may be associated with one particular carton die station, it may be associated with a damaged flight on a packaging-machine conveyor or it may be due to a damaged carton-forming mandrel. Sometimes the cause of a problem is related to some aspect of either the packing machine or the carton, or to some obscure interaction between the two; but it cannot be observed because the speed is too fast or the suspected position is difficult to access. In these cases, high-speed video should be used to observe the features in slow motion.

In his summary of carton–machine interactions, Hine (1999, p. 182) lists the important paperboard and carton properties in relation to the efficiency of the various machine functions and carton movements.

The properties and features listed are porosity, smoothness (roughness), friction, adhesion, dimensions (accuracy thereof), cut quality, paperboard stiffness, fold stiffness, carton opening force and flatness (Hine, 1999, p. 182: Table 8).

These paperboard and carton properties relate to the efficiency of the following carton machine and operational settings and procedures:

- loading the carton feed magazine
- extracting the cartons from the magazine
- erecting the carton
- conveying the carton through the machine
- filling the product
- closing the flaps
- collating the cartons as they leave the machine.

10.6 Distribution and storage

Cartons are usually collated or grouped together and packed in secondary packaging for distribution and storage. Typical distribution packs, are as follows:

- Unsupported blocks of cartons are stretch- or shrink-wrapped – the cartons may be packed in shallow-depth paperboard trays prior to stretch- or shrink-wrapping.
- The blocks of cartons may be protected with a wrap-around corrugated fibreboard sleeve and then stretched or shrink-wrapped. There are other designs which make use of corrugated fibreboard in this way. One of the objectives is to ensure that some cartons are visible so that the pack has good visual appeal when displayed in cash-and-carry-type warehouses from which many small traders obtain their bulk supplies.
- Regular slotted containers (RSC), Fig. 10.39, are made from corrugated fibreboard. The cases may have tear tapes to facilitate opening and displaying the contents at the point of sale. The case can be a wrap-around blank erected in situ at the end of the packaging line.

All these examples may be accompanied by the use of automatic equipment to collate the cartons and erect/pack the trays or corrugated fibreboard packaging. When the transit packing is completed, the packs are palletised, sometimes automatically.

The specification of all packaging components must take account of any special environments involved in the distribution, storage and merchandising at the point of sale.

Typical examples of special environments are those for frozen food (−40°C to −20°C) and chilled food (0°C to +3°C). An aid to monitoring that these products are not exposed to higher temperatures exists in the form of temperature monitors. At its simplest, this comprises a colour patch which can change colour if the temperature rises above specified limits.

In practice, there are several ways of ensuring a satisfactory carton performance in distribution. One of the considerations is the level of moisture resistance required. Moisture condenses

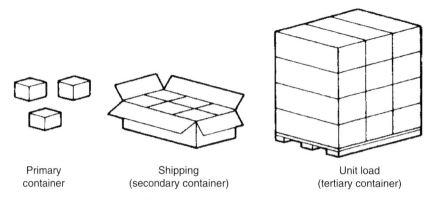

Figure 10.39 Packaging for distribution and storage. (Courtesy of the Packaging Society.)

on the surface of cartons when they are removed from the cold environment. Moisture will also affect cartons of chilled foods as they are stored in an environment with high RH.

Paperboard absorbs moisture on the surface and through cut edges. The effect is visual and results in distortion of the surface and loss in strength. Surface treatments, such as printing and UV varnishing, slow down moisture adsorption. Hard sizing also slows down moisture adsorption. Plastic coatings can be applied to one or both sides of the paperboard. Low-density polyethylene (LDPE) is the most commonly used plastic. High-density polyethylene (HDPE) has a better moisture vapour barrier. If additional performance needs are required, then there are additional choices. PET or PETE provides additional product resistance and can also be used where the product is reheated in microwave and conventional radiation-heated ovens. PP is satisfactory for reheating in convection steam-heated ovens. PET (PETE), PP and HDPE have good grease/fat resistance. All these plastics are heat sealable, and this property is frequently made use of by forming trays with the plastic on the inside. These trays can be lidded with peelable film or plastic-coated paperboard.

Care must be taken with exported goods which may be containerised and which pass through extremely warm conditions during transit. In these instances, hot-melt adhesives used to erect and/or close cartons must be replaced on the packaging machines because hot-melt adhesives can soften and the adhesion fails at high temperatures.

The hazards associated with distribution comprise:

- shock, for example due to dropping
- compression – both static, slow rate of loading, and dynamic, fast rate of loading
- vibration in transport causing destabilisation of the pallet and product damage (FOPT, 1999).

These factors can be studied in the laboratory. Safety factors are applied to static-compression loading results because, in practice, the box compression of the carton depends on:

- the structural design
- the direction of loading
- whether contents support the carton, as with a bottle or jar, or not, for example breakfast cereals in pouches
- the type of transit pack, for example corrugated fibreboard case

- the storage, palletisation, stacking and climatic conditions
- paperboard properties, such as grammage, thickness, moisture content, stiffness and short-span compression strength.

Complaints of carton damage need to be investigated carefully to find the real cause of the problem, because some examples of transit damage would not have been avoided even if the paperboard used to make the carton had been twice as thick and therefore much stronger. Everyone wants to reduce packaging, and clearly a balance must be achieved where the carton is strong enough without being open to criticism of overpackaging. This is best achieved with practical tests and by considering the whole packaging system, as it may be better to improve a situation by changing the specification of the transit pack or pallet arrangement rather than the specification of the carton.

A vast amount of work has been reported on the subject of relationships between paperboard properties and observed box compression strength (Hine, 1999, pp. 111–139). An example of this research is that carried out by Fellers et al. (STFI, 1983). This research showed that the compression of a panel, minimum size 60 mm × 90 mm, is described by the relationship

$$F_p = c\sqrt{F_c \sqrt{S_{MD} \times S_{CD}}}$$

where F_p is the panel compression strength; F_c is the compression strength (short span) in the direction of loading; S_{MD} is the MD stiffness; S_{CD} is the CD stiffness and c is a constant. $\sqrt{S_{MD} \times S_{CD}}$ is known as the geometric mean stiffness. It is recognised as an important strength-related feature of paperboard.

Under certain conditions and for a range of panel sizes, it was found that the constant c has a value 2π or 6.28. On the basis of that, for a complete box of four panels, the compression would be: $F_B = 4 \times F_p$

$$F_B = 4 \times F_P = 8\pi\sqrt{F_c \sqrt{D_{MD} S_{CD}}}$$

This shows that the box compression strength is dependent on paperboard stiffness and the short-span compression strength. Figure 10.40 shows the relationship between measured panel compression strength (N) and the predicted value based on stiffness and short-span compression using this equation, and the agreement is very good.

The importance of this work, in the author's opinion, is not that it is predictive, within limits, with respect to box compression, but that it shows that stiffness and short-span compression are important paperboard performance-related properties.

The short-span compression is measured by compressing a sample which is only 0.7 mm long. If a longer sample length is chosen, as would be the case with a tensile test, then under compression it would merely bend and buckle. When 0.7 mm is compressed, the failure point occurs when the fibres slide in relation to each other. This is an interesting phenomenon, bearing in mind the fibrous structure and interfibre bonding. The range in length of the fibres present in any given sample is also relevant, bearing in mind that the thickness of hardwood fibres is around 1.0 mm and softwood 3.0–3.5 mm.

The compression strength is two to three times lower than the tensile strength, and prior compression in this way does not affect the tensile measurement. This discussion explains how creasing and folding can occur, given the internal stresses withstood by fibres in tension and compression in close proximity. The various factors which are involved in studying compression strength are shown in Fig. 10.41.

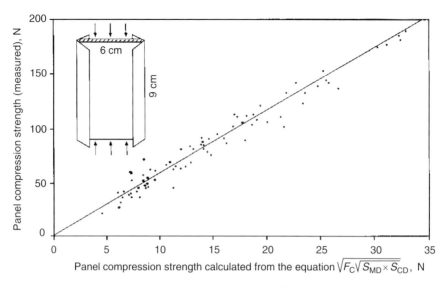

Figure 10.40 Panel compression strength, measured versus calculated (Fellers, de Ruvo, Htun, Carlsson, Engman and Lundberg). (Courtesy of Iggesund Paperboard.)

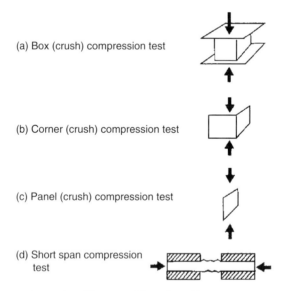

Figure 10.41 Compression testing. (Courtesy of Paperboard.)

10.7 Point of sale, dispensing, etc.

The consumer eventually takes possession of the cartonned product. As noted in Section 10.1, this can take place in a number of ways depending on the product and the intended market. In some situations, for example self-service retailing, the appearance of the pack is extremely important – a damaged or faded carton is unlikely to be purchased and an attractive carton may result in an impulsive purchase. In some situations, such as in the dispensing of a prescription medicine, the carton may play no part in the transaction, but nevertheless the brand owner would still want the carton to have a hygienic, quality image. Hence the printed appearance should be maintained, and the carton should be strong enough with an adequate box compression strength, good rub resistance, etc.

In the supermarket, the transit pack should be a conveniently handleable unit, easy to open and recycle. The cartons should be easy to merchandise, i.e. easy to stack and display, at the point of sale. The integrity of the pack is important to demonstrate that the product has not been tampered with.

An important consideration for the consumer today is that the product is genuine, i.e. not a counterfeit product. Several techniques are available to detect this (*Tobacco Asia*, 2010):

- Printing an identification on the carton using a clear, transparent varnish containing an ingredient which is only visible under UV or infrared (IR) illumination.
- Use of hot-foil stamping to give cartons a special visual characteristic.
- Use of holographic designs which show different images at different angles of observation.
- Use of microtext fonts digitally printed as small as 0.01 in. or 0.25 mm high requiring magnification to be read.
- Use of a non-toxic colour change coating which can be applied to a product or packaging item. When exposed to a low-power CO_2 laser this can make appropriate text apparent.
- Use of a heat-inducible colour change ink – in this the original impression/text changes colour when heated and may in fact disappear altogether.
- Incorporating a clear mark in the paperboard that is similar to a watermark. A recently published system can provide a mark which may be either visible to the naked eye or only visible under UV illumination. Other techniques which can be applied to paper and paperboard to produce a unique observable feature include the incorporation of coloured synthetic fibres.
- 'Fingerprinting' the approved paperboard using near infrared (NIR) spectroscopy (*Tobacco Reporter*, 2002). Every paperboard has an individual spectrum depending on the ingredients used in its manufacture and which will remain the same unless any of the ingredients are changed.
- Use of radio frequency identification (RFID) labels on pallet loads and transit packs (*Paper Technology*, 2003a). RFID labels are discussed in Section 4.5.3. This is a rapidly developing area mainly depending on reduction in the cost of the electronic components.

10.8 Consumer use

Quality aspects of packaging which are particularly noticeable to the consumers generally relate to such features as the ease of handling, any form of damage including security against pilferage, ease of opening and ease of reclosure (where relevant). If the reclosure uses a tuck-in-flap, then the durability of the lid-hinge crease is critical. The print should not be rubbed when handled – wet rub is a critical requirement for cartons used for frozen food, chilled food and ice cream. The printed instructions should be easy to read and unambiguous.

Convenience in the use of packaging is a feature that consumers respond to and manufacturers seek to provide. One of the more technically innovative developments which provide convenience in several ways has been the development of the PET (or PETE) extrusion-coated paperboard tray for microwaveable and radiant oven reheating of frozen and chilled

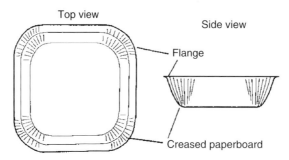

Figure 10.42 PET-lined paperboard deep-drawn tray with flange for applying lid. (Courtesy of Iggesund Paperboard.)

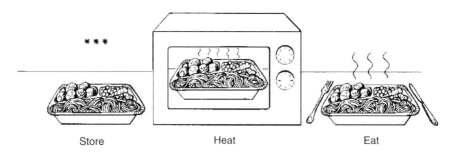

Figure 10.43 PET-lined paperboard trays stored in deep freeze and reheated in microwave oven. (Courtesy of Iggesund Paperboard.)

convenience ready meals at temperatures up to 200°C. To achieve this, several technologies have been brought together in a 'system'.

The extrusion coating process has already been described. For best results, the coating is applied to the reverse side of SBB which can be formed into a tray by one of two methods. In using the first method, a tray format can be cut and creased and formed on a tray erector, by either heat-sealing or interlocking corners. A leakproof tray-erected heat-sealed, web-cornered style is also possible (Fig. 10.5). Alternatively, the tray can be formed by deep drawing using metal tooling (Figs 10.6 and 10.42). Depths up to 25 mm can be formed in one operation and if the paperboard is moistened, a second draw to a depth of 45–50 mm can be achieved with heat and pressure, as noted in Section 10.3.2.

These tray designs can incorporate flanges to which plastic film or plastic-coated paperboard can be applied with peelable seals. Initially, microwave-heated foods provided convenience and rapid reheating (Fig. 10.43). The system could not brown or provide a degree of crispness, i.e. until a way was found to overcome the deficiency. This was achieved by including a susceptor inside the pack. Susceptors work by absorbing microwave energy which is made available to the food in the vicinity, causing localised browning and crispness. The susceptor is made from aluminium-metallised PET, (PETE), film. It is also made using Inconel (nickel/chromium), which can induce even higher temperatures (ASTM, 2003).

Another example of innovative paperboard packaging is provided by the following 'intelligent packaging' application (*Paper Technology*, 2003b). In this example, an SBB, chosen for its high security against cracking when the creases are folded, was printed with a conductive ink containing an embedded microchip, antenna and electronic circuitry. The application was a pharmaceutical blister pack. The microchip detects the removal of a pill, records the time of the event and can be programmed to bleep when the next pill is due to be taken. In addition, there is a row of buttons which enable the patient to enter feedback on the side effects of the pill. The information, which is encrypted, is stored in the chip, which includes the internet address to which the information can be sent. When the pack is empty, it can be scanned and the information downloaded to a PC. Alternatively, the data can be retrieved on a doctor's computer and viewed on screen.

Procedures are operated by paperboard manufacturers, carton makers and end-users to ensure that the consumer's needs in terms of product safety are met, particularly where the paperboard is in direct contact with, or in close proximity to, food, or other flavour- or aroma-sensitive product.

Two of the best-known authorities in this field are:

- USA, Food and Drugs Administration (FDA)
- Germany, Bundesgesundheitsamt (BgVV) (German Federal Institute for Consumer Protection and Veterinary Medicine Regulations).

In the USA, there are also the CONEG (Confederation of North Eastern Governors) Regulations which set limits for the content of lead, mercury, cadmium, chromium and nickel in packaging materials, such as paper and paperboard, in direct content with food.

There are particular regulations which apply to certain products, for example in Europe EN71 Safety of Toys Part 3 ISO 8126 sets limits for the migration of certain elements which are applied to packaging. There are requirements for plastics in the EC Plastics Food Packaging Directive 2005/79/EC and subsequent amendments.

The Confederation of European Paper Industries (CEPI) Food Contact Group includes CITPA (The International Confederation of Paper and Board Converters in Europe), CITPA (Paper and Board Converters) and CEPIC (Suppliers of Chemicals). This group developed a report which was published in March 2010 by CEPI and CITPA entitled *Industry Guidance for the Compliance of Paper & Board Materials and Article for Food Contact*. This is a voluntary route to obtaining compliance with current European food contact legislation (CEPI, 2010). (The substances permitted for use in paper and board conforming with this Guideline are given in BfR (Bundesinstitut für Risikobewertung) Recommendation XXXVI 'Paper and Board for Food Contact'.) This is an evolving subject and readers are advised to seek the latest position at any given time.

The protection of products from loss of taste, flavour or aroma is critical for some products. Chocolate confectionery, tea and tobacco are all particularly sensitive in this respect, and paperboard and paperboard cartons are regularly evaluated to ensure that they meet customer specifications in respect of odour and taint. There are several potential sources of contamination. These comprise:

- synthetic binders used in mineral-pigmented coatings
- pulp – chemical, mechanical and recycled: whilst the fibres comprise cellulose fibre which is tasteless and odourless, some fibres contain additional materials; there is the possibility of the oxidation of residual organic fatty acids to aldehydes and internal microbiological activity
- ink and varnish residues, such as residual solvents, products arising from oxidation–polymerisation of drying oils and unreacted components of radiation curing
- contamination as a result of pallets and conditions in transit and storage.

310 Handbook of Paper and Paperboard Packaging Technology

The human sensations of taste and smell are most sensitive. These faculties are used to test paperboard and printed cartons organoleptically. Panels of observers whose taste and smell faculties are normal test the samples for odour and taint. There are procedures for choosing panel members, and there are comprehensive standards issued by the main Standards organisations, for example International Organization for Standardization (ISO), ASTM International (formerly known as the American Society for Testing and Materials), European Committee for Standardization (CEN), British Standards Institution (BSI) and German Institute for Standardization (DIN). Topics covered include sensory testing and analysis. Many new and updated standards have been issued since 1995, and an up-to-date search is recommended. The main methods involve Triangular and Pairs testing for difference in both taint and odour evaluation as well as methods which allocate scores quantitatively, for example DIN 10955 for which a new standard was issued on 1 June 2004 (also known as the 'Robinson' test).

In addition, gas chromatography is carried out. Samples of the headspace of jars containing fixed amounts of paperboard conditioned for a fixed period at a fixed temperature are passed through a gas–liquid chromatograph (GLC). This separates the various components, indicates their relative volumes on a chart recorder (Figs 10.44 and 10.45) and measures

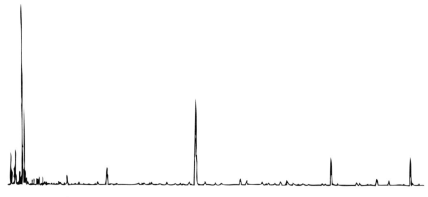

Figure 10.44 Chromatogram of unprinted paperboard. (Courtesy of Iggesund Paperboard.)

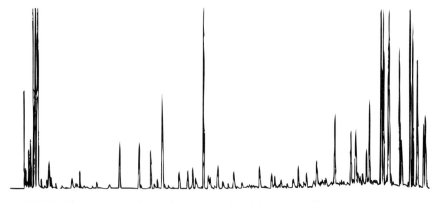

Figure 10.45 Chromatogram of paperboard printed with low-odour offset litho ink. (Courtesy of Iggesund Paperboard.)

their concentration in the material. It is possible to split the contents of the column and have an observer check the smell against the indications on the chart recorder. It is also possible to pass the eluted material from the gas chromatograph to a mass spectrometer (MS) and thereby identify the actual volatile materials arising from the original sample.

Meeting consumer packaging needs is the responsibility of packer/fillers, carton manufacturers and the manufacturers of the raw materials used in carton manufacture.

10.9 Conclusion

The folding carton has been around from the 1880s, and whilst traditional in concept it is the subject of constant innovation with respect to:

- increased productivity at all stages of manufacture and use
- manufacture and specification of paperboard, inks, varnishes, plastic coatings, adhesives, etc.
- surface and structural design
- printing and all conversion processes
- packaging machinery and methods, including product handling
- meeting new market needs and opportunities
- taking account of societal needs in respect of the environment, product and consumer safety.

References

ASTM, 2003, *Standard Test Methods for Temperature Measurement and Profiling for Microwave Susceptors*, ASTM F874-98, American Society for Testing and Materials International, West Conshohocken, Pennsylvania.
Atlas Die, 2012, *Reverse Cut Scores*, visit http://www.atlasdie.com/prod_result.php?name=Reverse Cut Score
Bernal, 2012, *Bernal Die Systems*, visit http://www.bernalrotarydies.com
Bobst, 2010, *Bobst Accubraille System for Folder-Gluers, for the Application of Braille Embossing to Folding Cartons*, visit http://www.bobstgroup.com
CEN (European Committee for Standardisation) 2010, EN 15823:2010 entitled *Packaging – Braille on Packaging for Medicinal Products*, visit http://esearch.cen.eu/
CEPI, 2010, *Industry Guidance for the Compliance of Paper & Board Materials and Article for Food Contact*, visit www.cepi.org
ECMA, (European Carton Makers publication) 2005, *Braille on Folding Cartons*, visit http://www.ecma.org
ECMA, 2009, *4th Edition of the ECMA Code of Folding Carton Design Styles*, visit http://www.ecma.org
FCI, 1988, Computer-based carton testing made easy, *Folding Carton Industry*, January.
FCI, 1996, Carton blank testing, *Folding Carton Industry*, January/February.
FOPT, 1999, *Fundamentals of Packaging Technology*, Institute of Packaging, Herndon, Virginia, pp. 361–380.
Hanlon, J.F., Kelsey, R.J., and Forcino, H.E., 1998, *Handbook of Package Engineering*, 3rd edn., Technomic Publishing Co., Lancaster, Pennsylvania, p. 175.
Hine, D., 1999, *Cartons and Cartonning*, Pira International. Indocomp, 2004, ACT II Carton testing, visit www.indocomp.com
Indocomp Testing Systems, 2012, *Automatic Carton Blank Testing System III*, visit http://www.indocomp.com
Packaging, 2004, Rotary screen printing, *Packaging Magazine*, 7(3), 22.
Packaging News, 2008, Iggesund launches bioplastic coated invercote, 5 December, visit http://www.packagingnews.co.uk

Paper Technology, 2003a, RFID radio tagging, 44(10), 3–4, December.
Paper Technology, 2003b, An SBB with ideal folding properties for ePackaging, 44(10), 5, December.
Pfaff, M., 1999, Die styles, cutting methods give converters options, *Paperboard Packaging*, November.
Pfaff, M., 2000, Choosing the method of die style, *Paperboard Packaging*, November.
PharmaBraille, 2009, *PharmaBraille News*, G. Steel 15/04/2009, visit http://www.pharmabraille.com
PharmaBraille, 2010, *PharmaBraille News*, G. Steel 21/08/2010, visit http://www.pharmabraille.com
PharmaBraille, 2011, visit http://www.pharmabraille.com
Pro Carton, 1999, *Carton packaging Fact File*, p. 5.
Roth, L. and Wybenga, G.L., 1991a, *The Packaging Designer's Book of Patterns*, 1st edn., John Wiley & Sons, Inc., New York.
Roth, L. and Wybenga, G.L., 1991b, *The Packaging Designer's Book of Patterns*, Van Nostrand Reinhold, New York, pp. 34, 71, 143, 279.
STFI, 1983, *Carton Board – The Profitable Use of Pulps and Processes*, Fellers, C. *et al.*, Swedish Forest Products Laboratory, Stockholm, Sweden.
Tobacco Asia, 2010, Allen Liao, Chinese Manufacturers adopt multiple anti-counterfeiting strategies, Q3 2010, visit http://www.tobaccoasia.net
Tobacco Reporter, 2002, Identification of paperboard using NIR, December.
Tobacco Reporter, 2008, David Willians, A look at who is who in the cigarette making and packing machinery business, October, visit http://www.tobaccoreporter.com

Suggested further reading

Boardtalk Trouble Shooting, published by M-real Corporation.
Cartons and Cartonning by Dennis Hine, published by Pira International.
Folding Carton Industry, visit http://www.brunton.co.uk/html/publishing/fci/fci_editorial.asp
Fundamentals of Packaging Technology by Walter Soroka, revised UK edition by Anne Emblem and Henry Emblem, published by The Institute of Packaging.
Paperboard Packaging, visit http://www.packaging-online.com/
Paperboard Reference Manual, published by Iggesund Paperboard.
Pfaff, M.R. Folding carton production in paperboard packaging in January, March, July and November 2002, visit http://www.packaging-online.com/packaging-online-author/michael-r-pfaff
Pharmaceutical Braille, 2010, G. Steel, *Pharmaceutical Braille News*, 21 August 2010, visit http://www.pharmabraille.com
The Carton Packaging Fact File, published by Pro Carton, UK.
The Packaging User's Handbook, ed. F.A. Paine, published by Blackie & Son Ltd. under the authority of The Institute of Packaging.

Websites

Atlas dies, http://www.atlasdie.com/
Bernal dies, http://www.bernalrotarydies.com/
British Carton Association, www.britishprint.com/sigs/bca.asp
Chesapeake corporation, http://www.chesapeakecorp.com/
European Carton Makers Association (ECMA), www.ecma.org
International Association of Diecutting and Die Making, http://www.iadd.org/
National Paperbox Association (USA), www.paperbox.org
Packaging Machinery Manufacturers Institute (USA), www.ppmi.org
Paperboard Packaging Council (USA), www.ppcnet.org
Pira International, now SmithersPira, at www.smitherspira.com
Pro Carton, www.procarton.com
Processing & Packaging Machinery Association (UK), www.ppma.co.uk

11 Corrugated fibreboard packaging

Arnoud Dekker

InnoTools Manager, Smurfit Kappa Development Centre, Hoogeveen, The Netherlands

11.1 Introduction

11.1.1 Overview

Corrugated fibreboard is the most popular packaging in the world for the storage and distribution of goods.

Corrugated fibreboard is two outside papers with a corrugated shaped paper in between. The outer papers are known as liners and the corrugated shaped paper is known as fluting or medium. Corrugated board can also be produced with two or three layers to increase its strength.

A large variety of packaging can be made from corrugated fibreboard. In this chapter, an overview of the mainstream applications will be given. Traditionally, corrugated fibreboard is used for the secondary or transport packaging. With the introduction of microflutes and high-quality printing, it is also used for the primary or product unit packaging.

Primary and retail display corrugated packaging is also used for advertising and brand promotional support. The package is a communication medium carrying information and artwork, and printing quality has been developed to meet these promotional needs. The attractiveness of the print can be decisive in catching the customer's eye.

11.1.2 Structure of corrugated fibreboard

Corrugated board is a light, high-performance material, with a structure that can be compared with modern materials used in the aerospace, automotive and construction industries. All these materials are designed to have a higher bending resistance than their individual components due to their created thickness. This is also the reason why it is worthwhile to form the flute and glue the two liners to the tops of the flute instead of simply gluing the three papers together.

Handbook of Paper and Paperboard Packaging Technology, Second Edition. Edited by Mark J. Kirwan.
© 2013 John Wiley & Sons, Ltd. Published 2013 by John Wiley & Sons, Ltd.

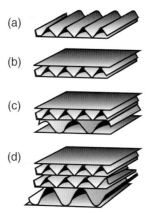

Figure 11.1 Corrugated boards: (a) single face, (b) single wall, (c) double wall and (d) triple wall. (Source: Smurfit Kappa.)

Table 11.1 Typical values for flute profiles and resulting board thickness

Flute type	Flute pitch (mm)	Take-up factor	Board thickness (mm) incl. papers
K	12	1.6	6.5
A	9	1.5	5.0
C	8	1.45	4.2
B	6.5	1.35	3.0
E	3.5	1.25	1.7
F	2.4	1.25	1.2
G	1.8	1.25	1.0
O	1.25	1.15	0.7

Corrugated fibreboard is two outside papers with a corrugated shaped paper in between. As already noted the outer papers are known as liners, the corrugated shaped paper is known as fluting or medium and this construction can be produced in two or three layers to increase its strength.

If one liner is used, the product is known as 'single faced' (Fig. 11.1a). If two liners are used, one on either side of the fluting, the product is known as 'single wall' (Fig. 11.1b). The combination of two flutings and three liners is called 'double wall' (Fig. 11.1c), and the combination of three flutings and four liners is called 'triple wall' (Fig. 11.1d).

The fluting shape is formed using heat, moisture and pressure by corrugated shaped rolls. This shape defines the flute profile.

Technically, any flute profile size can be used for making corrugated fibreboard. Historically flute profile sizes are indicated with a single letter to define their pitch, the number of flutes per unit length and the take-up factor. The pitch is the distance between two fluting tips. The take-up factor defines the length of the fluting (medium) paper used in a corrugated fibreboard structure compared with the length of the liners (Table 11.1).

Corrugated board is normally made in one of the following flute types, i.e. K, A, C, B, E, F, G and O. The most commonly used flute types are B, C and E.

It is generally known that with the same three papers, C flute gives a higher stacking strength than B flute and that B flute gives a higher stacking strength than E flute. More

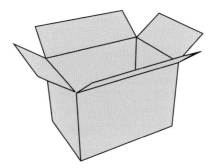

Figure 11.2 Regular slotted container. (Source: Smurfit Kappa.)

overall board thickness gives a higher stacking strength. Higher flutes are far more effective with respect to paper consumption than low flutes. On the other hand, the print quality is better for flute with a lower pitch than with a higher pitch. Hence E flute has a better print quality than B flute and B flute has a better print quality than C flute.

Corrugated fibreboard is a structure which can be designed and optimised for its intended use. There is a wide choice of papers for each liner and fluting. Also the flute type or flute types and their combinations can be chosen. Like so many aspects of design, there is no one single best solution. It is a matter of optimising the often conflicting requirements and prioritising them for a particular set of circumstances.

11.1.3 Types of corrugated fibreboard packaging

Due to the enormous difference in board properties between lightweight microflute single wall board and very heavyweight triple-wall board, the packaging is available in a wide variety of designs and sizes. The smallest packaging designs will be used for the primary packaging of small items, such as perfumes, electronics, hamburger clamp shells, etc. The largest packaging designs will be for the bulk packaging of granulates, liquids (bag-in-box), etc. containing up to a tonne of product.

The most commonly used corrugated fibreboard package is the case (box or carton) with a rectangular cross section together with top and bottom flaps. All flaps are the same length from the score (crease) to the edge of the flap. Typically, the major flaps meet in the middle and the minor flaps do not. This is known as a regular slotted container (RSC) as shown in Fig. 11.2.

The introduction of die-cut equipment gave complete freedom of packaging design. Considerable ingenuity is possible in pack design to produce one-piece or multiple-piece packaging to meet specific market needs. Several examples are shown in Fig. 11.3 wrap-around case, Fig. 11.4 corrugated fibreboard tray, Fig. 11.5 carry-home pack for six bottles of wine, Fig. 11.6 full depth divider for bottles in case and Fig. 11.7 point-of-purchase display stand. Corrugated fibreboard is also used in the form of pads and fittings to locate and protect vulnerable product components.

The most widely known box style library used in Europe is the European Federation of Corrugated Board Manufacturers and the European Solid Board Organisation (FEFCO-ESBO) code. Box styles are defined by a four-digit code specified by FEFCO: e.g. an RSC is coded 0201. FEFCO styles are normally the basis for more complicated special designs that incorporate, for instance, locking tabs or internal fittings.

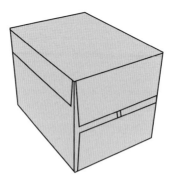

Figure 11.3 Wraparound case. (Source: Smurfit Kappa.)

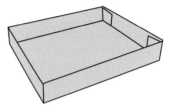

Figure 11.4 Corrugated fibreboard tray. (Source: Smurfit Kappa.)

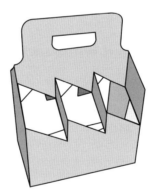

Figure 11.5 Carry-home pack for six wine bottles. (Source: Smurfit Kappa.)

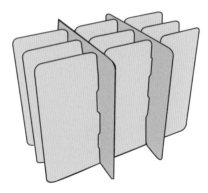

Figure 11.6 Full depth divider for bottles in case. (Source: Smurfit Kappa.)

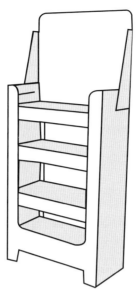

Figure 11.7 Point-of-purchase display stand. (Source: Smurfit Kappa.)

Generally, the most common printing used with corrugated fibreboard packaging is flexo – either as post-print, i.e. after the corrugated board has been made, or as pre-print where it is printed, reel to reel, prior to use on the corrugator. However, other print processes are used appropriately, e.g. offset litho for the high quality required with some, usually branded, higher value, retail packaging, or silk screen for short runs of point of purchase (POP)-display packaging. High-quality printed self-adhesive labels may also be used.

11.1.4 History of corrugated fibreboard

In the mid-nineteenth century, an ingenious concept enabled flimsy sheets of paper to be transformed into a rigid, stackable and cushioning form for the packaging of delicate goods in transit.

Corrugated (also called pleated) paper was patented in England in 1856 and used as a liner for tall hats, but corrugated boxboard was not patented and used as a shipping material until 20 December 1871. The patent was issued to Albert Jones of New York City for single-sided (single face) corrugated board. Jones used the corrugated board for wrapping bottles and glass lantern chimneys. The first machine for producing large quantities of corrugated board was built in 1874 by G. Smyth, and in the same year Oliver Long improved upon Jones' design by inventing corrugated board with liner sheets on both sides. This was corrugated board as we know it today.

The Scottish-born Robert Gair invented the pre-cut paperboard box in 1890 – flat pieces manufactured in bulk that folded into boxes. Gair's invention came about as a result of an accident: he was a Brooklyn printer and paper-bag maker during the 1870s, and one day, whilst he was printing an order of seed bags, a metal ruler normally used to crease bags shifted in position and cut them. Gair discovered that by cutting and creasing in one operation he could make prefabricated paperboard boxes. Applying this idea to corrugated boxboard was a straightforward development when the material became available in the early twentieth century.

The corrugated box was initially used for packaging glass and pottery containers. Later, the case enabled fruit and produce to be brought from the farm to the retailer without bruising, improving the commercial return to the producers and opening up export markets.

Corrugated fibreboard was born from a new way of using paper and from the increasing necessity to pack and protect goods. Thanks to its basic raw materials, and despite considerable changes, modern corrugated packaging is not so different than that used by our great grandfathers. This ingenuous construction is and will remain profitable, modern and innovative.

Since the end of the nineteenth century, many changes have occurred and remarkable progress has been made in the improvement of raw materials, in the equipment, in the production processes and the printing techniques of corrugated packaging (source: www.fefco.org, www.wikipedia.org).

Some examples are listed in the following:

- The number of paper grades used for the production of corrugated fibreboard is continuously increasing. The choice and the quality of all the different liners and fluting is improving constantly.
- Production speed increased dramatically with the developments in equipment. This is also true on the packaging user side, thanks to high speed case erectors, fillers and closers.
- The use of computers has revolutionised the industry by permitting continuous runs and avoiding machine stoppages. Its impact was also considerable on packaging design and order processing. Information technology will continue to drive these applications in computer use.
- In the last decade, new printing techniques have brought the biggest changes. The changing role of packaging from logistics to marketing functions and the use of bar codes for product identification have both required an enhancement in the quality of graphics for corrugated packaging.
- Small flutes and high-quality paper allow for a very high standard of graphics, which leads to new opportunities in end-use markets.

11.2 Functions

11.2.1 Overview functions

Packaging engineers have to design a corrugated package that has to meet the particular needs of the whole package cycle. An example for retail is given in Fig. 11.8. The product is packed in a corrugated package, the package is stacked on a pallet, and the pallet goes via a distribution centre to the supermarket. The product will be unpacked and placed on the shelf or the product will be shown in a display box or ready for the consumers to buy. The package is disposed of and collected. From the recovered paper and virgin fibres, paper is made. This paper is then used for the production of a new package. This completes the cycle.

This is only one example. Other uses and distribution chains are different and have to meet different requirements, e.g. business-to-business sales, postal service and mail order.

Every step in the cycle places its requirement on the package.

11.2.2 Corrugated fibreboard packaging production

The corrugated board package should be designed so that is can be produced cost-effectively. There is a large variety of printing and converting equipment available.

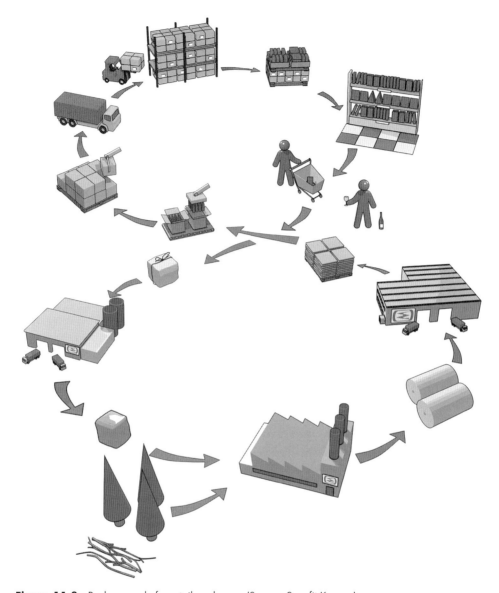

Figure 11.8 Package cycle for retail packages. (Source: Smurfit Kappa.)

11.2.3 Packing lines

There is a very large choice of equipment which can be used for erecting corrugated board packaging and filling and closing it. If a packaging end-user wants to use existing erection and closing equipment, its specific demands must be taken into account at the package design stage.

11.2.4 Palletisation and logistic chain

The main function of a corrugated paperboard box is to contain and protect packed goods during storage and distribution. The growing use of palletisation in warehousing and

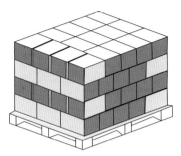

Figure 11.9 Structure of the pallet. (Source: Smurfit Kappa.)

transportation requires that corrugated boxes have good stackability. Specific commercial or in-house developed software is available to design the best layout/position of the corrugated boxes on the pallet with respect to utilisation of the volume and overall stability. The resulting palletisation plan can then be specified and implemented as shown in Fig. 11.9.

11.2.5 Communication

Most corrugated fibreboard boxes require printing to identify the contents, provide legal and regulatory information, and bar codes for routing. Boxes that are used for marketing, merchandising and point-of-sale often have high graphics to help communicate the contents. Some boxes are designed for display of contents on the shelf. Others are designed to help dispense the contents.

11.2.6 Retail-ready

A large proportion of the corrugated fibreboard packages will be sent to the retailers. Hence, the packaging has to meet the requirements of retailers, especially for in-store logistics. These are mostly specific for the retailer, due to their positioning, e.g. hard-discounter and hypermarket. Several guidelines have been developed to specify the functionality of the package, like publications of the Efficient Consumer Response Community (ECR). These focus on the easy recognition, opening, handling and disposing of the package in the supermarket and assisting sales of the packed product. These have been stated as the 'Five Easies':

1. Easy to identify.
2. Easy to open.
3. Easy to shelf.
4. Easy to dispose.
5. Easy to shop.

11.2.7 Product safety

The basic function of corrugated board packaging is the same as for any packaging, namely to protect products during distribution until the product is removed from the package. It may also protect the environment from the product – e.g. in the distribution of dangerous goods. The shipping hazards depend largely upon the particular logistics system being employed. For example, boxes unitised into a unit load on a pallet do not encounter individual handling whilst boxes sorted and shipped through part of their distribution cycle as mixed loads or express carriers can receive severe shocks, drops, etc.

Food safety and quality are also of paramount importance to consumers, manufacturers and regulators. Food safety is a scientific discipline describing handling, preparation and storage of food in ways that prevent potential health hazards. For all these packages, there are clear national and international regulations, e.g. EC (2004), Council of Europe (2002), CEPI (2010), BfR (2011) and French Decree (1992).

11.2.8 Recycling and sustainability

All corrugated fibreboard is recyclable, additions like staples, tape, tear strips, etc. are removed during the pulp preparation in the paper mill. All additions to the boxes should be checked, if they would cause problems for the recycling process.

11.3 Board properties and test methods

11.3.1 Overview of board properties and test methods

Most board properties can be measured with defined test methods. These test methods are present in the International Organization for Standardization (ISO), FEFCO or Technical Association of the Pulp and Paper Industry (TAPPI) standards. ISO is the world standard, where FEFCO and TAPPI are European and American standards. A comparison table can be found on the FEFCO website.

11.3.1.1 Board weight

The board grammage is the weight of one square metre of corrugated board (unit is gm^{-2} and measured under standard conditions: 23°C and 50% RH). It can be calculated from the paper grammages using a simple linear formula. For a single wall corrugated fibreboard, the formula is:

$$\text{Board weight} = \text{outer liner} + (\text{take} - \text{up factor} \times \text{fluting}) + \text{inner liner} + \text{glue}$$

with all the papers in $g\ m^{-2}$. In North America, weight per unit area is reported in pounds per 1000 ft^2 for corrugated board.

11.3.1.2 Thickness

The thickness is measured between a plunger and an anvil with 20 kPa. ISO 3034:2011 specifies a method for determining the single sheet thickness of corrugated fibreboard.

11.3.1.3 Flat crush strength

The main function of the fluting is its ability to retain its structure and its geometry. The traditional flat crush test (FCT) makes it possible to evaluate the performance of the fluting. Flat crush values depend on the shape of the flute and the quality of the fluting. Normally, semi-chemical fluting has a higher performance than normal recycled fluting for a similar grammage. The board sample is tested according to ISO 3035 for flat crush, as illustrated in Fig. 11.10.

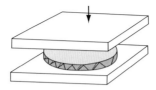

Figure 11.10 Flat crush corrugated board testing. (Source Smurfit Kappa.)

As a first approximation, there is a good relationship between the FCT of the board and the CMT (Concora Medium Test – ISO 7263) strength of the flutes.

The force displacement curve of the FCT has a typical shape. There is a first peak, a second and then there is the maximum force. The first peak is the most important of the curve. It is the hardness of the board which is needed for the support of the liner. The maximum force is independent of the first peak. There is no direct relationship with box performance.

11.3.1.4 Bending stiffness

Rigidity or bending stiffness relates to the force required to deflect a flat specimen of corrugated board. There are several methods for measuring bending stiffness, referred to as the two-point, three-point and four-point method. All these methods are described in the ISO 5628:2012. Bending stiffness is the main property which relates to the performance of the use of corrugated board packaging.

11.3.1.5 Edge crush test (ECT)

The edge crush test (ECT) is used to evaluate the compression strength of the corrugated board. The ECT of a corrugated board sample is tested according to ISO 3037-2007 or TAPPI 811 om-11. The board sample is mounted vertically, with flutes running vertically, between horizontal platens.

Different sample sizes, cutting shapes, cutting techniques and knife blades can change the measured value significantly. The main problem with the ECT test is the sensitivity to differences in the cutting quality of the samples. Large differences (>20%) between laboratories are not uncommon.

The ECT test is not dependent on the thickness of the board but depends mostly on the compression strength in the cross direction (CD) of the papers used and the gluing quality of the board. A combination of paper strength parameters can be made and used to predict ECT. Several formulas are found in literature, all based on combinations of SCT_{CD}, RCT or CCT.

SCT_{CD} is the short compression test in the CD of the paper.
RCT is the ring crush test in the CD of the paper.
CCT is the compression crush test strength in the CD of the fluting. (Note: CCT is not Concora Compression Test – a similar shaped sample is used as for Concora Medium Test (CMT) but the load is applied totally differently. The load is applied at the edge of the paper. The test is not used very often, but it is similar to the ECT test.)

11.3.1.6 Bursting strength

The measurement of burst strength is described in ISO 2759:2001. Burst strength is a commonly used measurement for classifying the quality of the corrugated board. Depending on

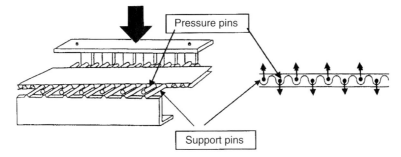

Figure 11.11 Pin-adhesion test. (Source: Smurfit Kappa.)

the grammage and the nature of the liners, i.e. whether they are made from virgin, recycled or blends of virgin and recycled fibres, a large range from 800 to 8000 kPa can be measured.

11.3.1.7 Puncture

Corrugated cases can be damaged by puncturing during distribution both internally by movement of the contents and externally by impact with sharp objects. Puncture resistance of the corrugated board is a requirement specified by some national specifications (France, Germany and Spain). The puncture test measures the energy required to penetrate the corrugated board. Standards for this test method are ISO 3036:1975, FEFCO TM5 and TAPPI T803.

11.3.1.8 Pin-adhesion test (PAT)

The pin-adhesion test (PAT) is used to evaluate the bonding between the fluting and the liners. It measures the force, in newtons per metre, required to separate the fluting from the liner. Pins are inserted between the flutes and the force required to break down the adhesion is measured, as indicated in Fig. 11.11. There are two different test methods, TAPPI T821 om-12 and FEFCO TM11, which can be used to evaluate the adhesion between the fluting and the liners.

11.3.1.9 Moisture content

The sensitivity of the paper to moisture variations is very important. The procedure for moisture measurement is by a gravimetric moisture analysis. This procedure is used to determine the moisture content of a sample of the corrugated board. It is achieved by the removal of water from a pre-weighed sample in an oven. The calculation of moisture content is described as follows:

$$\text{Moisture content} = \frac{\text{wet sample} - \text{dry sample}}{\text{dry sample}}.$$

11.3.2 Box tests

11.3.2.1 Internal dimensions

The size of a box can be measured for either internal (for product fit) or external (for handling machinery or palletising) dimensions. Boxes are usually specified and ordered by the internal dimensions.

11.3.2.2 Box compression test (BCT)

Stackability is best measured using the box compression test (BCT). BCT is the top-to-bottom compression strength. Several standards have been published, e.g. FEFCO, ISO, AFNOR, etc. In ISO 12048:1994, BCT measurement is made on filled boxes. In the FEFCO method, the measurement is made on an empty box. The application of a method for measuring BCT, mean and standard deviation is defined in FEFCO Testing Method No. 50.

The results depend on the method used due to the differences in empty versus with content, flaps fixed or loose and differences in testing speed.

11.3.2.3 Drop test

Drop testing is carried out according to ISO 2248:1985 standards. The height of the drop is increased incrementally during the test and the percentage of boxes damaged at each height is evaluated. The evaluation of corrugated board packaging is performed with 8 and 15 kg of packed sand, and there are different modes of drop. The case can either be dropped horizontally or on an edge. The potential energy of the box in drop testing is the product of the weight of contents and the drop height.

For other properties, such as resistance to high humidity and damage due to toppling of the box, other test methods are available.

11.3.3 Pallet tests

The complete pallet with all the boxes can be evaluated for the requirements of the logistic chain. Several testing schemes are available in the ISO, International Safe Transit Association (ISTA) and ASTM standards.

Depending on the logistics chain and logistics requirements, the pallet can be subjected to several test procedures, such as vibration on a vibration table, impact in an inclined plane test (Fig. 11.12), subjection to a sustained compression load (Fig. 11.13), rotational flat drop, etc.

11.3.4 Predictions

The prediction of the box performance is to reduce the time and effort to determine the specification of the corrugated board to be used and the box design to meet a required performance. Several approaches are used for these predictions, e.g. the empirical, statistical and finite element methods.

The most widely known formula which links the box compression strength with the mechanical properties of its components is the one devised by McKee (McKee et al., 1963):

$$\text{Box compression strength} = K \times ECT^a \times FS^b \times D^c$$

where D is the dimension of the box perimeter; FS is the flexural stiffness of the board; ECT is the ECT of the board and K, a, b and c are empirical constants.

For given box dimensions, the BCT of the corrugated board package depends both on the ECT and the FS of the combined board. Although this is an empirical approach, the FS term in the McKee formula is obtained from classical thin plate theory. (Thin plate theory is the classic approach for evaluating raw material performance in mechanics in order to predict the bending stiffness of composite materials, i.e. sandwiches.)

Corrugated fibreboard packaging **325**

Figure 11.12 Inclined plane test. (Source: Smurfit Kappa.)

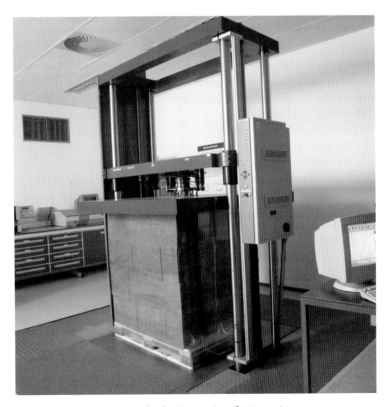

Figure 11.13 Sustained compression load. (Source: Smurfit Kappa.)

The calculation or measurement of the FS has been found difficult and hence a simplified formula has been made:

$$\text{Box compression strength} = K' \times ECT \times \text{thickness}^d \times D^c$$

where D is the dimensions of the box (perimeter); ECT is the ECT of the board; thickness is the thickness of the board and K', d and c are empirical constants.

11.4 Manufacturing

11.4.1 Overview

There are four main stages in the production of corrugated fibreboard packaging.

It starts with the production of the papers. Then the corrugated fibreboard is made after which the sheets of corrugated fibreboard are printed and converted into the packaging, which is delivered to the end-user product manufacturer. The final step is at the end-user product manufacturer's site, where the product is packed manually or with (semi-)automatic packing lines.

The paper is manufactured in paper mills. For the corrugated fibreboard and packaging production, there are two main types of production, i.e. either in integrated plants or in sheet plants. The integrated plant has both the corrugated board and packaging production. Sheet plants are restricted to the manufacture of corrugated board sheets, which are then transported to sheet-converting plants where the sheets are printed and converted into packaging.

The demand of high-quality printed corrugated fibreboard packaging has led to other production processes, e.g. pre-printing and litho-laminating.

Pre-printing of paper is an additional production step before the corrugated board is produced. This provides a higher quality of printing, although post-printing, i.e. the printing of corrugated board, has improved continuously by, e.g. the installation of infrared drying equipment. The difference between pre-print and post-print quality has been largely reduced.

Corrugated fibreboard boxes are manufactured in a corrugated board plant or in a sheet feeding plant. The corrugated board plant consists of the corrugator which produces flat sheets of corrugated fibreboard and of the converting equipment where the corrugated sheet is converted into corrugated board boxes by printing, cutting, scoring (creasing) and gluing (or possibly, taping or stitching). These latter processes are known as the converting operations. Corrugated fibreboard in sheet form is also sent to smaller factories known as sheet feeder plants, where they are converted into packaging for short run length orders, quick deliveries and for very specific local markets.

11.4.2 Paper production

Papermaking is a very capital-intensive industry driven with a large focus on productivity. The largest papermaking machines are over 10 m wide, 200 m long and can produce over 1500 km of paper a day. The papermaking process is explained in Chapter 1.

The four basic paper types used for corrugating (recycled and virgin, liner and fluting) all have their unique selling points. The virgin fibre-based paper is a world market and the recycled-based paper is a regional market.

The main function of liners is:

- to absorb and deflect static and dynamic forces such as tension, impact, pressure and bending
- to have a smooth surface for obtaining good printing
- to have coarse reverse side which helps the fluted webs and liner papers to adhere firmly together.

Liners can be white or brown.

Kraftliner is mainly produced in the grammage range of 115–440 gm^{-2}. Kraftliner contains at least 70% wood pulp, the rest is recycled fibres. The long fibres in the kraftliner give it good all round performance characteristics, ideal for packaging that is subject to rigorous demands (high levels of moisture, hazardous goods, bulk containers, etc.).

Testliner is mainly produced in the grammage range of 100–300 gm^{-2}. It contains shorter fibres and is built up with a thicker lower layer (bearing layer), couched with a high-quality liner web. Testliner is best suited for supply chains with no extreme demands ('dry' conditions, average strength, etc.).

Fluting has to be stiff and elastic at the same time, because the main function is to sustain the structure of the corrugated board. It provides cushioning from vertical pressure but has to have sufficient elasticity to be shaped into waves.

Semi-chemical fluting is produced in the grammage range of 115–275 gm^{-2}. The long virgin fibres give great rigidity and resistance to impact. This paper is ideal for packaging that is subject to rigorous demands (high levels of moisture, hazardous goods, bulk containers, etc.).

Recycled-based fluting is produced in the grammage range 75–175 gm^{-2}. Only good-quality recycled raw materials are used to achieve the required performance with these shorter fibres. Recycled-based fluting is best suited for supply chains with no extreme demands ('dry' conditions, average strength, etc.).

Papers are the building blocks of corrugated fibreboard. The properties of corrugated board are determined by the quality and features of the paper:

- mechanical properties
- optical properties
- production properties.

The mechanical properties of paper are strongly oriented. The properties in the machine direction are higher than the properties in the CD. Properties of paper are strongly dependent on fibre type and length. Additives are used to further increase the properties (especially for paper made from recycled fibre).

Papers based on virgin fibres have better performance at high environmental moisture levels.

For liners the following properties can be specified:

- SCT_{CD} (or RCT)
- burst
- moisture
- friction
- porosity

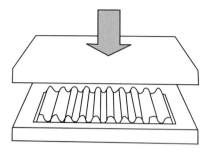

Figure 11.14 Concora Medium Test apparatus. (Source: Smurfit Kappa.)

- Dennison wax
- Cobb
- colour
- brightness
- roughness.

For fluting, the following properties can be specified:

- SCT_{CD}
- CMT liner
- moisture
- friction
- porosity.

SCT_{CD} is the short CCT strength in the CD. This test has been developed to overcome the problems with the RCT at low grammages. This test becomes unreliable at low grammages <150 gm^{-2}. Also the SCT_{CD} test is less labour intensive than the RCT.

The CMT is described in ISO 7263. It is a test of the compression of the paper after fluting in a fluting apparatus (A flute), shown in Fig. 11.14. The measurements made are CMT_0, i.e. the compression strength immediately after fluting, and CMT_{30}, which is the compression strength after 30 min conditioning at 23°C and 50% RH. The CMT_{30} give a more reliable lower value. The lower value is due to the longer reconditioning of the paper after the heating in the fluting apparatus. The CMT_0 is less reliable due to the impact of small time differences between the flute formation, tape applications and compression testing.

11.4.3 Corrugated board production

The production of the corrugated board is carried out in several stages in-line:

- Production of the single face corrugated board (Fig. 11.15): The fluting paper is conditioned with heat and steam to make it pliable enough to accept and retain the shape of the flute profile. The flute profile shape is pressed into the fluting paper either using two profiled corrugator rolls or, in the 'fingerless' process, by forming the fluting medium on one profiled roll using vacuum. After the corrugation of the fluting paper, starch adhesive is applied to the tips of the corrugations and the fluting paper is combined with the liner

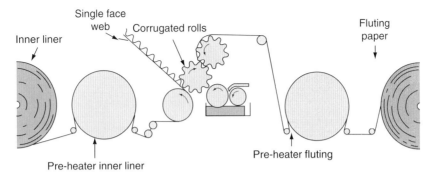

Figure 11.15 Production of single face corrugated board. (Source: Smurfit Kappa.)

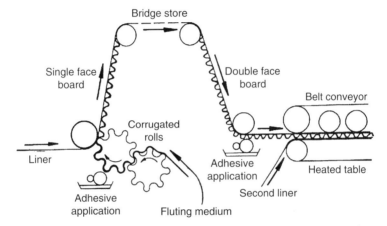

Figure 11.16 Production of single wall (double face) corrugated board. (Courtesy of the Packaging Society.)

which has also been conditioned to bring it to the same temperature and moisture content as that of the fluting paper.
- A second liner is applied at the double backer to produce single wall, or double face, corrugated board (Fig. 11.16).

After passing through a drying section, the board is matured and cooled before being slit to the required width and cut to the required length. Scores (creases) may also be applied to the board in the machine direction of the corrugator.

The speed of the single facer is more than that of the double backer, and the excess board accumulates in a bridge system between the two lining stations balancing the difference in speed.

In order to make a double-wall corrugated board, the machine would incorporate additional sections as shown in Fig. 11.17. In this process, two single face corrugated webs are formed in Zone A. The flutes of Single Face 1 are glued to the liner of Single Face 2 in Zone B. A liner for the flutes of Single Face 2 is applied to the flutes of Single Face 2 and the combined board is dried between heating plates in Zone C.

Starch adhesive preparation is an important element for the orderly running of the process. Essentially, a corrugating starch adhesive is a four-component system. It consists of a

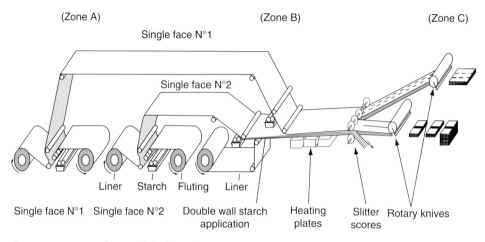

Figure 11.17 Production of double-wall corrugated board. (Source: Smurfit Kappa.)

carrier or cooked starch component, a raw starch component, caustic soda and borax, all prepared in water. All ingredients are mixed together in a 'kitchen'.

Most corrugators are two knife corrugators, which means that they can produce two different sheet lengths side-by-side. This leads to an optimisation problem, known as the cutting stock problem.

11.4.3.1 Flatness of corrugated fibreboard

Corrugated board sheets often exhibit curvature, also referred to as warp or curl, which can cause great difficulty in subsequent converting operations and in box set-up in the customer's facility. Hence, flatness or the avoidance of warp is a major consideration in the corrugating industry. Since warp varies inversely with board thickness, thin boards like F and E flute are much more prone to warp than thick boards such as C flute. As the production of these thinner flutes and corrugator speed have both increased dramatically, warp has become a much more important issue. There are different forms of warp, normal warp or curl in both directions and twist warp.

Warp reduction at the end of the corrugator exit, or delivery, is one of the main preoccupations of the production staff. Warp can have a serious effect on the printing, converting and use of corrugated board. Many studies have been made to investigate the parameters in the manufacture of corrugated fibreboard which are responsible for warp phenomena.

11.4.3.2 Wash-boarding

Using low grammage papers in the outer liner creates in some cases, but not consistently, a wash-boarding phenomena, as indicated in Fig. 11.18. In post-print, i.e. printing on corrugated board, it is difficult sometimes to obtain good results, especially when it is required to print an illustration in solids and halftone or a bar code.

11.4.4 Corrugated fibreboard converting

The flat sheets of corrugated board coming off the corrugator are transferred to the converting operations where the main machines are printer/slotters, die-cutters and folder

Figure 11.18 Wash-boarding effect of print on corrugated board surface with low grammage white top double face. (Source: Smurfit Kappa.)

gluers, depending on the corrugated product being produced. The printing is explained in Section 8.3.5

11.4.4.1 Flexo folder gluers

Blanks for RSCs from the corrugating machine are further converted on flexo folder gluers. Special design profiles are cut and scored on a die-cutter/scorer. Machines are also available, which incorporate many typical features of corrugated box printing and conversion in one in-line machine (Fig. 11.19).

The following features are indicated in Fig. 11.19:

- A is the feeder which can handle a wide range of board thickness, from thin microflute to double-wall corrugated board.
- B is a four-colour flexographic printing unit.
- C is a rotary die-cutter. A die-cutter is necessary for the production of sophisticated box designs, i.e. boxes with features other than simple straight line scores and slots.
- D is a slotting unit with waste removal.
- E is a folder-gluer unit where the side seam is completed.
- F indicates two-process touch-control screens which are used to set up and operate the machine.
- G stacks the boxes in bundles and ejects them from the machine in a controlled way.

Features of such a machine are the automation of the adjustments, i.e. the set-up, required when changing from one box design to another, attention to the removal of board dust generated during cutting and noise reduction.

The manufacturer's joint is most often joined with adhesive but it may also be taped or stitched.

11.4.4.2 Die-cutting

There are two types of die-cutting: flat bed and rotary. For the flat bed die-cutting, a sheet is gripped by several clamps and pulled under a flat die. The flat die is pressed down with several tonnes of pressure. After the die is lifted, the die-cut sheet is pulled to the next position, where trim is separated from the packaging design. The advantage of using a flat die is the accuracy of the cutting and creasing operation.

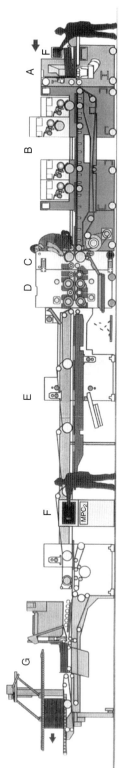

Figure 11.19 Martin combined in-line rotary die-cutter, slotter, flexo folder gluer. (Reproduced with permission from Bobst S.A.)

With rotary die-cutting, the die is a drum with another drum acting as anvil. After the sheet is passed through the die drum and anvil drum, the sheet is vibrated to eject the trimmings. The accuracy in the CD of the die is good, but in the length direction slip of the sheet reduces the accuracy. Developments in modern rotary die-cutters have increased the accuracy of cutting in the length direction.

Die-cutting has provided design freedom and is the main driver for the increase of die-cutting compared with the flexo folder gluers with their limited range of possible designs.

11.4.5 Corrugated fibreboard printing

In addition to its primary role to protect and transport products, corrugated board packaging is increasingly being used as an inexpensive advertising medium. Therefore an increasing proportion of boxes are printed and also the number of colours used in printing is increasing.

The printing techniques used for corrugated board are:

- Flexographic printing
  - Basic post-print flexo
  - Standard post-print flexo
  - High-quality post-print.
- Pre-print (flexographic printing).
- Litho-laminated offset.
- Digital printing.
- Screen printing.

The application of the different techniques is dependent on the graphical requirements, on the print quality and on the order quantity (see Fig. 11.20).

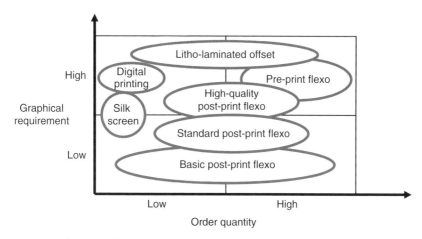

Figure 11.20 Relationship between print quality, printing technique and order quantity. (Source: Smurfit Kappa.)

11.4.5.1 *Flexographic post-print*

Flexographic printing is a 'letterpress' process. This means that the printing sections of the plate are slightly higher than the non-printing sections. The ink is transferred by a rubber roll to the Anilox roll, which distributes the ink evenly. The Anilox roll transfers the ink to the printing plate and the printing plate transfers the ink to the corrugated sheet.

11.4.5.2 *Flexographic pre-print*

A roll of paper is printed before being laminated on different board grades. Pre-print flexo is cost effective when printed on large and medium runs of corrugated boxes with demand for high-quality printing. It is possible to print on coated papers with varnish. A roll of paper is printed in a separate factory before being laminated on different board grades in a 'normal' integrated corrugated factory.

11.4.5.3 *Litho-laminated offset printing*

Litho-laminating is the process of laminating offset printed sheets of papers on single face material. Afterwards the sheets are mostly die cut to make the packaging. All the process steps are done in one factory and used to produce all types of consumer packages.

Offset printing is a method which reproduces documents and images on sheets of coated paper which are subsequently pasted onto a sheet of corrugated board. This printing system consists of applying an oil-based ink on a metallic sheet, generally made from an aluminium alloy. This takes up the ink in the areas where there is a hydrophobic (water rejecting) composition, and the rest of the sheet is covered with water as a repellent against the ink; the image or text is pressed onto a rubber blanket and is finally pressed onto the paper. The fact that the ink goes directly onto the paper is the main reason for the high quality of this printing system.

Offset gives photographic print quality, but it is a more expensive technique, though it has lower initial costs. The printing is done on heavy coated papers like GD2, where UV varnish is also possible.

11.4.5.4 *Digital printing (inkjet)*

Digital printing is a direct printing process which prints directly using digital data, i.e. from texts and images created with electronic design or auto-editing programmes. Unlike conventional printing processes, digital printing does not require intermediate processes such as films, plates, phototypesetting, etc. One of the main advantages of digital printing is that there is no limit to the minimum order quantity but high volume output may be a problem.

11.4.5.5 *Screen printing*

Screen printing is a technique which reproduces documents and images on a sheet of corrugated board. The technique consists of transferring an ink through a gauze or mesh stretched on a frame. The ink is prevented from entering areas where there is to be no image through an emulsion or varnish, leaving free the area where the ink will be transferred. The printing system is repetitive: once the first model has been achieved, the printing can be repeated hundreds and even thousands of times without losing definition. Screen prints can differ as each colour has a certain relief and in addition the colours have a high degree of tonality.

11.4.6 Customer packing lines

Corrugated board packaging (RSCs and die-cut blanks) are packed manually, semi-automatically or fully automatically. The selection of loading method will depend on the product, the package, the line speed and the capacity needs. The packages can be either erected and then filled and closed or formed around the product and closed. In general, the filling line speed increases from 5 to 10 packages per minute to 30 packages per minute in a semi-automatic line and on an automatic line at around 1 pack per second. To ensure packing line efficiency, the corrugated board blanks and cases have to meet certain requirements of flatness, structural stability and suitability for closure.

11.4.6.1 *Closure of corrugated cases*

There are different closure possibilities:

- cold glues, e.g. polyvinyl acetate (PVA) emulsion
- hot-melt
- tape (paper, film or reinforced paper with various adhesive systems)
- strapping (plastic or metal)
- stitching (metal wire based).

11.4.6.2 *Cold glue application*

The important characteristics of a synthetic cold glue, such as a water-based dispersion of PVA, are the solids content, open time and setting time.

11.4.6.3 *Hot-melt application*

Corrugated board gluing is very often carried out using hot-melt glues. Heat is used and the adhesive is applied to the package by extrusion. As compared with vinyl adhesives where the paper absorbance of water is very important, the quality of corrugated board glued with hot-melt depends on the paper surface porosity in order to achieve micro-penetration of the adhesive into the paper surface.

11.4.7 Good manufacturing practice

The International Good Manufacturing Practice Standard for Corrugated and Solid Board was published in October 2003 by the FEFCO and the ESBO. This is concerned with the manufacturing of corrugated and solid board in a controlled environment as far as quality, hygiene and traceability are concerned. For details see FEFCO website.

11.5 Corrugated fibreboard and sustainability

Sustainability becomes more and more important. A decade ago, only recycling was on the agenda, whereas today sustainability is about social equity, environmental protection and economic prosperity.

The best-known description of sustainability is 'Development that meets the needs of the present without compromising the ability of future generations to meet their own needs' (Bruntland report, 1987).

But what does this mean for secondary packaging, mostly corrugated fibreboard, in the modern supply chain? How can it be translated into packaging language, meaningful figures and everyday operations? The purpose of this paragraph is to show the implications for and discussions within the supply chain around these questions and to give a brief overview of activities and challenges in the area of sustainability.

Many different stakeholders are involved. Retailers, with the products they choose to sell and promote, are seen by governments as an ideal platform to reach millions of consumers, enabling them to have more sustainable consumption. Likewise, retailers, by means of the products they choose to sell, are also in a position to communicate sustainability requirements to producers and even further down the supply chain to producers' suppliers. To create trust and credibility towards consumers, many retailers and producers ask Non-Governmental Organisations (NGOs) to support them in measuring their sustainability performance (Smurfit Kappa, 2011).

Why is secondary packaging on the agenda? Packaging is a relatively easy target – it is one of the most visible, specific products for consumers to easily understand. Consumers see recycling of packaging as key for sustainability and they are also concerned about what is perceived as too much packaging and overpacking.

The direct impact of secondary packaging on the environmental footprint is only 4%. Yet it is very high on the agenda today, as it is relatively easy for packaging users and specifiers to change and control.

The aim of a company must be to minimise their total impact per unit of product delivered to the customer. Recommended reading is the publication from ECR, (Efficient Consumer Response) Europe entitled *Packaging in the Sustainability agenda: a guide for corporate decision makers*, (ECR, 2009).

Key packaging sustainability issues are:

- Minimising total impact → focus on all packaging functionalities and materials
- Per unit of product → look at the interaction of package (secondary and primary) and product
- Delivered to the consumer → consider the total supply chain

This means that the key focus points for sustainable secondary packaging in the current supply chain are:

- Reduction of over packing
- Increasing recycling rate
- Reduction of carbon emission, water and energy consumption
- Protection of natural resources: forest, water
- Reduction of waste going to landfill
- Improvement of the supply chain

There are several elements in focus.

11.5.1 Sustainable sourcing of raw materials

It should be ensured that the raw material for corrugated fibreboard packaging is sourced in a controlled and sustainable way. Paper, which is the raw material for corrugated boards, is produced from two types of fibres:

1. Virgin fibres – virgin mills can obtain their timber from Forest Stewardship Council (FSC) or Programme for the Endorsement of Forest Certification (PEFC) Chain of Custody certified sources. This provides a credible guarantee that the produced virgin papers are originating from well-managed forests or controlled woodland.
2. Recycled fibres – produced from the used packaging recovered from the closed loop between retailers and industrial users of corrugated packaging. For the European paper packaging industry, the recycling rate is over 75% – the highest rate of all packaging materials. Fibres are reused seven to eight times.

The raw material for corrugated boards is the ultimate sustainable resource. These recycled paper mills can also be FSC recycled certified.

11.5.2 Sustainable production

As well as fibres, energy and water are also needed to produce paper:

- Energy – paper production is energy intensive but not carbon intensive. A substantial share of the total energy consumption comes from green energy (biomass fuel).
- Water – more than 90% of water used is cleaned and reused many times on-site.

Besides paper, many valuable by-products are utilised for other applications, leading to an amount of waste going to landfill of less than 1%. A significant part of these by-products are used to create biomass fuel. The core of the industrial activity is the re-circulation of used material into a valuable new raw material.

11.5.3 Sustainable packaging design

The goal of the design process is to develop packaging solutions that are both functional and attractive whilst keeping the impact on the environment as low as possible. The graph in Fig. 11.21 shows the direct impact of secondary packaging on the environmental footprint per unit of product delivered to the consumer. Overpackaging is a waste of packaging

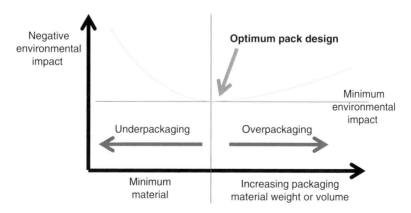

Figure 11.21 Optimal pack design. (Source: Smurfit Kappa.)

material, and underpackaging will lead to product loss, which mostly has a much larger environmental impact than the secondary packaging.

11.5.4 Sustainable supply chain

Sustainable packaging alone is not the only target. We work on the role of packaging to make the supply chain more sustainable. The primary and most vital function of secondary packaging is to protect the products it contains. Without proper corrugated packaging, a huge proportion of products would never reach the consumer – the wastage would be enormous with corresponding negative impact on the sustainability. Besides that, corrugated secondary packaging contributes in reducing the environmental impact related to:

- distribution chain
- retailing
- collection of post-used packaging
- reusing, recycling and recovering
- disposal.

The way forward is by close cooperation with all the stakeholders to understand the entire supply chain of any specific product, and that will enable the creation of sustainable transport packaging that minimises total impact per unit of product delivered to the consumer.

References

BfR, 2011, German Recommendation BfR XXXVI on papers and board materials and articles to come into contact with foodstuffs, the German Federal Institute for Risk Assessment, 1 January 2012, pp. 1–16.

CEPI (2010), Industry guideline for the compliance of paper and board materials and articles for food contact (developed by the European paper and board food packaging chain). Published by CEPI (Confederation of European Paper Industries) and CITPA (International Confederation of paper and Board Converters in Europe), Brussels, Belgium, March.

Council of Europe, 2002, Resolution AP (2002) 1 and its technical documents. Documents part of the Council of Europe's policy statements concerning paper and board materials and articles intended to come into contact with foodstuffs.

EC, 2004, Regulation (EC) No 1935/2004 of the European Parliament and of the Council on 27 October 2004, published in the *Official Journal of the European Union*, November 13, pp. L 338/4–L 338/14.

ECR, 2009, *Packaging in the Sustainability agenda: a guide for corporate decision makers*, ECR Europe & EUROPEN (2009), visit Publications, ECR EUROPE at http://www.ecr-all.org

French Decree, 1992, French Government Decree no 92/631 related to material and objects intended to come into contact with foodstuffs, 8 July, pp. 4–8.

McKee et al., 1963, Compression strength formula for corrugated boxes, *Paperboard Packaging*, 48(8), 149–159.

Smurfit Kappa, 2011, KMPG Sustainability presented the *Independent Assurance Report* in the *Smurfit Kappa Sustainability Report 2011*, published Dublin and London, June, p. 88.

Websites

www.smurfitkappa.com
www.fefco.org
www.fefco-esbo-codes.com (International Case Code)

www.wikipedia.com
http://en.wikipedia.org/wiki/Corrugated_box_design
http://www.ecr-all.org (see *Packaging in the Sustainability agenda: a guide for corporate decision makers*)
www.brc.org.uk (see *BRC Global Standard for Packaging & Packaging Materials (Issue 4)* – Interpretation Guideline)

Suggested further reading

Boxes, corrugated in Yam, K.L., 2009, *The Wiley Encyclopedia of Packaging Technology*, John Wiley & Sons, Inc., Hoboken, New Jersy.

12 Solid board packaging

Mark J. Kirwan
Paper and Paperboard Specialist, Fellow of the Packaging Society, London, UK

12.1 Overview

Solid board is primarily used for distribution and shelf-ready packaging. It is a rigid, puncture-resistant and water-resistant material which varies in thickness from 0.8 to 4.0 or 4.5 mm and in grammage from around 550 to around 3000 g/m^2.

Solid board is either made on a multi-ply paperboard machine forming with vats, Fourdrinier wires, or is made by a combination of forming methods. It is also made by multi-ply lamination. The higher thicknesses are always made by laminating two or more layers of paper or paperboard together to provide the strength necessary for the intended use. Additional materials, including polyethylene (PE), may be incorporated to meet specific performance needs. The board is cut into sheets, printed and converted into boxes, trays and other paperboard structures.

Solid board is used in a wide range of packaging and display applications, such as:

- packaging of meat, poultry, fish and horticultural produce in boxes and trays
- promotional displays in self-service merchandising
- games and games packaging
- shelf-ready packaging
- pallet interleaving slip sheets, pallet trays and caps for use in bulk distribution and storage
- fitments and divisions for glass and plastic bottles in corrugated cases
- pads to stabilise layers of stacked containers
- packaging of engineering products
- export packaging
- bag-in-box liquid containers.

In Europe, the largest market for solid board packaging is for poultry, which together with meat accounts for around 60% of the usage. Table 12.1 indicates the percentage of overall European market usage. However, the actual percentages of the usage of solid board in specific countries for each market will reflect the importance of the various markets in each country. Hence the proportion used, for example, in horticulture may be significantly higher than 15% in some countries and lower in others.

Table 12.1 Percentages for typical European market usage of solid board production

Markets	%
Horticulture	15
Meat	25
Poultry	35
Fish	10
Others	15

Source: Smurfit Kappa Corby.

Standard quality solid board is more resistant to water and damp conditions than standard corrugated board. The water resistance of solid board can be significantly improved during manufacture to meet the needs of differing wet environments. Performance needs can be met for packing and storage in frozen, chilled, ice-packed, wet or humid conditions. Higher levels of moisture resistance are achieved by internal treatment (sizing) of the board during board manufacture and by the use of kraft facing, or liner, extrusion coated with PE on one or both sides.

Inevitably, solid board packaging is compared with corrugated board packaging. At equal *grammage*, a case made from corrugated board will have a higher box compression strength as a result of its higher thickness.

At equal *box compression strength*, the solid board box will be heavier and as the weight of the material has a major influence on the cost, the corrugated board container is preferred.

The solid board container will, however, be specified where its strength, toughness, puncture and water resistance are essential for a satisfactory packaging performance in specific conditions of use, which can include rough manual handling in cold and wet environments.

Though both solid and corrugated board containers are normally thought of as one-trip packages, where it is feasible to specify multi-trip usage, solid board is a better choice. This is because it is less easily crushed than corrugated board, and where it is damaged, it is easier to repair with self-adhesive tape.

Solid board packaging is manufactured primarily from recycled material. It is both recyclable and biodegradable at the end of its useful life. It can be either collected from households and businesses, or from 'bring' systems, to be returned to a mill which uses recovered paper and paperboard to be recycled in the manufacture of solid boards.

12.2 Pack design

A wide range of packaging designs has been published in the International Fibreboard Case Code (www.fefco/ESBO.org). The European Solid Board Organisation (ESBO) has collaborated with the European Federation of Corrugated Board Manufacturers (FEFCO) in the preparation of the Code. It is a structured presentation of existing box designs with a code number assigned to each design. The Code is used worldwide and has been adopted by the United Nations. It has also been adopted by the International Corrugated Case Association (ICCA).

In addition to the designs in the Code, individual solid board packaging manufacturers can extend the design range with customised packaging solutions to meet specific market needs.

Solid board packaging is supplied flat to save space in storage and distribution. It can be erected by the packer manually or with mechanical assistance. Where product volume is

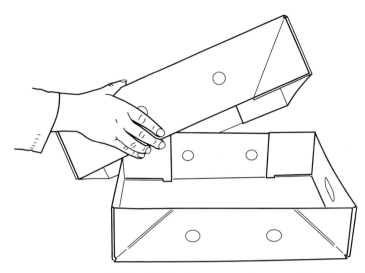

Figure 12.1 Separate base and full depth lid, both based on a four-point glued blank. (Reproduced with permission from Smurfit Kappa Corby.)

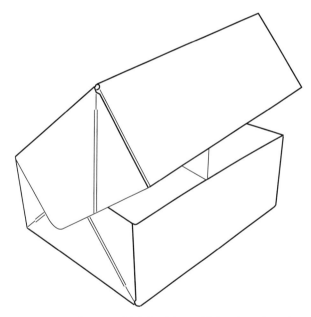

Figure 12.2 One-piece box with tray and full depth hinged lid and tray based on a six-point glued blank. (Reproduced with permission from Smurfit Kappa Corby.)

high, and a high packing speed is required, fully automatic machinery is used for erecting, packing and closing. The solution in any given application depends on the volume of usage and the packaging environment.

Typical packaging designs are as follows:

- separate base and full depth lid both based on a four-point glued blank (Fig. 12.1)
- integral full depth hinged lid and tray based on a six-point glued blank (Fig. 12.2)

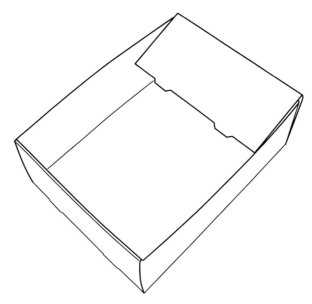

Figure 12.3 Double end wall self-locking tray based on a die-cut blank. (Reproduced with permission from Smurfit Kappa Corby.)

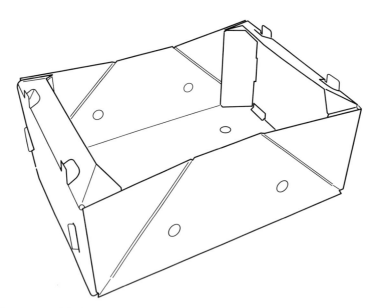

Figure 12.4 Tray with ledge at each end with stacking and holding features, based on an eight-point glued blank. (Reproduced with permission from Smurfit Kappa Corby.)

- double end wall self-locking tray (Fig. 12.3)
- tray with ledge at each end with stacking and holding features, based on eight-point glued blank (Fig. 12.4)
- stackable tray with clipped-in plastic corner supports (Fig. 12.5)
- tray with reinforced corners (Fig. 12.6).

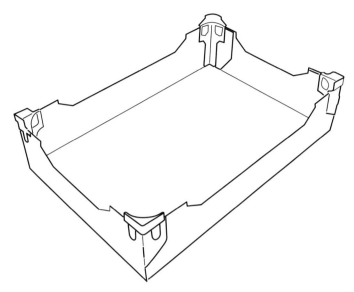

Figure 12.5 Stackable tray with clipped-in plastic corner supports based on a die-cut blank. (Reproduced with permission from Smurfit Kappa Corby.)

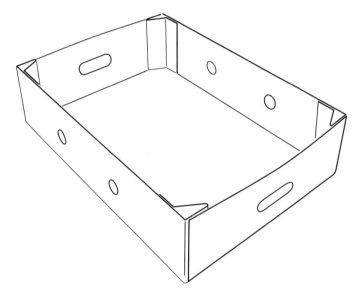

Figure 12.6 Tray with reinforced corners which is glued up at point of use based on a die-cut blank. (Reproduced with permission from Smurfit Kappa Corby.)

12.3 Applications

12.3.1 Horticultural produce

Fruits, vegetables and flowers are packed by the grower, or in cooperatives, for supply to the markets on a daily basis. The water-resistant qualities of solid board are important where the product is packed wet in damp conditions and to keep the product fresh during storage and transportation.

Flowers are frequently packed in lidded trays, i.e. cartons comprising a base and a lid. Fruits and vegetables are packed in machine-erected trays that include stackable features, which enable packs to be stacked 10 high on a pallet. As already noted, some products may be packed wet.

12.3.2 Meat and poultry

These products, frozen and vacuum packed, are packed in trays with full depth lids. Where PE-lined solid board is used, it must not be corona discharge treated as this would cause wet products to stick to the PE. Such designs can be glued using hotmelt adhesives. These packs can be used for a product weight range of up to 20 kg weight.

12.3.3 Fish

Fish may be packed wet, chilled or iced for distribution in a wide range of sizes, ranging from 3 kg (half-stone) to 25 kg (4 stone), base and lid. The trays need to be in a leak-proof style, in which the corner design is described as 'webbed'. Two-side PE-coated solid board with additional hard sizing provides the maximum protection for this type of packaging.

Double-walled designs give increased stacking strength to provide a real alternative to polystyrene fish boxes. Fish may be frozen in the box and the freezing time is less than it would be with a corrugated board box.

12.3.4 Beer (glass bottles and cans)

The wrap-around litho-printed multipack is the standard specification for this application.

12.3.5 Dairy products

Compartmentalised machine-erected trays are typically used for multipacks of plastic yoghurt pots. The solid board design is preferred as the corrugated board equivalent has a larger area on account of the thickness of the sides. This difference can be critical in supermarket shelf display.

12.3.6 Footwear

Solid board is used in shoe box packaging, alongside folding cartons and corrugated board packaging.

12.3.7 Laundry

In some European markets, large solid board cartons are used for detergent and soap powders.

12.3.8 Engineering

Products which are heavy, liable to be shifted in transportation, require moisture protection and possibly with protruding parts need the protection of a puncture-resistant solid board. These boxes can provide rustproof packaging to the Ministry of Defence (UK) Standard.

12.3.9 Export packaging

Solid board packaging is robust and both puncture and moisture resistant. Where boxes are to be shipped in containers, the box dimensions are designed to ensure optimum use of the available volume.

12.3.10 Luxury packaging

Litho-printed solid board is used for luxury packaging of products, such as chocolates, cosmetics and whisky. (Solid board as a material in sheet form is also used by rigid-box manufacturers, see Chapter 9.)

12.3.11 Slip sheets

Slip sheets can be used in place of wood or plastic pallet bases. The items are assembled on a sheet of solid board, known as a slip sheet, shown in Fig. 12.7. In order for a fork-lift truck to move such a unit load, it has to operate with a system, known as a Quick Fork Mount (QFM), which grips a protruding edge of the slip sheet so that it can pull, and subsequently push, both the sheet and the stacked load onto a platform attached to the truck (Walter, 1996).

For a unit load size of 1000 mm × 1200 mm, the protruding edge needs to be between 75 and 100 mm wide. This flap is pre-creased so that it can be easily folded against the side of an adjacent load. Depending on customer requirements, protruding edges can be specified on one side, two sides opposite, two sides adjacent, three sides or four sides.

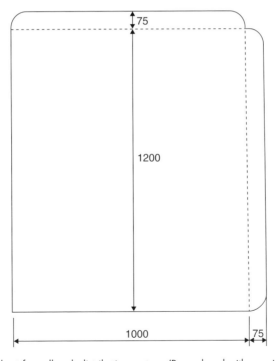

Figure 12.7 Slip sheet for pull-push distribution system. (Reproduced with permission from Smurfit Kappa Corby.)

Slip sheets can be made from three plies of kraft with a combined grammage of 960 g/m^2. They are also made using 100% recovered fibre at a grammage of 1200 g/m^2 and with kraft outer liners and recovered-fibre middle plies. The use of kraft provides a stronger sheet and one that is possible to reuse.

PE-lined board and hard-sized board can be specified for use in wet/damp conditions. The presence of PE improves load stability by increasing the coefficient of friction of the surface of the slip sheet.

An anti-skid varnish can be applied to the slip sheet to assist load stability, but this is only effective where a similar varnish is applied to the cartons forming the unit load.

Similarly, thinner sheets, without flaps, are used as layer sheets to assist the sweeping of cans, plastic and glass containers at the container in-feed on packing lines. It is superior to corrugated board for this purpose as it does not indent with the shape of the upturned container. Solid board sheets are also used as pallet top boards and as caps for pallets.

12.3.12 Partitions (divisions, fitments and pads)

These are interlocking pieces, divisions or partitions of solid board, which are used to create cells in a case or a box wherein individual items of packaging and packaged products can be placed. The solid board is cut and the fitments or divisions are interlocked and folded flat for storage and shipment.

Divisions used in this way protect the products by restraining them from movement rather than by cushioning. Typical examples, including sheets which fit over bottles to restrain movement, are shown in Fig. 12.8. Restraining in this way is particularly important for glass packaging, for example glass containers and bottles of wines, spirits, liqueurs, beer, etc. Divisions also protect labels on glass and plastic bottles from damage due to scuffing during transportation.

Divisions are also used in the packaging of pharmaceuticals, cosmetics, ceramics, confectionery, fresh produce and electronic and engineering components. Divisions can be erected

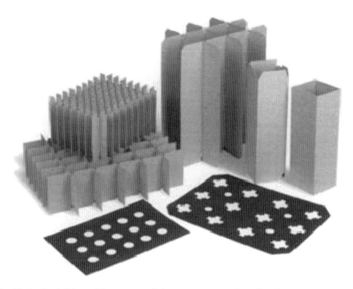

Figure 12.8 Typical solid board fitments and divisions. (Reproduced with permission from Smurfit Kappa Lokfast.)

Solid board packaging **349**

Figure 12.9 Divisions being inserted by hand. (Reproduced with permission from Smurfit Kappa Lokfast.)

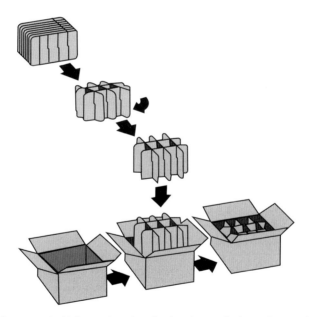

Figure 12.10 Sequences in high-speed packing for the glass and other industries. (Reproduced with permission from Smurfit Kappa Lokfast.)

and inserted by hand as shown in Fig. 12.9. They can also be erected automatically and inserted on high-speed packing lines running at 600 bottles/minute in a sequence shown in Fig. 12.10.

Whilst corrugated board partitions and dividers can also be used, they are less preferred as they are thicker and require larger dimensions in transit cases. Another important advantage of solid board dividers is that they can be cleanly cut and are, therefore, free from excessive edge dust or board slivers and trimmings.

Die-cut pads are made with solid board to locate containers, such as tubs of vegetable oil/fat spreads, stacked in layers in multilayer stacks.

12.3.13 Recycling boxes

Waterproofed boxes, with lids, can be used for the collection of waste paper and paperboard. They are used by some local authorities in the UK instead of plastic boxes. The boxes can be attractively printed to encourage their use and display the collection calendar dates. They can be used in a variety of situations, for example kerbside collection schemes and in offices, homes and at supermarkets. The boxes themselves are recyclable.

12.3.14 Bag-in-box liquid containers

Liquid containers comprising solid board packs with flexible plastic liners are available in sizes up to 10, 15 and 20 L.

12.3.15 Shelf-ready packaging

Solid board retail-ready packaging fulfils two roles. Firstly, it provides secondary packaging for the safe storage and distribution of multiples of primary packs of branded goods which are easily opened and, secondly, it is placed in self-service shelf displays. It is high quality printed to compliment the primary packs providing greater product identification and visual presentation. It improves the display of packs on the shelf making the individual packs more easily available to the shopper. Retail-ready packaging assists the identification of the product as it arrives in the store and prior to movement into the sales environment.

12.4 Materials

As already noted, solid board can be made on multi-ply paperboard machines. The higher thicknesses are made by lamination on a paster. This is a machine which takes reels of board, usually up to six reels in line, glue laminates them and cuts them into sheets at the end of the paster machine.

The middle layers are made from recycled paper fibre and are grey in colour. The moisture resistance is increased by hard sizing. Starch is used to increase strength.

In addition to the use of hard-sized middles, one or both outer surfaces can be laminated with PE-coated kraft to achieve the best water and water-vapour resistance. The typical weight of PE applied in this way is 15 g/m^2. White lined folding boxboards and solid bleached (white) boards can also be used as required.

12.5 Water and water-vapour resistance

Figure 12.11 shows the superior wet resistance of water-resistant solid board compared with standard solid board and corrugated board measured over a 24-h period.

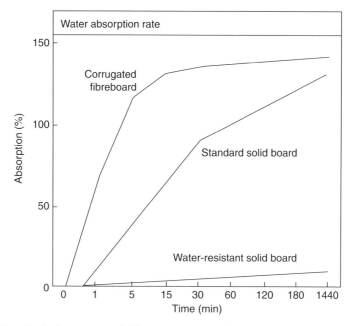

Figure 12.11 Graph showing typical differences in water absorption over 24 h for water-resistant solid board compared with standard solid board and corrugated board. (Source: Papermarc Merton Packaging.)

Table 12.2 Effect of humidity and water on the strength of typical solid and corrugated boards

	Corrugated board	Standard solid board	Water-resistant solid board
Effect of relative humidity on rigidity at 90% RH	60% loss of strength	40% loss of strength	20% loss of strength
Effect of water immersion on strength	80% loss after 5 min	80% loss after 60 min	10% loss after 60 min
Effect of 15 min water immersion on puncture strength	Reduced from 6 to 2.5 J	Reduced from 6 to 5 J	—

Source: Papermarc Merton Packaging.

Other key data demonstrate that water-resistant solid board is significantly better than alternative materials as shown in Table 12.2.

As already noted, the water resistance of standard solid board is increased by hard sizing the middle plies and by PE coatings on one or both outer surfaces.

12.6 Printing and conversion

12.6.1 Printing

Depending on the quality required, printing is carried out on sheets by either letterpress, offset litho or flexo.

12.6.2 Cutting and creasing

Three methods are available:

1. Rotary bending machine bends and slots simple designs, such as four-flap cases and slotted bases and lids.
2. Flatbed cutting and creasing forms using a make-ready.
3. Rotary cutting and creasing forms have the advantage of incremental cutting and creasing and hence do not need the high pressures used with an auto platen.

12.7 Packaging operation

Solid board packaging can be supplied either for easy make-up by hand or by machine.

12.8 Waste management

As noted, solid board cases are usually one-trip containers after which they can be recovered and the fibre recycled. Where special conditions apply, such as box cleanliness and appropriate return transport, solid board cases can be folded flat after use and returned to the packer for reuse.

12.9 Good manufacturing practice

The International Good Manufacturing Practice Standard for Corrugated and Solid Board was published in October 2003 by FEFCO and ESBO. This is concerned with the manufacturing of corrugated and solid board in a controlled environment as far as quality, hygiene and traceability are concerned. For details, visit the FEFCO website.

Reference

Walter, S., 1996, *Fundamentals of Packaging Technology*, revised UK edition, Anne, E. and Henry, E. (eds), Institute of Packaging, Lincolnshire, UK, p. 347.

Websites

http://www.fefco.org (for the International FECO ESBO Case Design Codes)
http://www.smurfitkappa.co.uk/

13 Paperboard-based liquid packaging

Mark J. Kirwan

Paper and Paperboard Specialist, Fellow of the Packaging Society, London, UK

13.1 Introduction

The story of paperboard-based packaging and systems for dairy products, food and beverages over the last century is a classic business case study. It contains all the elements from the concept to a successful worldwide business based on vision, motivation and entrepreneurial drive, utilising skills in science, technology and commerce.

Technically, the concept perceived around 1900 sought a leak-proof paper-based container to match the glass and metal containers in common use and to replace the traditional distribution of milk, whereby it was ladled out into the consumers' own ceramic containers from churns in the street. In this instance, according to Gordon Robertson's account of the early development of paperboard-based beverage cartons (Robertson, 2002, pp. 46–52), there appears to be no dispute about where the credit lies for the first successful solution. Though earlier attempts to solve the problem have been recorded, it was in 1915 that a US patent was granted to John Van Wormer of Toledo, Ohio, for what was described as a 'paper bottle' and which he called Pure-Pak®. It was a folded carton blank, which would be supplied flat to dairies for the packaging of milk. The concept offered several important advantages which still apply today – those of savings in delivery, storage and weight compared with glass bottles.

Glass packaging for milk was well established by this time with equipment for washing, filling and capping. There were systems for the distribution and recovery of the bottles. The machinery for forming, filling and sealing the paper bottle still had to be invented.

The carton blank had to be made into a container with the help of adhesives. It was then made leak-proof by dipping it in molten paraffin wax. The container was then filled and the top closed by heat, pressure and stapling, as described in the articles in *Modern Packaging* and *Popular Science Monthly* from the early 1930s. Clearly, considerable engineering development work was necessary to accomplish the forming, filling and sealing and to do this at reasonable speeds added to the complexity of what was required. Ex-Cell-O Corporation emerged as the owner of Pure-Pak® and the successful machinery producer. Milk was packed in Pure-Pak® cartons at 24 quart, i.e. 2-pint, cartons per minute in bulk from 1936.

Handbook of Paper and Paperboard Packaging Technology, Second Edition. Edited by Mark J. Kirwan.
© 2013 John Wiley & Sons, Ltd. Published 2013 by John Wiley & Sons, Ltd.

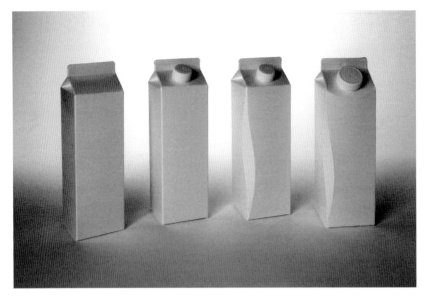

Figure 13.1 (Left to right) Pure-Pak® carton, Pure-Pak® with cap, Pure-Pak® curve, Pure-Pak® Diamond® curve. (Reproduced with permission from Elopak.)

Elopak (European Licensee of Pure-Pak®) was formed in 1957 and, with an agreement signed with Ex-Cell-O, started producing cartons in Norway. The first order was to supply milk cartons to the US forces in Europe.

In 1987, Elopak bought the Ex-Cell-O Packaging Systems Division and the Pure-Pak® Licence, and with developments in carton design has improved functionality, convenience and product protection, (Fig. 13.1).

There were alternatives launched in the USA – some ten companies had launched a paper-based package for milk, by the 1930s. In Germany, Jagenberg had introduced Perga, which was also waxed. It had a circular base and a square top. Perga developed quickly and by 1939, there were 26 factories in eight countries including Germany, England, Sweden, the USA, Canada and Australia. Jagenberg founded PKL in 1958 and introduced a package known as Blocpac.

Meanwhile, from 1943, a different approach was in development in Sweden. The company of Åkerlund and Rausing had been formed in 1929 to manufacture consumer packaging. This company developed Satello during World War II. This was a wax-coated cylindrical container produced during the time when glass and tinplate containers were in short supply. Satello was used to pack jams and marmalade. The company then set out to develop a milk package which would be formed from the reel by cutting and sealing a tube of moisture-proofed paperboard at alternate right angles below the level of the liquid. This produced a tetrahedral-shaped package and a patent was applied for in Sweden in the name of R. Rausing in 1944 (Leander, 1996a).

It did not have an auspicious beginning. Ruben Rausing afterwards said that it was 'a package (shape) which you only ever saw in geometry classes, to be manufactured by a machine of whose appearance nobody had even the faintest idea, made from a material which did not exist and intended for one of our most vulnerable foodstuffs'. Of the person who invented the package, Rausing said that 'this was what you get for hiring people who

Figure 13.2 Tetra Classic®. (Reproduced with permission from Tetra Pak Group.)

have no idea how a package is made and what it ought to look like' (Leander, 1996b), clearly, a justification of lateral thinking!

As with Pure-Pak®, the development of a forming, filling and sealing machine together with that of a suitable packaging material took several years. The name Tetra Pak was registered in 1950 and a subsidiary company of Åkerlund and Rausing, AB Tetra Pak, formed in 1951. Tetra Pak became an independent company under the ownership of Ruben Rausing in 1965.

Several plastic materials had been tried by Tetra Pak and a polystyrene blend had been successful. The first packaging machine was installed in a dairy in Lund, Sweden, in 1952. This produced a tetrahedral-shaped package. Today, it is known as Tetra Classic® (Fig. 13.2).

Polyethylene (PE) had been invented in England in 1933 by ICI and was given the brand name 'polythene'. Its use as a packaging material was not envisaged. Subsequently, DuPont with a licence to manufacture PE encouraged the firm of Frank W. Egan to develop an extrusion coating machine in 1954–1955. Tetra Pak began coating paperboards with PE in 1956 (Leander, 1996c).

A brick-shaped PE-coated paperboard pack was introduced by Zupack (subsequently bought by Tetra Pak in 1982) in 1959 and, slightly earlier, PKL had introduced Blocpac with a similar shape. The Pure-Pak® gable top from Elopak was being sold in Europe since the 1950s. These developments prompted Tetra Pak to develop a brick-shaped pack, and this was achieved by 1963. The main advantage of a brick-shaped pack called Tetra Brik®, with a square or rectangular cross section, is that it makes the best use of storage volume and is easier to handle mechanically in the dairies and in distribution (Fig. 13.3).

In addition to developments in machinery, materials and pack shapes, it was also recognised that it would be a major benefit to develop an *aseptic* paperboard-based packaging system for milk. Aseptic packaging requires the separate sterilisation of the package and the product for them to be brought together in an aseptic environment, i.e. one from which micro-organisms are excluded, and for the pack to be hermetically sealed. Sterilising food, particularly liquid food, in this way achieves its objective with a shorter exposure of the food to heat, thereby minimising the loss of nutritional content and flavour. This process would also result in an extended product shelf life. It was demonstrated by Tetra Pak on tetrahedral-shaped packs in 1961 and on the Tetra Brik® in 1969.

In 1970, Tetra Pak bought an Austrian company, Selfpack, which had also introduced an aseptic system. PKL replaced Blocpac with Quadrobloc and offered an aseptic system. About the same time, this company changed its name to Combibloc and became SIG (Swiss Industrial Company) Combibloc. Since 2007, SIG Combibloc has been owned by Rank

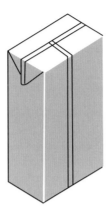

Figure 13.3 Tetra Brik®. (Reproduced with permission from Tetra Pak Group.)

Group Ltd., New Zealand. Later, the aseptic Pure-Pak® was launched using aseptic processing licensed from Liquipak.

In 1989, an expert panel on food safety and nutrition at the Institute of Food Technology, Chicago, ranked the aseptic process and packaging as the most significant of 10 food science innovations of the previous 50 years (AIChE, 2009).

The leading companies in both retail aseptic and non-aseptic paper-based liquid packaging today are Tetra Pak, Elopak (with Pure-Pak® cartons) and SIG Combibloc.

All three companies not only supply the carton packaging but they also provide the erecting, filling and sealing machinery. Where necessary the machinery is also involved in processing and this is most importantly demonstrated where the packaging is processed and filled aseptically. As will be seen subsequently, the packaging machinery is fully integrated with the packaging in what is described as a packaging system.

Despite the fact that Pure-Pak® was initially the leading paperboard-based liquid package for many years (1915–1955), Tetra Pak is the dominant company worldwide today. The market expanded significantly from 1960 onwards. Initially, only milk and cream were packed in paperboard-based packaging; then juices were launched and from the 1980s, soups, sauces, cooking oil and, more recently, water. The range of pack sizes is extensive – from the Tetra Classic Aseptic® range of 65, 100, 150 and 200 mL to the popular range of packs with square or rectangular cross sections from 200 mL to 2 L. Even larger 4 and 5 L pack cartons are available in the Pure-Pak® range. Pack shapes have expanded beyond the brick and gable top cartons to include multifaceted, wedge, pouch and hexagonal designs.

The range of products and the shelf-life requirements have extended the types of packaging materials to include aluminium foil, high barrier and ionomer plastics – the latter providing easier heat sealing, product resistance where required and excellent adhesion to aluminium foil.

A major feature of pack development in recent years has been the attention paid by the manufacturers to open, pour, reclose and demonstrate tamper evidence. In 2001, a fully retortable paperboard-based carton called Tetra Recart® was launched to compete with the processed food can. SIG Combibloc subsequently launched *combisafe* also a retort sterilisable high-barrier carton.

A major component of the success of these forms of liquid packaging has derived from the attention paid to logistics – the science of material flows. Various designs of brick-shaped *transit* packaging have been developed to fit the pallet bases in common use. This in turn fixed the dimensions of the individual package to minimise the volume required in storage and distribution. Mini pallets and roll cages were developed to minimise handling and provide suitable displays at the point of purchase.

In addition to the liquid packaging so far discussed, the bag-in-box pack which comprises a corrugated fibreboard outer box and a high-barrier plastic inner bag is a well-established liquid packaging system (see Fig. 13.20).

It was invented by W. R. Scholle in the USA in 1955. The first application was for battery acid. A wine pack was introduced in Australia by T. Angrove in 1965 but was not until 1967 that a pack was produced for Penfold Wines by 1967 Charles Henry Malpas with an integral air-tight tap.

Today bag-in-box serves retail, catering and industrial markets for the packaging of a wide range of liquid foods, beverages and non-food products in volume sizes which range from 2 L up to just over 1000 L. It is an established form of cost-effective packaging which is disposable, recyclable and efficient in transportation and storage (Yam, 2009).

A major feature of paper-based liquid packaging, in Europe and North America, has been the attention paid to environmental issues in terms of minimising the use of materials, energy savings in the packaging chain and to the recovery and recycling of the paperboard, plastics and aluminium.

In 1996, the aseptic package received the Presidential, US, Award for Sustainable Development – the first package to receive this environmental award (Presidential award, 1996).

13.2 Packaging materials

13.2.1 Paperboard

Paperboard provides strength, structure, a hygienic appearance and a good printing surface for liquid packaging cartons. The basic construction is multi-ply paperboard made from virgin fibres to ensure a high standard of odour and taint neutrality. The outer layer is always made from bleached, i.e. white, chemical pulp. (*Note*: This is also referred to as bleached sulphate pulp as the chemical separation of the cellulose fibres from wood is carried out by the *sulphate* process.) This outer layer may be white pigment coated to give the best print reproduction. The other layers may comprise either bleached or unbleached, i.e. brown, chemical pulp.

It is also possible to incorporate chemically treated thermomechanical pulp (CTMP) which has a lighter shade than ordinary thermomechanical or refiner mechanical pulp but is not as white as bleached chemical pulp. Where it is used, it is sandwiched between layers of bleached chemical pulp in the paperboard. CTMP provides more bulk and hence more stiffness than chemical pulp of the same grammage, or basis weight, and is lower in cost. The thickness and hence grammage or basis weight used depends on the size of carton and whether CTMP is used in the construction.

The main difference between liquid packaging board and board used for folding cartons is in the type of internal sizing applied at the stock preparation stage. Even though the pack design ensures that neither the edges nor the flat surfaces of the paperboard are exposed to the liquid product, because dairies have high humidity and wet environments, it is necessary to ensure that raw edges of board which are exposed to that environment do not readily absorb water or product.

Liquid packaging paperboard has developed into a mature product with capital-intensive large-scale production from relatively few producers (SPCI, 2002).

The bag-in-box package is a larger pack where the size starts around 2/3 L for retail packs of wine, increasing to around 10–30 L for bulk catering packs whilst industrial sizes can be very much larger. Hence the strength requirements of the box will vary depending on the size of the pack. Single wall, double wall and triple wall in various grammages and specified liners and fluting will be used.

13.2.2 Barriers and heat-sealing layers

The basic construction requires compatible heat-sealing polymers on both the face, top or print side, and the reverse side. This is usually provided by extrusion coatings of low-density polyethylene (LDPE). In some designs, as we will see later, PE is sealed from the outside surface to the inside surface and hence the PE on the two surfaces must be heat-seal compatible; and in other designs, the sealing is inside to inside. Most designs also require outside to outside sealing. Figure 13.4 shows a two-side PE-lined paperboard.

This construction will provide liquid tightness and humidity protection. It is mainly used for fresh products requiring a relatively short shelf life in chilled distribution.

The thicknesses of PE used depend on the size of carton. The outer layer may be either 14 or $26\,g/m^2$ and the inner layer 26, 41 or $56\,g/m^2$, the latter two being applied in two or three consecutive applications to achieve the total PE thickness required.

PE, as with most plastics, should not be thought of as a single material in the way, for instance, that we think of a specific inorganic chemical compound. There is a large family of PEs of different densities and molecular structures, which are controlled by the conditions under which the ethylene was polymerised – the conditions referred to concern pressure, temperature and the type of catalyst used. The recent introduction of metallocene (cyclopentadiene) catalysts has had a major impact on the properties of PE and other plastics. Hence this is an active area of development for the manufacturers of PE. The LLDPE (linear low-density polyethylene) is superior to LDPE in most properties such as tensile strength, impact and puncture resistance.

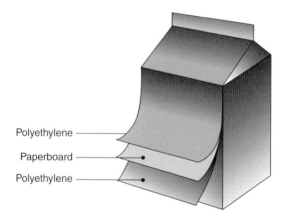

Figure 13.4 Two-side polyethylene-lined paperboard. (Reproduced with permission from Elopak.)

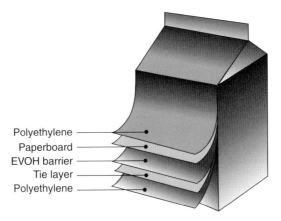

Figure 13.5 Two-side polyethylene-lined paperboard incorporating ethylene vinyl alcohol (EVOH). (Reproduced with permission from Elopak.)

Another development of interest is the use of bioplastic extrusion coatings, i.e. of non-fossil origin which have been made available. They meet EN 13432 for compostability.

It is also possible to blend the various PEs with other compatible polymers to enhance or vary the properties of the coating. In this connection, ethylene vinyl acetate (EVA) should be mentioned. As an extrusion coating, EVA improves heat sealability. In films, such as those used in the laminates and extrusion coatings in the bag-in-box applications, EVA improves strength and flexibility as well as heat sealability.

For higher barriers in liquid packaging, it is necessary to use other materials in addition to PE. Ethylene vinyl alcohol (EVOH) and polyamide (PA) are good oxygen barriers, and they also provide flavour protection and oil/fat resistance. EVOH and PA are not particularly good barriers to moisture vapour, the barrier for which is provided by the PE. These structures are used for medium and long-term shelf-life products in ambient and chilled distribution and provide an alternative to the use of structures which include aluminium foil (Fig. 13.5).

Aluminium foil is a well-established high-barrier material, which provides a barrier to light, oxygen and moisture vapour. It provides excellent protection to flavours and has oil/fat resistance. It may be laminated to either side of the paperboard. When laminated to the outside (top side) of the paperboard, it provides the carton with a metallic finish. In the majority of the cases, aluminium is used solely for barrier properties and an additional layer of PE is applied as indicated in Fig. 13.6. These laminates are used for aseptic and hot-filled products, requiring a long shelf life in ambient distribution and fresh juices in chilled distribution.

Ionomer (Surlyn™) is a moisture vapour barrier like PE and, additionally, has oil/fat resistance. It has very good hot-tack and heat-sealing properties and is used as a tie layer (process aid) on aluminium foil on which PE can be extruded. Ionomer can be extruded onto aluminium foil at lower temperature than PE and thereby avoid potential odour from PE extruded at the temperatures necessary to achieve good adhesion to aluminium foil.

The plastic bag used in bag-in-box packs will incorporate the plastics already mentioned such as PE, EVA-modified PE, metallised polyester (PET or PETE) and EVOH depending on the barrier and strength required.

An alternative tie layer for PE/PA would be one of the Bynel® range of co-extrudable adhesion promoters from DuPont (Tampere, 2000; DuPont website).

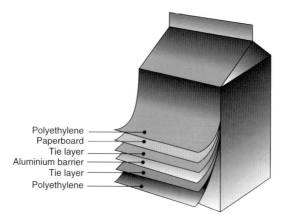

Figure 13.6 Two-side polyethylene-lined paperboard incorporating aluminium foil. (Reproduced with permission from Elopak.)

PET, (PETE), film vacuum metallised with a thin layer of aluminium can be a good barrier to oxygen and moisture vapour. It can be laminated to paperboard and is used in bag-in-box laminates. Other films such as polypropylene (PP) and PA can also be metallised. Another high-barrier film construction has a plasma-coated layer of silicon oxide (SiO_2) on polyester film which also provides a completely non-metallic barrier which can be extrusion laminated to paperboard.

There are many possible combinations of extrusion coating and laminating. For example, PP, with resistance to essential oils, is extrusion coated onto paperboard for cartons used to pack orange juice. The PP is overcoated with a blend containing PP plus 20% LDPE and 4% tackifier to improve the heat sealing of the inner layer (TAPPI, 2001). PP also has higher heat resistance than PE.

In practice, the product, the shelf life required and commercial considerations will all have a bearing on what is used in any specific liquid packaging application.

Another possible approach to barrier coating which may find a place in future developments is by way of size-press application on the paperboard machine of polymers in water-based dispersions. Both Surlyn™ and Nucrel® are thermoplastics from DuPont with oil/grease resistance and low seal-initiation temperature properties, which are available as dispersion coatings. Lower costs are favoured by the treatment which takes place on the paperboard machine (Tampere, 2000; SPCI, 2002; DuPont website).

Another reported development involves the production and use of polymer nanocomposites in which nanoscale particles of surface-modified clay are dispersed in the polymer. Enhanced barrier properties are claimed, making the material suitable for the packaging of oxygen-sensitive beverages such as fruit juices (Krook & Hedenqvist, 2001).

13.3 Printing and converting

13.3.1 Reel-to-reel converting for reel-fed form, fill, seal packaging

The paperboard is printed reel to reel by flexo or gravure using conventional, water-based or ultraviolet (UV) curing inks. The printed reels are then either PE extrusion coated on both sides or, if an additional barrier layer is required, the reels are PE extrusion laminated to aluminium foil. With the heavier PE extrusion coatings on the inside of the package, a second PE application is required. As noted earlier, an alternative barrier layer may be applied – sandwiched between the paperboard and the outer coating of PE. The reels are then creased and slit to size and sent to the packer where the reels are formed, filled and sealed. An exception to this is Tetra Classic® which is not creased during conversion.

13.3.2 Reel-to-sheet converting for supplying printed carton blanks for packing

There are options possible for producing flat carton blanks from extrusion coated and laminated paperboard. These materials are printed, cut and creased in-line. The printing process would most likely be gravure or flexo. In order to achieve good print adhesion to the PE, it is necessary to treat the surface of the PE after extrusion coating with an electric corona discharge, or direct gas flame, which oxidises the surface of the PE, making the surface molecules more reactive. Where printing is applied to the outside surface of the PE, the inks must have good wet and dry rub resistance.

PE and barrier coatings can also be cut into sheets before printing, usually by offset litho. The printed sheets are then cut and creased to produce individual carton blanks. The print must also be product resistant.

The design includes a narrow fifth panel on the side of the blank about 15 mm wide. The edge of this panel is then 'skived'. In this process, most of the paperboard from a 4 mm wide strip on the edge is removed. This leaves a narrow strip, 4 mm wide, of the inner layer of material, i.e. PE + aluminium foil + residual fibre, which is then folded over through 180°, so that it is then in contact with PE on the remaining 10 mm wide panel. As this is completed, the carton panels are folded over, and the 10 mm wide side seam of the carton is formed using hot air or gas flame. Skiving enables efficient folding and a side seam which is restricted to two thicknesses of the paperboard. This construction seals the inner surface of the narrow sealing panel to the inner surface of the joining panel and ensures that when the carton is filled, the liquid does not have any contact with a raw or exposed edge of paperboard, as shown in Fig. 13.7. The side seam sealed carton blanks are then ready for dispatch to the packer.

13.3.3 Sheet-fed for bag-in-box

The box is made by manufacturers of corrugated fibreboard. Where special openings are required for the taps, the boxes are die cut. Printing may be done by pre-printing the liner, or by standard flexo or offset litho printing.

13.4 Carton designs

There is a wide range of carton shapes, openings and closures. It is a constantly developing area. Readers are advised to consult the websites and the companies offering paperboard-based packages and systems for dairy products, foods and beverages. This section gives an indication of the wide range of carton designs which are available. Opening and closure features are discussed in Section 13.5.

Figure 13.7 Heat sealed side seam. (Reproduced with permission from Tetra Pak Group.)

Figure 13.8 Tetra Rex® gable top. (Reproduced with permission from Tetra Pak Group.)

13.4.1 Gable top

The original Pure-Pak® (from Elopak) carton was a gable top as shown in Fig. 13.1; today this is most likely to incorporate a screw cap as shown in Fig. 13.1. Gable top cartons have also been available from other suppliers for many years. Tetra Rex®, as shown in Fig. 13.8, is a gable top carton produced by Tetra Pak. These cartons are produced from side seam sealed carton sleeves.

The traditional gable top cartons now available incorporate additional design features. Additional curved side panels are eye catching and improve handling and pouring. Increasing the front gable panel and incorporating a curved front crease enables the use of higher diameter plastic closures which improve pouring particularly for higher viscosity liquids, e.g. Pure-Pak Curve®, Pure-Pak Diamond® and Pure-Pak Diamond Curve® as shown in Fig. 13.1. Elopak also has Slim® with a smaller cross section and hence is taller for the same volume.

These gable top designs are made from carton sleeves and used for liquid products which are distributed chilled or in ambient conditions after aseptic processing and filling. Tetra Gemina® Aseptic is a reel-fed gable top package, and as the name implies, it is also for use in aseptic packaging for ambient distribution (see Fig. 13.9).

13.4.2 Pyramid shape

This is the shape of the first liquid pack launched by Tetra Pak. It is known today as Tetra Classic Aseptic® and is shown in Fig. 13.2. The board is supplied uncreased in reel form.

Figure 13.9 Tetra Gemina® Aseptic. (Reproduced with permission from Tetra Pak Group.)

The pack is formed over a shoulder and around a vertical tube where the side-seam is heat sealed. Prior to sealing, a thin ribbon of PE is fed and heat sealed across the raw or exposed edge of the longitudinal seal. This ensures that the liquid is not in contact with the raw edge of the paperboard. The product is fed into the tube where it enters the folded side seam sealed tube. The horizontal seal is made intermittently through the product. As succeeding heat seals are made at right angles to each other, the resulting pack is shaped like a pyramid.

13.4.3 Brick shape

The Tetra Brik® is made from reels and the Combibloc (SIG Combibloc) is made from carton sleeves. In the case of Tetra Brik®, the side-seam is also covered with a strip of PE film to prevent the product from contacting the raw edge of the overlapping seam. This style results in ears formed when the horizontal heat seals are made. They are folded flat against the pack and lightly sealed. There is Tetra Brik® Aseptic for ambient distribution and Tetra Brik® for chilled distribution – there are differences in the material specifications and in the filling machines used.

Most of the brick-shaped packages have a rectangular cross section, as shown in Fig. 13.3, but it is possible to have a square cross section, as shown in Fig. 13.10. There are in fact five formats of Tetra Brik® – slim, mid, base, square and edge – in a volume range of 200, 250, 500 and 1000 mL.

The *combifit* carton is a square cross section carton with a slanted top to improve handling in use (see Fig. 13.11). The c*ombibloc* carton is also available with a rectangular cross section or can also have a square cross section and is also shown in Fig. 13.11. There are nine different base formats for *combibloc* and a volume range of 125–2000 mL. There are five base formats for *combifit* with filling volumes from 150 to 1500 mL.

Figure 13.10 Tetra Brik® – square. (Reproduced with permission from Tetra Pak Group.)

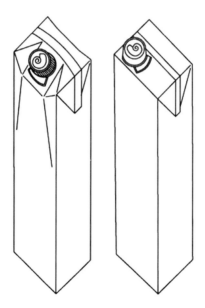

Figure 13.11 *Combifit* (left) and *Combibloc* (right), both in the square cross-section format and with plastic caps. (Reproduced with permission from SIG Combibloc.)

13.4.4 Pouch

Tetra Fino® is a paperboard-based pouch (Fig. 13.12). This pack and the associated packaging machinery provide a low cost and low investment system. It also places a lower demand for personnel, training and spare parts compared with other systems.

13.4.5 Wedge

The wedge shaped package is a stand-up pouch (Fig. 13.13). An example is the Tetra Wedge® which, with a straw, is a convenient-to-use drinks pack.

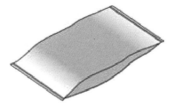

Figure 13.12 Tetra Fino® – pouch-shaped liquid package. (Reproduced with permission from Tetra Pak Group.)

Figure 13.13 Wedge®-shaped paperboard-based package. (Reproduced with permission from Tetra Pak Group.)

13.4.6 Multifaceted design

Tetra Prisma Aseptic® is an eight-faceted design, as illustrated in Fig. 13.14, which has a distinct image with an easy to handle shape.

13.4.7 Bottle shapes

Tetra Top® (Fig. 13.15) is a container with a square cross section modified with rounded corners and a plastic top.

This design is differentiated by the range of tops which are discussed in Section 13.5 and shown in Figs 13.22 and 13.23.

Tetra Evero® Aseptic is also bottle shaped (see Fig. 13.16). Pre-printed reels are converted into sleeves on the filling machine, and injection moulding technology is used to fuse the capped neck to the sleeve. The containers are then sterilised using heat and hydrogen peroxide and finally with sterilised air in preparation for product filling.

Figure 13.17 shows the difference in the folded and heat sealed bases of Tetra Top® and Tetra Evero® Aseptic.

13.4.8 Round cross section

The Lamican 250 Aseptic is a round cross section container manufactured by Lamican Oy. It is a 'drink from the can' container, with a foil ring-pull or peel-tab opening (Fig. 13.18). An aluminium-free specification is also available.

Figure 13.14 Tetra Prisma®. (Reproduced with permission from Tetra Pak Group.)

Figure 13.15 Tetra Top®. (Reproduced with permission from Tetra Pak Group.)

Figure 13.16 Tetra Evero® Aseptic. (Reproduced with permission from Tetra Pak Group.)

Paperboard-based liquid packaging **367**

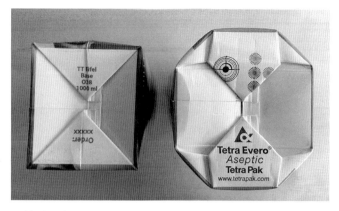

Figure 13.17 Folded and heat sealed bases of Tetra Top® and Tetra Evero® aseptic. (Reproduced with permission from Tetra Pak Group.)

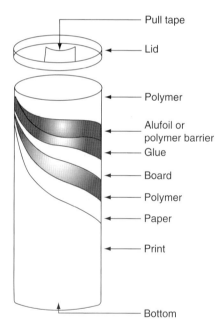

Figure 13.18 Lamican 250 aseptic. (Reproduced with permission from Lamican Oy.)

13.4.9 Bottom profile for gable top carton

This is not a carton design as such, but a modification to an existing design with a technical or performance-related benefit. This is the Sahara bottom which reduces water absorption by the raw edge in the bottom design of the Tetra Rex® carton.

It is a feature which can be applied on the forming and filling machine. It has the effect of raising the uncoated paperboard edge in the sealed bottom of Tetra Rex away from contact with the surface, which may be wet, on which the bottom of the carton is resting. The raised

Figure 13.19 Modified carton base profile – Sahara design for Tetra Rex®. (Reproduced with permission from Tetra Pak Group.)

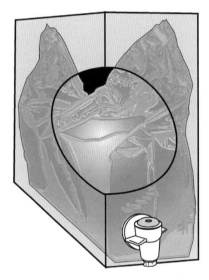

Figure 13.20 Schematic drawing of bag-in-box with tap. (Reproduced with permission from Rapak.)

area of the base is shown in Fig. 13.19. It is claimed that raising the exposed edge by 1 mm reduces the uptake of water by the raw edge of paperboard by 80% and prevents the bottom of the carton from becoming saturated with water, i.e. soggy (Packaging, 2000).

13.4.10 Bag-in-box

The box for bag-in-box usually has a rectangular shape (see Fig. 13.20). A 2 L round pack is used for olive oil (Rapak, 2010). For the really large volume packs, an octagonal shape set on a standard 1000 × 1200 mm pallet and with a separate lid may be necessary to provide adequate strength for handling in storage and distribution.

The bags are usually supplied singly and separately to the filler or on reels where each bag is separated as required from the next by breaking perforations between the bags. It is also possible for the film to be supplied to the filler on reels and for the bags to be made, filled, tap inserted and the bag sealed in-line.

13.5 Opening, reclosure and tamper evidence

The main disadvantage of liquid food and beverage cartons has, until comparatively recently, been the absence of convenient and safe ways of opening them. Though many packs, particularly the smaller ones, were considered to be one-shot packs, reclosure was a poorly addressed feature, particularly for the larger pack sizes, and this became a barrier to developing the use of these containers in some product markets.

The early gable top cartons from Ex-Cell-O were to be opened 'with a knife or scissors'. Clips were used to reclose gable tops. In 1955, the 'Pitcher Pour' built-in pouring spout concept replaced the perforated tabs and openings covered by paper patches (Robertson, 2002, p. 48).

Subsequently one was advised to open a gable top carton by 'pressing back two of the wings and compressing them to form a spout'.

It is also likely that the advent of aseptic liquid packaging cartons actually set back concerns about opening liquid food and beverage cartons, as a balance has to be struck between the opening convenience and the integrity of the contents. There were also the questions of what sort of closure is needed, how it would be applied and what would it cost.

One was invited to open brick-shaped packs with the aid of scissors. In another example, a pull tab is sealed across the access point, as shown in Fig. 13.21.

The simplest approach to drinking the contents of paperboard-based liquid packaging is by means of a straw, wrapped in film, attached to the body of the carton. A hole is punched in the paperboard prior to applying aluminium foil and PE. This provides an easy entry point for pushing the straw into the carton. The position of the hole is covered with a pull tab. The straw is packed in a film sachet and attached to the carton with a spot of hotmelt adhesive (Fig. 13.22).

The Tetra Prisma®, described in Section 13.4.6, is fitted with a pull-tab closure, which when removed reveals a large drinking aperture that also is excellent for removing the contents by pouring (Fig. 13.23).

In recent years, a wide range of reclosable injection-moulded plastic closures has been launched with built-in tamper evidence.

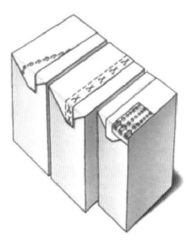

Figure 13.21 Opening indications on brick-shaped cartons. (Reproduced with permission from Tetra Pak Group.)

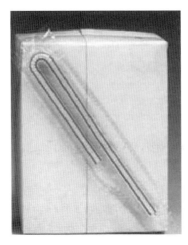

Figure 13.22 Drinking straw attached to carton. (Reproduced with permission from Tetra Pak Group.)

Figure 13.23 Pull-tab closure and wide drinking aperture. (Reproduced with permission from Tetra Pak Group.)

Figure 13.24 shows Recap 3 from Tetra Pak. This has a flexible aluminium foil-based diaphragm seal over the carton opening. This has to be removed in order to access the product and is therefore a tamper-evident feature. The hinged overcap can be used to reclose the container with a sharp click.

There are several types of screw-type closure. combiTop from SIG Combibloc was the first, launched in 1994. There are designs which are applied over an area of the carton from which the paperboard was removed prior to extrusion coating and laminating with aluminium foil. When this cap is removed for the first time, the foil/plastic material is perforated and pushed away from the opening. Others have external tamper-evident security plastic rings, which have to be broken in order to remove the cap, as shown in Fig. 13.25.

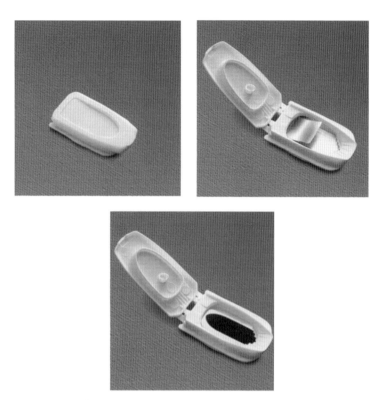

Figure 13.24 Tamper-evident closure with reclosure, Recap 3. (Reproduced with permission from Tetra Pak Group.)

Figure 13.25 Screw cap with security ring. (Reproduced with permission from Tetra Pak Group.)

Improved pourers and wide aperture closures have been incorporated where they are considered appropriate.

Tetra Top®, the pack with the square cross section and rounded corners, is offered with a family of alternative plastic closures integrated with the paperboard-based body of the

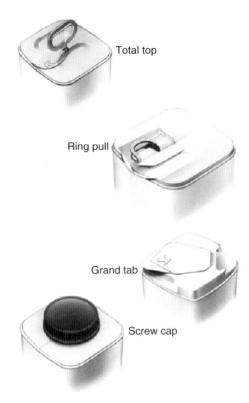

Figure 13.26 Four closure designs for Tetra Top®. (Reproduced with permission from Tetra Pak Group.)

container. Each closure is designed to meet specific market needs. The closures are described in the following and are shown in Fig. 13.26:

- grand tab – a lift-off closure for a container, which is easy to drink from directly
- total top – fitted with a ring pull which removes the entire top of the container making it suitable for foods which are eaten from the container such as soups, yoghurt and other desserts
- screw cap – with a tamper-evident seal
- ring pull – which is easy to open and close effectively.

The combiSwift cap is a screw closure that can be applied to the *combibloc* and *combifit* cartons. It is lightweight, weighing 2.7 g and very easy to open. As has already been described for the straw opening, a hole of the correct size is punched into the printed cartonboard before it is laminated and extrusion coated with aluminium foil and PE.

After the pack has been aseptically sealed inside the filling machine, the three-part plastic closure is applied over the paperboard hole to the carton from the outside (see Fig. 13.27). The cap consists of a flange with an integrated cutting ring and screw cap. When the cap is twisted the ring slices easily and cleanly through the thin layers of aluminium foil and PE.

The combiSmart cap is a screw closure designed to be more suitable for pouring products such as evaporated milk and cream and for drinking directly from the container. This

Paperboard-based liquid packaging **373**

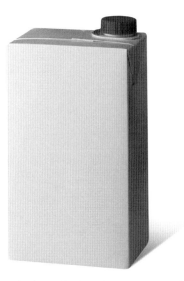

Figure 13.27 Combibloc*Standard*, 1000 mL, with screw cap combiSwift. (Reproduced with permission from SIG Combibloc.)

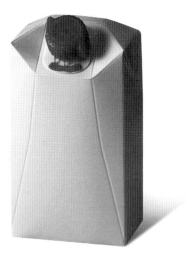

Figure 13.28 Combifit*Small* 250 mL, with screw cap combiSmart. (Reproduced with permission from SIG Combibloc.)

is achieved by the cap having a raised 14.3 mm flange compared with 12.6 mm for the flange on combiSwift. CombiSmart is used for the lower volume cartons compared with combiSwift which is used on cartons for medium and high volume. With combiSmart, the paperboard is not pre-punched before the cap is applied. When the customer wants to open the carton, a simple twist enables a ring in the cap to cut through the paperboard and the other layers. The cap is designed to be more easily unscrewed but in other respects is similar to combiSwift with respect to the way it is applied and opened (see Fig. 13.28).

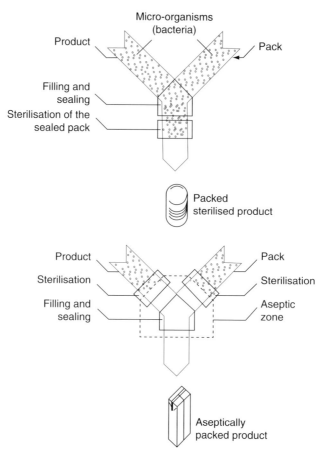

Figure 13.29 Comparison between sterilising a packed product and aseptic filling technology. (Reproduced with permission from SIG Combibloc.)

13.6 Aseptic processing

In the aseptic packaging process, ultraheat treatment (UHT), in-line food sterilisation thermal processing is followed by packing in a sterilised container whilst still *within* the sterile environment, also known as the aseptic zone. The liquid dairy product, food or drink is sterilised or pasteurised in a continuous process as it travels through a heat exchanger, after which it is cooled and filled. Meanwhile, the packaging machine shapes the container and sterilises the packaging material.

This is significantly different to the procedure adopted when sterilising food *after* it has been packed in a *sealed* container, such as a processed food can. This difference is demonstrated in Fig. 13.29.

Typical temperatures and holding times in a UHT process are of the order of 140°C for a few seconds, for example 3–15 s (Tucker, 2003). Sterilisation of the packaging is carried out using hydrogen peroxide, UV and sterile hot air. The system is purged with an overpressure using a sterile air system. A typical flow line is illustrated in Fig. 13.30.

Aseptic filling of pureed tomatoes in cartons
UHT heating: 125° for about 4 min.* (HTST High Temperature – Short Time)
*Temperature and duration depend on the product

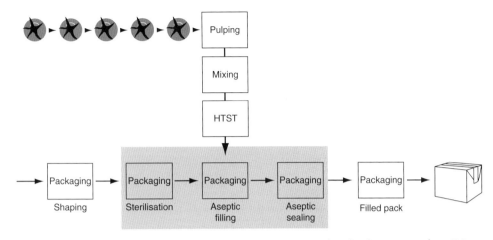

Figure 13.30 Aseptic filling of pureed tomatoes in cartons. (Reproduced with permission from SIG Combibloc.)

The main benefits of aseptic packaging are:

- food safety – harmful bacteria and other microbiological contaminants are eliminated from liquid food and drinks
- no refrigeration necessary – cost and environmental impact reduced
- convenient – lightweight, shatterproof and convenient for distribution
- better taste and nutrition – retains natural taste and colour, less damage to heat-sensitive constituents, such as vitamins, and reduced use of chemical preservatives
- longer product shelf life.

13.7 Post-packaging sterilisation

Tetra Pak launched a retortable brick-shaped carton known as Tetra Recart® in 2001 (Packaging, 2001) (see Fig. 13.31). The material is a six-layer laminate including paperboard, aluminium foil and polyolefin, i.e. PP, with heat sealing adequate to withstand retorting at temperatures around 130°C. Opening instructions for Tetra Recart® are shown in Fig. 13.32. A similar package, *combisafe*, has been introduced by SIG Combibloc.

These packages are an alternative to the heat-processed can, glass jar and retort pouch. They are suitable for the packaging of a wide range of foods, such as vegetables, fruits, soups, ready meals, sauces, pastas, salsas and pet food. They are capable of handling food in particulate and chunky form.

A form, fill, seal machine erects cartons from flat sleeved blanks and, after filling, seals the bottom. The filling machines are capable of handling 24 000 cartons per hour, i.e. similar to that of a modern canning line.

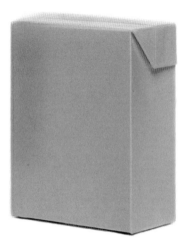

Figure 13.31 Tetra Recart® retortable carton package. (Reproduced with permission from Tetra Pak Group.)

Figure 13.32 Opening instructions Tetra Recart®. (Reproduced with permission from Tetra Pak Group.)

The main benefits are:

- food safety – harmful bacteria and other microbiological contaminants destroyed in the retorting process
- long shelf life comparable with processed food cans, glass jars and retort pouches
- no refrigeration necessary – cost and environmental impact reduced
- convenient to handle and open, lightweight and shatterproof
- more efficient distribution, storage and merchandising compared with cylindrical cans, glass jars and retort pouches.

13.8 Transit packaging

The needs of transit packaging, storage and merchandising at the point of purchase have all been addressed by the suppliers of liquid packaging materials. Packs are collated or grouped for sleeving, placing in trays and shrink or stretch wrapping. Special packaging formats are used for boxing the tetrahedral, wedge and pouch-shaped packs. Plastic crates have been designed which also serve as merchandising units, or retail-ready packs, at the point of purchase. Figure 13.33 shows three presentations of stackable distribution packs.

Paperboard-based liquid packaging **377**

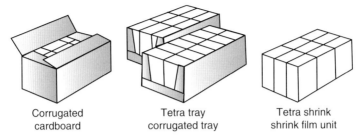

Figure 13.33 Distribution packs. (Reproduced with permission from Tetra Pak Group.)

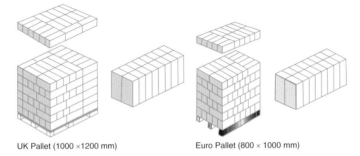

Figure 13.34 Efficient palletisation in use of space and stability of the pallet. (Reproduced with permission from Tetra Pak Group.)

Figure 13.35 Roll cage container for use in distribution and at point of purchase. (Reproduced with permission from Tetra Pak Group.)

These packs are designed for efficient stacking, i.e. best use of the pallet volume and with alternate rows having a different pattern to give pallet stability (Fig. 13.34).

For the large retail groups selling large volumes of products in paperboard-based liquid packaging, the usual package for distribution is the roll container as illustrated in Fig. 13.35.

Delivery lorries have been designed to handle crates and facilitate loading/unloading.

13.9 Applications for paperboard-based liquid packaging

The main application area in volume terms up to 2 L retail market is for the packaging of dairy products, such as milk and cream, and utilising both refrigerated non-aseptic and aseptic packaging. This has been extended to a wide range of juices. Wine and mineral water have been packed in liquid packaging cartons. Ice lollies are packed in Tetra Classic® and Tetra Classic Aseptic® packaging.

Other examples are as follows:

- cooking oil
- liquid egg
- soups, broths and stocks
- ice cream mix
- baby food
- pet food
- beans and other vegetables
- fruit-based desserts
- tea
- yoghurt-based drinks
- soya bean-based products.
- In the non-food area, softener products used in washing clothes have been packed in cartons as a refill pack for plastic containers.
- Gable top cartons can be used for dry food. Freeze-dried vegetables have been packed in large Pure-Pak® cartons for the catering and institutional markets.

Bag-in-box packaging is used for larger volume packs for many of the products mentioned earlier. It is available in an extensive range of volumes from 2 to 1100 L with many intermediate sizes. The volumes packed depend on the markets being served. They can be retail markets from 2 to 5 L, catering markets, including those for drinks dispensers, from 5, 10 and up to around 20/30 L, and industrial packs at say 200, 500 and up to 1000/1100 L.

The products packed fall into three main categories.

Beverages which include water, wine, juices, tea and coffee. Even beer is packed in this type of container, but this requires a system whereby the carbonation is reduced at the brewery and is recarbonate activated during dispensing.

Liquid foods include milk and other dairy products, fruits and purées, syrups for beverage preparation, edible oils, liquid eggs, sauces and soups. Interesting new products packed include cocktails for restaurants and bars. These are products demanding time skill in preparation on-site, and hence there are several benefits in providing ready-to-serve packs containing up to 67 servings per pack (Rapak, 2011).

Liquid yeast is supplied to bakers in bag-in-box packs.

Other products supplied to industry include adhesives, chemicals, coatings, pharmaceuticals, motor oils and household products.

13.10 Environmental issues

The liquid-food packaging industry has invested much effort into promoting the environmental benefits of liquid packaging cartons through industry-wide associations. In general, this has concentrated on aspects which apply to paper and paperboard packaging as a whole and as such will not be discussed here.

Environmental issues which are specific to this form of packaging concern are discussed in the following sections.

13.10.1 Resource reduction

The material used in beverage cartons has been reduced significantly. 'Over the last 20 years technology and best practices have made it possible to produce 40% more cartons from the same amount of wood. With carton manufacturers who are members of the Alliance for Beverage Cartons and the Environment, (ACE), using around 2.5 million tonnes of wood fibres annually, this has meant a significant resource saving. It also helps decouple environmental impact from economic growth. Worth noting too is that the European forest industry does not even use the annual growth of forests, a renewable natural resource' (ACE, 2011).

An average beverage carton for milk or juice weighs around 5% of the total pack. (A 1-pint glass milk bottle in UK weighs approximately 220 g, which is roughly 26% of the total pack weight.). On the journey to the filler's plant, 600 000 empty 1 L beverage cartons fit into one truck, which results in a significant energy saving compared to transporting the same number of empty glass bottles.

After filling, beverage cartons are easily stacked and take up less space. The overall effect in transportation is significant in reducing emissions of nitrogen oxides, carbon monoxide and carbon dioxide.

Manufacturers continually study the impact of their products on the environment. This can be monitored by calculating the carbon or CO_2 footprint. One important way of reducing the CO_2 footprint is by resource reduction. The material used to make the packaging can be investigated though it is necessary to ensure that product protection is maintained. This could involve the use of a lighter grammage of paperboard, changing the type of paperboard or modifying the weight of a plastic closure. This has been applied by Elopak using lighter grammage paperboard and a lighter plastic closure for the Eco™ cartons used to pack fresh and ESL (extended shelf life) milk and by SIG Combibloc where their EcoPlus aseptic cartons have achieved a 28% reduction in carbon footprint with a new material specification (see websites).

The weight of the plastic caps has been reduced since they were originally introduced. One example quoted is a reduction from 4.9 to 1.9 g. This is the combiCap closure which is available on the combibloc *Slimline* (SIG Combibloc, 2011b).

13.10.2 Life-cycle assessment

Life-cycle assessments (LCAs) have been conducted on beverage cartons in comparison with other forms of food and beverage packaging since the 1990s. The potential environmental impact of products during their entire life cycle is recorded and evaluated. The factors which have to be taken into account have been identified. They are conducted according to ISO 14040 and are subjected to critical reviews by independent supervisory project groups.

The following factors are taken into account:

- CO_2 emissions (greenhouse effect)
- total energy consumption
- non-renewable energy consumption

- use of sustainable resources
- mineral resources
- water consumption
- generation of municipal waste
- generation of hazardous waste
- acidification of the environment
- eutrophication of the environment.

LCAs have been commissioned by SIG Combibloc at the Institute for Energy and Environmental Research (IFEU), Heidelberg, Germany. The IFEU is one of the most reputable environmental research institutes in Europe. The LCA work was carried out in accordance with ISO 14040 and independently verified. The results of three projects have been published.

13.10.2.1 Life-cycle assessment of retortable cans, glass jars, pouches and cartons

An LCA for food metal cans, glass jars, retort pouches and carton packs was published by IFEU in 2009. The work showed that using carton packaging had significantly lower CO_2 emission equivalents and a significantly lower demand of fossil resources compared with the metal, glass and plastic retortable packs. The reductions are shown in Table 13.1.

13.10.2.2 CO_2 emissions of a standard aseptic carton compared with those of an aseptic carton designed to result in a lower CO_2 emission

In July 2010, IFEU completed a Europe-wide LCA for the SIG Combibloc EcoPlus aseptic carton. The LCA study showed that the new paperboard specification and changes in other layers in the laminate for this carton resulted in a reduction in CO_2 emissions of 28% compared with a standard SIG Combibloc carton pack of the same format and product protection requirements (SIG Combibloc, 2010).

13.10.2.3 Reductions in the consumption of energy and fossil fuel and the CO_2 emissions of a soft drinks carton compared with bottles made from glass and PET (PETE)

An LCA for PET (PETE) bottles, disposable glass bottles and carton packs for the packaging of non-carbonated soft drinks (NCSD) was published by IFEU in 2011. The work

Table 13.1 The reductions in CO_2 emission equivalents and use of fossil resources for a 400 mL aseptic carton compared with glass, metal and plastic retortable packs

	400 mL aseptic carton, 13.9 g	
Alternative retortable packs	Fossil resource reduction (%)	CO_2 emission equivalents reduction (%)
Glass jar and metal closure, 211.3 g	60	60
Three-piece metal food processing can, 52.5 g	41	55
Retort pouch, 10.1 g	45	36

Source: SIG Combibloc (SIG Combibloc, 2009).

Table 13.2 The *reductions* in CO_2 emission equivalents and in primary energy and fossil resource usage for a 1 L NCSD carton compared with disposable glass bottles and PET bottles

	One litre non-carbonated soft drink (NCSD) carton		
	Primary energy reduction (%)	Fossil resource reduction (%)	CO_2 emission equivalents reduction (%)
Disposable glass bottle	56	68	70
Monolayer disposable PET bottle	24	51	28
Multilayer disposable PET bottle	34	59	39

Source: SIG Combibloc (SIG Combibloc, 2011a).

showed that the use of carton packs showed significantly lower figures for CO_2 emission equivalents, primary energy and fossil resource usage compared with the PET (PETE) bottles and disposable glass bottles (see Table 13.2).

LCA is used by many packaging material and packaging suppliers. It is ongoing work as there is a need for constant improvement in seeking to reduce the environmental impact of packaging. (Readers are recommended to study the websites of the leading liquid packaging companies for detailed comments on LCA.)

13.10.3 Recovery and recycling

Recycling of beverage cartons in Europe (27 EU countries + Norway and Switzerland) has grown steadily over the last 15 years with around 350 000 tonnes (over 13 billion cartons) recycled in 2010. This represents a rate of 36% of all cartons sold in Europe being recycled, with some countries like Belgium or Germany having rates twice this average or more. Combined recycling and energy recovery rates for the whole region reached roughly 650 000 tonnes (a 66% recovery rate) (ACE, 2012). The fibre recovered from paper-based liquid packaging is of high quality and therefore is useful to maintain the quantity and quality of recovered fibre. The separated plastic and aluminium material is used in one of several different ways.

13.10.3.1 *Fiskeby Mill*

Fiskeby Mill in Sweden invented a system for separating the fibre from the plastic and aluminium in waste packaging such as beverage cartons. The fibre was recycled in the manufacture of multilayer paperboard. The plastic and aluminium is incinerated in an energy recovery plant producing electricity and steam which is used in the manufacture of the paperboard (Fiskeby, 2012).

13.10.3.2 *Stora Enso Barcelona Mill*

Stora Enso Barcelona Mill. The fibre is separated from the plastic and aluminium and is recycled in the production of paperboard (white lined chipboard). The plastic and aluminium is processed by pyrolysis, i.e. heating in the absence of oxygen. This makes the long chains of PE divide into gases and light oils which are used to generate energy in the mill. The aluminium remains unoxidised in this process and can be recycled and remelted without problems. This process is unique for separating plastic and aluminium.

The process was refined in collaboration between Stora Enso's Barcelona Mill and Alucha Recycling Technologies. The innovation was selected as a winner of the 2010 'Best of the Best' LIFE Environment Projects award, granted by the European Union to recognise projects that have a positive impact on the environment (Stora Enso, 2012).

13.10.3.3 Lucart Group, Italy

Lucart Group, Italy, is able to separate the fibre from the PE and aluminium. The fibre is used to make tissue. The aluminium and PE is processed into an homogenous material (EcoAllene) with excellent injection moulding characteristics and is used in place of virgin plastic to make pens and similar stationery products for the Lecce Pen Company (Lucart Group, 2010).

13.11 Systems approach

Today, there are many examples of a total system approach involving one packaging company acting in partnership with manufacturers, particularly in the food industry, in developing, installing and maintaining the whole system from the point of packaging to the point of sale.

In the case of liquid packaging, this has tended to go one step further and includes the food processing. This is inevitable if, as with aseptic packaging, the processing and packaging is inevitably closely integrated. This has led to liquid packaging manufacturers being closely associated with food processing, and this in turn has led to packaging solutions which are highly cost effective.

In another way, the manufacturers of liquid packaging have also gone further than most packaging manufacturers in studying and getting involved with solutions for the logistics of distribution from the end of the packing line to the point of sale at the supermarket. This has resulted in maintaining, developing and expanding liquid packaging in paperboard-based packaging.

Solutions have been developed with the worldwide market bearing in mind the serious wastage of food which occurs in the developing world as a result of the lack of suitable technology that can be adapted to local needs.

References

ACE, 2011, *The Alliance for Beverage Cartons and the Environment*, Katarina Molin, Director General of ACE at Conference in Brussels, Belgium, 10 November, visit http://www.ace-uk.co.uk/news/article/resources_unlimited

ACE, 2012, *The Alliance for Beverage Cartons and the Environment*, Press Release, visit http://www.ace.be//en/working-with-nature-2/alias-2/recycling-in-europe

AIChE, 2009, *Chemical Engineering Innovation in Food Production*, The American Institute of Chemical Engineering and Chemical Heritage Foundation (AIChE), Philadelphia, Pennsylvania, visit http://www.chemicalengineering.org/docs/ChemE-Food.pdf

Fiskeby, 2012, Recovered fibre – an environmentally-friendly raw material on Fiskeby, visit http://fiskeby.com/en/

Krook, M. and Hedenqvist, M., 2001, Nanocomposites: one way to improve the barrier properties of degradable polymers, *European PLC Conference Proceedings*, 8th European polymers, films, laminations and extrusion coatings conference, Barcelona, Spain, TAPPI Press, Atlanta, Georgia.

Leander, L., 1996a, *Tetra Pak – A Vision Becomes Reality*, Tetra Pak International S.A., Pully, Switzerland, p. 27.

Leander, L., 1996b, *Tetra Pak – A Vision Becomes Reality*, Tetra Pak International S.A., Pully, Switzerland, pp. 25–27.
Leander, L., 1996c, *Tetra Pak – A Vision Becomes Reality*, Tetra Pak International S.A., Pully, Switzerland, p. 53.
Lucart Group, 2010, *Environmental Report 2010, Waste*, p. 35, visit http://www.lucartgroup.com/UserFiles/File/Environmental_Report_2010.pdf
Packaging, 2000, *Packaging News*, May, p. 2.
Packaging, 2001, New Tetra Pak carton challenges food cans, Hilary Ayshford, *Packaging Magazine*, 16 May.
Presidential Award, 1996, *Presidential Award for Sustainable Development to Aseptic Packaging Presented to Tetra Pak*, visit http://www.tetrapakrecycling.co.uk/tp_didyouknow.asp
Robertson, G.L., 2002, The paper beverage carton: past and future, *Food Technology*, 56(7), 46–52.
SIG Combibloc, 2009, Brochure, *Getting the Complete Picture*, Press Release, April, visit www.sig.biz at Environment Food LCA Europe.
SIG Combibloc, 2010, *Comparative LCA of Beverage Cartons*, IFEU July 2010, visit www.sig.biz at Environment, EcoPlus LCA Europe.
SIG Combibloc, 2011a, *Latest IFEU comparative analysis for PET, Glass and Carton Packs*, Press release May, visit www.sig.biz at Environment NCSD LCA Europe.
SIG Combibloc, 2011b, Getting the Complete Picture, Environmental Brochure, Press Release, January, visit www.sig.biz
SPCI, 2002, *7th International Conference on New Available Technologies*, Stockholm, Sweden, SPCI Swedish Association of Pulp and Paper Engineers, 4–6 June.
Stora Enso, 2012, *Award-Winning Recycling Process Inaugurated in Barcelona*, visit at http://www.storaenso.com Barcelona Mill pyrolysis.
Tampere, 2000, *Barrier Coatings in Packaging Conference*, Tampere University of Technology, Tampere, Finland, June.
TAPPI, 2001, R. Edwards, Polypropylene extrusion coating, in *Proceedings of 2001 Polymers, Laminations and Coatings Conference*, San Diego, California, 26–30 August, TAPPI: Press, Atlanta, Georgia.
Tucker, 2003, Food and beverage packaging technology, in: Gary S. Tucker (ed.), *Continuous Thermal Processing (Aseptic and Hot-fill): Food Biodegradation and Methods of Preservation*, paragraph 2.3.1.3, Blackwell Publishing Ltd., Oxford, UK, pp. 44–46.

Suggested further reading

Muthwill, F., 1996, Continuous aseptic packaging of liquid foodstuffs using a complex combination of paper, polyethylene and aluminium, in: Bureau, G. and Multon, J.-L. (eds), *Food Packaging Technology*, John Wiley & Sons, Inc., Hoboken. pp. 51–64.
Paine, F.A. and Paine, H.Y. (eds.), 1993, Aseptic packaging using flexible materials, in *Handbook of Food Packaging*, 2nd edn., pp. 278–284, Blackie Academic & Professional Publishing, London,.
Rapak, 2010, *Press Release September 2010*, visit http://www.rapak.com/pages/news/press39.asp
Rapak, 2011, *Press Release November 2011*, visit http://www.rapak.com/pages/news/press46.asp
Yam, K.L., 2009, *The Wiley Encyclopedia of Packaging Technology*, John Wiley & Sons, Inc., Hoboken, New Jersey.

Websites

Alliance for Beverage Cartons and the Environment (ACE), http://www.ace.be/
DuPont, www.dupont.com/packaging
Elopak, http://www.elopak.com/
SIG Combibloc, www.sig.biz
Tetra Pak, www.tetrapak.com
Rapak (bag-in-box), www.rapak.com
Scholle (bag-in-box), www.scholle.com

14 Moulded pulp packaging
Cullen Packaging Ltd, Glasgow, UK

14.1 Introduction

Amongst the largest volume of mass-produced moulded pulp products in Europe and North America which are instantly recognised by consumers are egg trays, shown in Fig. 14.1, and egg boxes. Other examples of typical moulded pulp packaging is shown in Fig. 14.3.

As the name implies, this type of packaging is made from pulp which is moulded to a shape designed to hold and protect the product to be packed.

The primary function of moulded pulp packaging is impact protection against breakage, chipping, etc. This is achieved in the design which locates and stabilises the product. The structural design can also provide a degree of springiness and, therefore, shock amelioration.

14.2 Applications

Moulded pulp packaging includes:

- Trays. In addition to those used for eggs and fruits, such as apples, similar trays with different cavities are used for ampoules and vials.
- Punnet-style trays with handles for mushrooms.
- Top and bottom trays as in Fig. 14.2 are designed to locate and protect bottles and jars.
- Clam shell style containers in which the product is enclosed as for 6/12 eggs or single or multiple bottles, as in Fig. 14.4.
- Corner or edge protectors for ceramics, radiators and furniture, as in Figs 14.5 and 14.6.

Recent new applications of moulded pulp packaging include trays for the location of collapsible tubes, e.g. foods and toiletries, and for electronic products, e.g. car radios, and computer associated products such as flat screens and laptop computers, as in Figs 14.7, 14.11 and 14.12.

Handbook of Paper and Paperboard Packaging Technology, Second Edition. Edited by Mark J. Kirwan.
© 2013 John Wiley & Sons, Ltd. Published 2013 by John Wiley & Sons, Ltd.

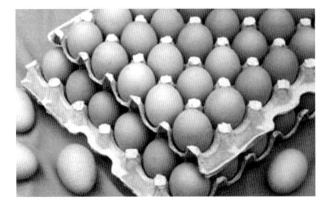

Figure 14.1 Egg trays.

Figure 14.2 Bottle protection with top and bottom fitments.

Figure 14.3 Typical tray shapes.

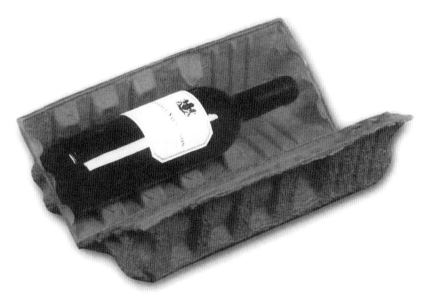

Figure 14.4 Glass bottle protection suitable for postage.

Figure 14.5 Complete kit of shower fittings.

The main end-use industries served are:

- food and drink
- chemicals
- electronics and IT equipment
- furniture
- ceramic wares and radiators.

Figure 14.6 Examples of edge and corner protection.

Figure 14.7 Protection of laptop computer.

Moulded pulp packaging is used in retail, wholesale, industrial and mail-order applications. It can be used for disposable containers in hospitals.

14.3 Raw materials

Every moulded pulp item is produced by mixing water with either wood pulp or pulp made from recovered waste paper/paperboard to a consistency of normally 96% water and 4% fibre. Where required a waterproofing agent such as rosin or a wax emulsion is added. Dye may be added to produce a specific colour.

The fibre used is predominantly made from specific grades of recovered paper and paperboard. However, where required, virgin fibre either chemical or mechanical, bleached or unbleached may be used. Baled recovered paper or pulp is hydropulped and diluted to the correct consistency.

14.4 Production

Initially there were two moulding processes: pressure moulding and suction moulding. In the former, the pulp mixture was fed into the mould and the product formed using hot air under pressure. This process was semi-automatic and therefore of lower output. Additionally, mouldings made by the pressure-forming process had a thicker and more variable wall thickness. The pressure-moulding process was also less suitable for producing more complicated designs and it has been superseded by the suction-moulding process.

In suction-moulding pulp is pumped into the perforated mould where water is removed by vacuum forming. The moulded item is then dried (Fig. 14.8).

The mould is essentially the 'shape' of the product required. All tool sets are two pieces of the 'male and female' type. This results in the moulded pulp product having one side which is smooth and one which is rougher. The mould is perforated to allow the removal of water by suction. It is covered, or lined, depending on the shape and which side is required to have a smooth surface finish, by gauze (see Fig. 14.9). This gauze is made from strands of stainless steel wire 50 μm thick and with a 50 μm gap, or pitch, between parallel strands. It imparts a smooth surface to the surface of the moulded product.

For an egg box produced by this process, the outside surface is required to be smooth so that a printed self-adhesive label can, subsequently, be applied.

The die set would comprise, for the outside, a smooth/female tool/mould and, correspondingly, an inside male/de-mould. This results in a rough finish on the inside of the egg box.

If we require a tray with a smooth *inside* surface, then the inside would be in contact with a smooth male mould and the outside would be in contact with a female de-mould. This imparts a rough finish to the *outside* surface of the moulded product.

Figure 14.10 shows all the components which are needed to make a full tool set, i.e. aluminium backplate, retaining plates, etc.

Figure 14.8 Forming tools.

390 Handbook of Paper and Paperboard Packaging Technology

Figure 14.9 Gauze used to mesh the mould tool surface.

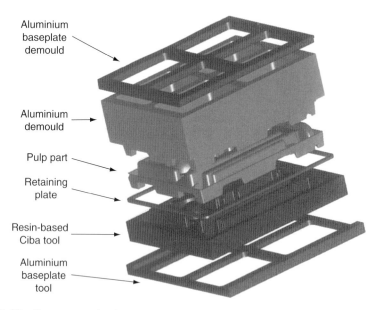

Figure 14.10 Components of a die.

The forming mould is a complex piece of engineering. It is expensive. It is designed by specialist engineers and is normally made from aluminium, though resin-based tooling, e.g. 'Ciba' indicated in Fig. 14.9, can also be used. The use of computer-aided design (CAD) has facilitated mould design and enabled much more complicated designs to be produced than hitherto.

The tool set is made on a milling machine under computer numerical control (CNC). This is based on the output from the CAD tool design and a computer-aided manufacturing (CAM) tool path.

After CNC machining and then manual drilling, the face of the mould used to form the product is covered with the fine mesh gauze. This is applied manually. It is a specialised skill acquired after years of training.

Once the product is formed by vacuum (suction) on the mould, it is transferred to the drying process using a transfer mould which is a mirror image of the forming mould and is made from aluminium or epoxy resin.

Reverse air flow is used to eject the formed piece of pulp from the suction-forming mould onto the transfer mould.

Machines used for pulp moulding range from inexpensive hand-operated machines to fully computer controlled automatic machines capable of producing thousands of tonnes of moulded pulp packaging per annum.

14.5 Product drying

The product is dried in one of two ways. It is either dried by the circulation of heat inside long aluminium gas burning driers or by in-mould thermoforming which uses additional heated moulds to further press and dry the product. This in-mould drying results in a very high quality finished product, which rivals vacuum and thermoformed plastic mouldings in both aesthetics and geometrics.

Along with its low environmental cost in real and life-cycle senses, this new in-mould pressed pulp packaging is now the most popular choice for packing in the electronics and mobile communication industries, e.g. the packaging of products such as modems, mobile phones and computer printers (Figs 14.11 and 14.12).

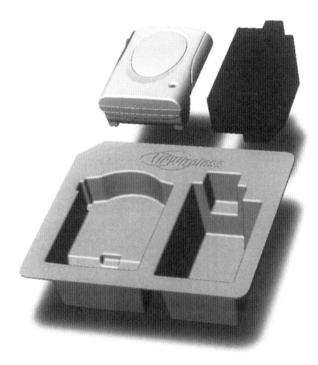

Figure 14.11 Electronics tray pack.

Figure 14.12 Wireless satellite modem packaging.

14.6 Printing/decoration

As already noted, coloured, moulded pulp packaging can be produced by using a dyed pulp. Decorative finishes can be applied by spray gun.

Text such as brand and end-user names, symbols such as the recyclable logo or trademarks and decorative patterns can be incorporated in the mould to produce an embossed or debossed effect.

Multicoloured self-adhesive labels provide the best option for high-quality printing.

Direct printing is also possible on moulded pulp surfaces and whilst the better result is achieved on the smooth side, it is also possible to print a small font size adequately on the rough side, e.g. inside of egg boxes.

14.7 Conclusion

Moulded pulp packaging provides a cost-effective solution for the packaging of a wide range of products, many of which are fragile. There are ancillary benefits in the sense that multiple cavities can be customised for the collation of difficult to handle products. The environmental benefits include features shared by all paper and paperboard-based packaging, namely those of being recyclable, biodegradable and by being based on a naturally renewable resource. Moulded pulp packaging provides an important use for recovered paper and paperboard and by the application of innovative design techniques, moulded pulp packaging provides protection with a minimal weight of material.

Whilst tooling costs are high and order sizes need to be relatively high to spread the cost over a large number of units of production, it is also the case that many of the products protected also have a high unit cost. With many of the other products where the item packed has a relatively low cost, e.g. eggs and fruit, the tooling cost is spread over vast numbers of packs and therefore the cost of the tooling is a lower issue. The use of CAD has significantly extended the complexity of pack design for which tooling can be produced.

Website

www.cullen.co.uk

Appendix: Checklist for a packaging development brief

Why this is needed?

This appendix highlights the reasons why packaging development can be a complicated multi-discipline procedure and suggests a structured outline of how the facts necessary can be worked through to ensure that all the important facts are considered.

The development of a package for a new product or for the relaunch of an existing product is a complex procedure for a package using company (package end-user).

There are several important reasons for this.

There is a wide choice of packaging. The packaging discussed in this book is restricted to that with a paper or paperboard content, and even with this restriction there are 12 different types of paper and paperboard packaging which can be used as a unit pack. Additionally, packaging may be based on plastics, metal or glass and there may well be choices for a combination of pack designs and materials within the overall specification for a product's packaging.

In addition to a unit pack, it is necessary to develop a secondary or bulk pack and to specify the unit load to be used in distribution (tertiary packaging).

Many appearance and protection needs have to be met:

- Protection, which may involve product preservation needs.
- Promotion of the brand for retail products.
- Communication of information to the customer and consumer concerning convenience-in-use safety.
- Logistical needs in transportation, storage and distribution.
- Environmental needs.
- Legal needs.

Coordination with the many company departments/disciplines involved, e.g. Marketing, Production, Engineering, R&D, Purchasing, Quality Control, Legal and Financial.

Coordination with suppliers and customers is necessary.

Coordination is also required to produce an overall time plan. Some activities may have to be completed before others can begin, e.g. market research, product research and the pack design.

The coordination of the design and development of a package therefore requires the cooperation and coordination of the efforts of many people and functions.

Hence, the identification of a comprehensive list of all the requirements is necessary to ensure that all concerned with the project work in an integrated way to achieve a successful result. This body of knowledge is referred to as the ***packaging brief***.

Handbook of Paper and Paperboard Packaging Technology, Second Edition. Edited by Mark J. Kirwan.
© 2013 John Wiley & Sons, Ltd. Published 2013 by John Wiley & Sons, Ltd.

To ensure that no important requirements are omitted, it is useful to define the packaging brief by working through a ***checklist*** of the required information.

....

The following suggested checklist for developing a packaging brief is based on a checklist prepared by 'SiebertHead – Brand Design Consultants', a leading international packaging design company (SiebertHead, 2011) whose permission for its use is acknowledged.

Suggested Checklist for a Packaging Brief

1 General information

1.1 The market

Describe:
   Total size, volume, value and trends in the market
   Brand fragmentation and how the brands share the market
   Regionality/seasonality
   Wholesale distribution analysis
   Retail distribution analysis

What is the consumer profile in terms of age, sex, socio-economic groups and any special characteristics?

1.2 The product

History of the product
Brand share
Description and usage
Frequency of purchase
Sizes and prices
Advantages and disadvantages compared with competitive products
Brand loyalty
Regionality and seasonality
Wholesale distribution analysis
Retail distribution analysis:
   By outlet type
   By counter versus self-service
   Direct to purchaser

Profile of the purchaser now and in the future by age, sex, socio-economic group and any special characteristics
Copy strategy
Planned advertising and promotion
Any features of the product which are too highly regarded to be changed, criticised or curtailed

2 Pack technical design

2.1 Specific design

New design – revolutionary/evolutionary
Redesign/updating – design continuity/analysis of existing elements
Imagery
Standout
Range integration/extension
Competitors
Countries of sale

2.2 Consumer requirements

Ideal quantities, sizes, weights
Size and value for money impression
Importance of style, shape, colour, texture, product visibility
Inspection and handling prior to sale
Ease of take home/delivery
Need for storage
Life required
Opening, closing, reopening requirements
Child-resistant
Tamper evidence
Dispensing/extraction
Measuring
After-use
Protection of the consumer against hazard from product

2.3 Environmental requirements

Appropriate pack for product
Energy use, renewable and non-renewable, in production, transport and distribution
CO_2 emission equivalents, targets
Raw material resources and their sustainability
Waste at each stage of manufacturing and processing
Impact of post-consumer waste
Ability to be reused, recovered, recycled, incinerated with energy recovery, compostability
Degradation in landfill
Absence of toxic materials
Compatibility with recycling infrastructure
Overpackaging
Single packaging material or easily segregated materials

2.4 Merchandising requirements

Size compatibility with own/other displays and fixtures
Use of planograms
Carded products:
   Hanging slot configuration
   Requirement for radius/square corners
   Requirement for price mark
   Product orientation
   Required product visibility
   Need for tactile product examination on display
   Shelf-ready packaging

2.5 Product requirements

Compatibility of packaging materials and product
Product life/deterioration characteristics
Shelf life
Product size, weight, volume
Protection against liquid/moisture, microbiological contamination, gases/odour, temperature extremes/thermal shock
Infestation
Light UV/artificial
Mechanical damage such as crushing, dropping, squeezing, puncture and additional transit hazards

2.6 Production requirements

Forecast throughput
Need for use of existing plant/machinery

Line speeds
Machinery, lines and operatives available
Opportunity for introduction of new processes and machinery
In-plant facilities available for material manufacture, package and pack component manufacture, printing or decoration
Forecast purchase quantity/call offs
Pack cost budget
Prearranged (tied) or favoured suppliers/suppliers not to be approached
Contract packaging

Cost analysis (breakdown) of existing packaging processes:
    Sterilisation/pasteurisation
    Pre-weighing/metering
    Filling
    Closing/sealing/capping
    Checkweighing/headspace/ullage, metal detection
    Labelling/printing/sleeving/other decoration
    Overprinting
    Inspection/quality control
    Collation
    Method of intermediate and outer packaging

Storage:
    Method
    Time requirements
    Stacking, weight requirements
    Pallet size/type
    Palletised height restrictions

Opportunities for:
    Standardisation
    Cost reduction
    Improved pallet utilisation

2.7 Distribution requirements

Method of distribution:
    Mail order
    Wholesaler
    Retailer
    Direct/internet

Estimated length of time stored at:
    Warehouse
    Wholesaler (including cash and carry)
    Retailer
    Consumer

Method of:
    Transportation
    Loading
    Warehousing
    Wholesaler storage (including cash and carry)
    Retail storage

Importance of divisibility at all stages
Transportation weight requirements
Requirements for standardisation of range

2.8 Legal requirements

National, regional and international trading requirements, e.g. European Union, USA and UN.
Transportation

3 Pack graphic design

3.1 Specific design objectives

New design:
   Revolutionary/evolutionary
Redesign/updating:
   Design continuity/analysis of existing elements
Imagery
Branding
Standout
Range integration/extension
Competitors
Countries of sale

3.2 Consumer requirements

Size impression
Appropriate pack for product
Importance of package appearance in use
Preferred or associated colours (brand or house colour)
Usage instructions

3.3 Special requirements

Anticipated future copy
Language variants/multilingual copy
Price flashes
Promotional flashes
Bar codes
Need to link with advertising/point of sale
Pack to incorporate counterfeiting feature

3.4 Retail display requirements

Shelf sizes
Estimated display life
Retail-ready display
Position of display:
   Shelf
   Counter
   Dispenser
   Check-out
   Window
Position in relation to purchaser:
   Eye level
   Above/below
Display by brand group/product group
Panels most often seen:
   Front/back/top/sides
Lighting levels at point of sale
Proximity of competitor/other products

3.5 Wholesale display requirements (including cash and carry)

Need for coding/simplification on outer (secondary) packaging
Need for sub-divisible sales units
Degree of branding necessary

3.6 Production requirements

Number of colours/varnish (matt or gloss)
Size of print area

Overprint areas
Print method
Materials
Quantity
Run length
Cost limitations
Artwork requirements

3.7 Legal requirements

Cautionary (safety) information to be included, including Braille for some products/markets

Legislation to be observed:
    Correct style and size of copy
    Content declaration, weight/volume
    Ingredients copy, order and size
    Manufacturer's name and address
    Illustrations/photography appropriate to product
    Special claims
    Hazard warning requirements
    Inclusion of special national or international symbols where required, such as the 'e' symbol, where appropriate within the EU
    Environmental symbols

Restraints:
    Materials
    Processes
    Inks
    Pack style/form

Reference

SiebertHead 2011, SiebertHead Ltd., visit http://www.sieberthead.com

Further reading

Coles, R., (2011), See paragraphs 1.5 packaging strategy and 1.6 packaging design and development, in *Food and Beverage Packaging Technology*, 2nd edn., R. Coles and M.J. Kirwan (eds), pp. 10–27, Blackwell Publishing Ltd., Oxford.

Index

acidification 51
acid rain 69
additives for strength 11
adhesion 31, 32, 45
air knife coating 18
air, manufacturing emissions 68, 69
alum size 11, 12
aluminium foil 5, 20–22, 30, 32, 46, 93, 97–9, 101–2, 106–7, 109, 129, 208, 228, 268, 319, 375
Anaconda sealing (composite cans) 195
anti-counterfeiting 307
appearance properties 26–34
aseptic packaging/processing 355–7, 374–5

bags *see* paper bags
bag in box 315, 341, 350, 361, 368,
bag papers 21, 175–7
barrier properties, (introductory summary) 5, 95–99
basis weight 35
beating 12, 13
biodiversity 50, 55, 56, 59, 86
bioplastics 5, 268
blade coating 18
bleaching process 10
brightness 11, 27, 28
brushing 19
burst resistance 38, 39, 223, 224, 322, 323

calendering 18, 21
caliper (calliper) 4, 35
carbon cycle 86
carbon 'fixing' 55, 81, 82
carbon footprint 81
carton profile testing 291

cast coating 19
cellulose 6–9, 46, 309
chemi-thermomechanical pulp, CTMP 9, 357
chemical pulping 9, 10
chemical recovery 67
chemicals used in paper manufacture 11, 12, 47, 48, 67–68
classifications of recovered waste paper and paperboard 10, 11, 75–7
climate change 51, 52
coated oriented polypropylene (OPP or BOPP) 103, 107, 130, 133–5, 268, 299
coated with an insect repellent 5
coating with mineral pigment 11, 18, 19
Cobb test 44, 45
colour 26–8
combined heat and power (CHP) 65, 66, 78, 85
composite cans
    Anaconda sealing of liner 195
    applications 189
    barrier properties 195, 196
    construction of sidewalls 195–196
    convolute winding 185
    definition 185
    design 190–194
    gas flushing 188, 197
    glossary of composite can related terms 201–3
    historical background 187–9
    label printing 197–199
    labeling and printing options 197–9
    labeling options 199
    linear draw 186
    manufacture 185–7, 194–7

composite cans (*cont'd*)
  opening/closing options 191–193
    single wrap 187
    skiving of body ply 196
    spiral winding 185
compression strength 40–41, 305–6, 324–6
consumption of paper and board 2
corrosion inhibition, vapour-phase inhibitor 5
corrugated fibreboard
  applications 313
  description 313–15
  pack designs 315–17
  product safety 320–321
  recycling 335–338
  shelf/retail ready packaging 320
  sustainability 321, 335–8
corrugated fibreboard manufacture 326–34
  adhesives 328–330, 335
  die cutting 331
  folder-gluer 331–333
  good manufacturing practice 335
  printing 333–335
  stitching 331
  taping 331
  wash boarding 330–331
corrugated fibreboard performance
  bending stiffness 322
  box and pallet tests 323–5
  burst strength 322
  cold glue 335
  compression crush test, (CCT) 322
  Concora Medium Test (CMT) 328
  edge crush test, (ECT) 322
  flat crush test 321
  flatness 330
  flute size 314
  grammage 321, 327
  hot-melt adhesives 335
  inclined plane impact test 324–325
  McKee formula 324
  moisture content 323
  pin adhesion test 323
  properties and test methods 321–3
  puncture 323
  thickness (caliper) 321
counterfeit prevention 307
creaseability 41

creasing 4, 41–2, 280–292, 329, 331–3
cross direction (CD) 16, 35
cylinder mould forming 17

debris, e.g. clumps of fibres, particles of coating etc. 33, 279–80
deforestation 58
de-inking 66, 69, 77
Dennison Wax Number 31
digital printing *see* printing processes
dimensional stability 35–7, 43–4,
Dow Jones Sustainability Indexes 53
Dust and debris *see* debris

electron beam ink drying, EB 32
elemental chlorine free, ECF 10
embossing 292–3
energy recovery 77–79
energy sourcing 62–66
equilibrium moisture content 35–7
ethylene vinyl alcohol (EVOH) 5, 99, 359
extrusion coating with polyethylene (PE) 5, 100–102

fibre drums
  applications 205–207
  barrier, functional 213–214
  cleanliness 214
  drum designs 205–208
  manufacture 198, 200, 205–12
  opening/closing systems 210–212
  performance 212–14
  printing/decoration 214–15
  standards 216
  waste management 215
fibre recovery 74–7
fibre separation 8–10
fibre, sources of 7–8
fillers in paper manufacture 11
finishing 19–20
flatness 37, 43–4
flexible packaging
  applications for foods, electrical, chocolate and sugar confectionary, DIY, horticulture etc. 95–99
  bags 92, 100, 109–10, 114–16

cap liners, wads and diaphragms 91, 118–19
direct seal paper 112
ethylene vinyl acetate, EVA 95, 101–2
horizontal f/f/s, flow wrap 116, 117, 119
lids, lidding including die-cut lids 92, 100, 109, 113, 117
manufacture 99–104
medical packaging 109–14
medium density PE 99
microcrystalline wax 98, 100
oriented polypropylene film, OPP 104, 107
overwrapping 91, 118
packaging machinery 114–18
paper cushioning 121–122
polyamide 104, 114
pouches 91, 93, 109, 113–114, 116
properties 92, 94–9
regenerated cellulose film (RCF) 94
release lacquer 104
roll packs 114
sachets 91, 93, 109, 114, 116
sealing tapes 121
solvent based coatings 100
strip packs 109, 115
Surlyn® 95, 99, 102, 114, 359
tea and coffee 119–121
Tyvek® 114
vertical f/f/s 115
waxed paper 100–101, 118
flexible packaging machinery 114–118
horizontal form, fill, seal machine 114–117
horizontal form/fill/seal pouch/sachet machine 116
overwrapping machine 118
roll wrap 114
thermoforming, labelling and lidding, in-line 114, 117
tea and coffee bag machines 119–121
twist wrap 118
vertical form, fill, seal machines 115, 116
flexible packaging manufacture 99–104
dispersion coating, paper/PVdC 100
dry bonding, lamination 106–107
extrusion coating with PE 100–102
extrusion lamination 107–108
hot-melt coating 103–104
lamination with water based adhesives 105–106
lamination with wax 108–109
metallization 102–103
printing 99–100
water based coatings 100
wax coatings 100
wax lamination 108–109
flexible packaging performance
barrier properties 95–99
cold seal 104–105
ethylene oxide, EtO, sterilisation 109–111
gas barrier e.g. oxygen, carbon dioxide and nitrogen 99
heat sealing 95
light barrier 99
moisture and moisture vapour 97–98
oil, grease and fat barrier 99
radiation, gamma, sterilisation 109–111
steam sterilisable medical packaging paper 109–111
flexography *see* printing processes
fluorescent whitening agents (FWA) 10–11
fluorocarbon 5, 12
foldability 41
folding boxboard, (FBB) 23–4
folding carton manufacture 277–96
carton dimension checking 290–291
computer aided design, CAD 277
creasing 280–292
creasing internal delamination 290
crush cutting 284
cutting and creasing 288–92
dimension checking 286–287
dust and debris 279–80
embossing 292–3
flatbed platen dies 281–284
folder-gluer 294, 331–3
folding 287–292
gluing 294–5
hot foil stamping 293–4
notches 282
pressure or shear cutting 285–286

folding carton manufacture (*cont'd*)
  printing 277–80
  rotary dies 284–287
  warming up times for printing efficiency 277–9
  waxing 296
  windowing 295
folding carton performance
  coefficient of friction 302
  compression strength 266–7, 281, 291, 305–6
  folding, foldability 287–292, 296–300
  high speed video 303
  hot melt gluing in packing 298–300
  microwaveable 307–8
  odour and taint testing 301–11
  ovenable 307–8
  packaging line efficiency 300–303
  packing operation 296–303
  product and consumer safety 309–10
  runnability, packing line efficiency 300–303
  storage 300
  susceptor 308
  timing, cartonning machine 301–2
  warming up times for packing efficiency 300
folding cartons
  applications 266–7
  carton erecting/closing 296–303
  coatings/laminations 268
  definition 265
  designs 269–77
  distribution and storage 303–6
  easy open features 276–81, 298
  grammage, basis weight 266
  paperboard grades 22–5, 267–8
  point of sale, purchase, dispensing 306–307
  storage *see* distribution
  tamper, pilfer evidence 276
  thickness, (calliper) 266
food waste during distribution 54
forest certification 59, 85
forests 54–61, 84–6
fossil based energy 52, 55–6, 62–6, 68, 71, 77–8, 82

forming using wires 15–16
forming using vats/cylinder moulds 16
Fourdrinier 16
fresh water 51–2, 66–7
friction glazing 19

gas chromatography 310–11
gas flushing 92, 95–99, 188, 197, 395
general information required for a new package 394
glassine 18, 21, 30
global warming (alleged) 52, 55, 68
gloss 30
gluability 45–6
grammage 3, 35, 48–49
gravure *see* printing processes
greaseproof 21–2
greenhouse effect 55, 68, 78

hand made paper 13–15
high speed video 303
hygrosensitive 35–7
hysteresis effect 36

IGT printability test 31
impregnating papers 22
incineration 64, 73, 77–9
ink and varnish absorption and drying 31–2
insect repellent treatment 5
intelligent paperboard packaging 308–9
internal sizing 11, 12
Inverform process 16

Kraft or Sulphate process 9, 10

label manufacture
  bronzing 156
  die cutting 156–159
  digital die-cutting 158
  digital printing 155–6
  direct thermal printing 154
  dot matrix printing 154
  embossing 156
  guillotining 156–157
  hot foil blocking/stamping 152–3
  lacquering 156

laser cut die boards 158
on-press slitting and sheeting
 157–159
printing 145–156
punching 156–157
variable information printing,
 electronic 153–5
labels
active, intelligent labels 142–144
adhesives 136–139
application 159–163
collar or neck labels 142
colour change sterilisation labels 144
diffractive optically variable image
 devices, DOVIDs 145
guilloches 142
hologram labels 142
hot melt adhesive 137
in-mould labels 133
label paper substrates 129–31
label testing methods 164–7
leaflet and book labels 142
linerless self-adhesive labels 132
logistics labels 141
optically variable devices, OVIDs
 142, 145
pre gummed labels 125
primary labels 117, 140
RFID labels 141
secondary labels 117, 141
security labels 142
self adhesive labels 126–129,
 130–132
smart labels 142–4
solvent-based adhesives 138
swing tickets 142
tactile labels 142
tags 142
tie on labels 142
types of labels 128–36
waste and environmental issues
water-based adhesives 137–8
wet glue labels 125–126, 128–129
laminating papers 22, 32
lamination with plastic films 22
land quality deterioration 51
landfill 79, 85, 86
landfill gas 63 4

life cycle assessment, (LCA) 79–81,
 379–381
lightfastness 34
lignin 8–9
liquid packaging, paperboard-based
applications 378–9
aseptic processing 374–5
barriers 358–60
distribution, transit packaging 376–7,
 382
heat sealing 353–60, 374–5
LCA, liquid packaging 79–81,
 379–81
materials 391–4
opening, reclosure and tamper
 evidence 369–74
pack designs 361–9
post packaging sterilisation,
 retortable 375–6
printing and converting 360–361
recovery 381–2
recycling 381–2
resource reduction 379
systems approach 282
transit packaging 376–7

machine direction, (MD) 15–16
machine glazed, MG, finish 17, 21. 30
machine glazing or 'Yankee' cylinder 17
machine, MF, finish 21
manufacture of paper and paperboard 1–2,
 13–19
Marbach Crease Bend Tester 292
mass spectrometry 47, 311
mechanical pulping 8–9, 23–4
microcreping 22
mineral pigment for surface coating 11,
 18–19
Modulus of Elasticity (E) or Young's
 Modulus 37, 40
moisture content 17, 35–6, 43
mould inhibitor 5, 22
moulded pulp packaging
applications 385–8
manufacture 389–92
pack design 385–8
printing and decoration 392
raw material 388–9

multi-ply, sheet forming 16–17
multiwall paper sacks
   applications 217–18, 246
   baling systems 247–8
   definition 217
   designs, open mouth 219–22
   designs, valved sacks 222–5
   environmental position 250–251
   filler (filter) cords 231
   handles 218, 223, 231
   manufacturing tolerances 250
   printing 232
   sack identification 245
   standards for paper sacks 249
   sewn closures 225–6
   tapes 221
   threads 221
   valve design 223–5
multiwall sacks, materials 226–30
   adhesives 231
   bitumen laminated kraft 230
   coated kraft 227–8
   creped kraft 227
   extensible kraft 227
   laboratory testing of paper sack materials 232–5
   laminated kraft, (aluminium foil, plastic film) 228, 230
   laminated kraft, (woven and non-woven plastics) 230
   PE coated kraft 229
   PVdC coated kraft 230
   scrim and cloth laminates 230
   silicone coated kraft 229
   use of PE liners 217, 220
   waxed kraft 230
multiwall sacks, performance
   filling, weighing and closing 237–245
   sack flattening and shaping 247
   sack testing, drop testing 236–7
   slip resistance, slip resistant agents 232, 235
   weighing, filling and closing systems 237–48

new product/package
   development 393–398

odour *see* taint and odour neutrality
opacity 11, 30–31
optical brightening agents (OBA) 10–11
organic recycling, composting 75, 77
ozone depletion 50

package technical design 394–396
package graphical/appearance design 397–398
packaging brief/specification factors 393–398
packaging development considerations 393–398
packaging needs 4, 47
packaging requirements 25–6
   appearance properties 26–34
   performance properties 34–47
packaging waste 62, 73–9
packaging waste recovery 73–9
palletizing 25
paper and paperboard
   environmental impact of manufacture and use 61–73
   fibre separation and processing 8–10, 12–13
   manufacture 13–20
   raw materials 6–8, 11–12
paper and paperboard packaging types
   composite cans 183–203
   corrugated fibreboard packaging 313–339
   fibre drums 205–216
   flexible packaging 91–213
   folding cartons 265–312
   liquid packaging 253–383
   moulded pulp packaging 385–292
   multiwall paper sacks 217–251
   paper bags 169–182
   paper labels 125–168
   rigid boxes 253–263
   solid fibreboard packaging 341–352
paper and paperboard packaging
   uses 3–4
   requirements 3–5, 25–26
paper bags
   applications, uses 169–70
   bag designs 170–175

bag papers 175–7
carrier bags 174–175
dust bags 179
flat bags 170–172
heat sealed paper bags 178
historical development 169, 181
manufacture 177–9
paper for paper bags 175–177
performance testing 179–80
printing 180–181
printing paper bags 180–181
satchel bags 170–172
SOS bags 170–175
steam sterilized hospital bags 179
stringing 179
strip window bags 172
testing paper bags 179–180
tin ties 179
types of paper bags 170–175
uses of paper bags 189–190
wicketting 179
paper labels *see* labels
Parker PrintSurf (PPS), Bendtsen and Sheffield instruments 28
performance properties 34–47
photosynthesis 55–6
pigment coating waste 73
Pira Cartonboard Creaser 291
Pira Crease Tester 292
plantation forestry 57–8
ply bond (interlayer) strength 12, 42–3
points, (thickness) 4, 35
pollution of rivers, lakes and seas 51–2, 69–71
polyethylene terephthalate (PET or PETE) 5, 68, 99,196
polymethyl pentene (PMP) 5
polypropylene (PP) 5, 68, 172, 272, 360, 375
polyvinylidene chloride (PVdC) 5, 95, 97, 99, 100
population 51, 52, 58
porosity 44
press section 10, 17
print mottle 32
printability 11, 30–31
printing processes

digital 31, 94, 155–6, 277–80, 333–4
flexography 31, 94, 133, 138–40, 148, 180, 197, 232, 277–80, 332–4, 351
gravure 31, 94, 133, 150–151, 180, 197, 277–80, 361
letterpress 31, 133, 146–7, 277–80, 351
lithography 30, 133, 149–50, 197, 277–80, 333–4, 351
silk screen 31, 151–2, 180, 214, 277–80, 333–4
printing variable information (VIP), electronic
dot matrix printing 154
direct thermal printing 154
laser printing 153–4
ink jet printing 154–5
ion deposition 153
thermal transfer printing 154
product safety 22, 24, 47, 216, 309, 335
pulping 3, 7–10
puncture resistance 39

quality standards 48

rain forests 57–9
raw materials for paper and paperboard manufacture 6–12
recovered fibre 10–11, 20, 24–5
recovered waste paper and paperboard classification 11
recycling 10–11, 73, 75–7, 84–6
refiners 12
regenerated cellulose film, (RCF), 94
relative humidity (% RH) 8, 34–7, 300
renewable energy 56, 62–4, 77–81, 83, 86
rigid boxes
adhesives 257
applications 256
definition 253
designs 254–5, 257–8, 261–2
manufacture 258–62
paper and board used 256–7
printing 257
strengths, benefits and weaknesses 253–4
roll bar 18
rub resistance 33

sack kraft 22, 226–30
sealing 4, 20, 45–6
    see also *the various types of packaging*
secondary fibre *see* recovered fibre
sheeting 20, 53
shelf/retail ready packaging
    corrugated fibreboard 320
    solid fibreboard 350
silicone 5, 21, 222, 229
size, internal 11–12, 44–5
size, surface 11–12, 17, 44–5
slitting 20, 33
solid bleached board (SBB or SBS) 22–3, 267
solid fibreboard packaging
    applications 341–2, 345–50
    bag-in-box liquid containers 350
    cutting and creasing 252
    cutting and creasing 352
    description 341
    manufacture 351–2
    materials 341, 350
    pack designs 342–5
    partitions, divisions and fitments 348–50
    printing 351
    recycling, waterproofed collection boxes 350
    shelf/retail ready packaging 350
    slip sheet 347
    waste management 352
    water and water-vapour resistance 350–351
solid unbleached board (SUB) 23, 267
solid waste residues 62, 70, 73
specifications 48
spots (hickies) causing print blemishes 33, 279–80
starch 12, 17
sterilisable, retortable packaging/processing 375–6
stiffness 17, 24, 30–40, 305, 322, 324
stock preparation 12, 20, 27, 39
stretch or elongation 22, 35, 37, 38, 41
substance 3, 35
sulphite process 9
supercalender 18
supercalendered, SC, finish 21

surface cleanliness 33, 279–80
surface pH 32
surface smoothness 28
surface strength 12, 31
surface structure, sheet 28–30
surface structure of fibre 12
surface tension 32
sustainable development 33–34, 53–54, 80, 85–6
sustainable forest management 58, 61, 86

taint and odour, neutrality/testing 46, 310–311
tea and coffee packaging 119–112
tearing resistance 38
temperature equilibrium 36, 43
tensile energy absorption, or TEA 38, 233
tensile strength 35, 37–8, 41
thermomechanical pulp, TMP 9
thickness (caliper) 4, 7, 17, 20, 34, 35, 40, 48–49
tissues (tea, coffee and wrapping) 1, 4, 20, 35, 46
totally chlorine free, TCF 10
transport, environmental concerns 68, 69, 73, 79
types of packaging 4

unlined chipboard 25
urea and melamine formaldehyde, (wet strength) 12, 39, 227

vapour phase inhibitor 5, 22
varnishability 30–31
vats and vat forming 13–14, 16–17
vegetable parchment 21
virgin, or primary, fibre 7–10, 22, 24, 46

waste management options 72, 74–9
waste minimization 74
water 51, 52, 54, 61, 66–7, 69–71, 85
water footprint 66
water absorbency 44–5, 350–351
water, manufacturing emissions 69–71
water, WF, finish 21
water usage 66–67

wax size 17
wax treatment 5, 12, 22, 98–100, 103, 108–9, 230, 296
wettability 31, 32
white lined chipboard, (WLC) 24–5, 27, 267
whiteness 11, 27–8
whitening (bleaching) 10, 11, 27

wicking tests 45
wire forming 13, 15, 16, 19
wood usage 56, 58–61
world concerns, population, resources etc. 51–53
world (apparent) consumption 1, 2
world production 1, 2

Food Science and Technology

GENERAL FOOD SCIENCE & TECHNOLOGY, ENGINEERING AND PROCESSING

Title	Author	ISBN
Organic Production and Food Quality: A Down to Earth Analysis	Blair	9780813812175
Handbook of Vegetables and Vegetable Processing	Sinha	9780813815411
Nonthermal Processing Technologies for Food	Zhang	9780813816685
Thermal Procesing of Foods: Control and Automation	Sandeep	9780813810072
Innovative Food Processing Technologies	Knoerzer	9780813817545
Handbook of Lean Manufacturing in the Food Industry	Dudbridge	9781405183673
Intelligent Agrifood Networks and Chains	Bourlakis	9781405182997
Practical Food Rheology	Norton	9781405199780
Food Flavour Technology, 2nd edition	Taylor	9781405185431
Food Mixing: Principles and Applications	Cullen	9781405177542
Confectionery and Chocolate Engineering	Mohos	9781405194709
Industrial Chocolate Manufacture and Use, 4th edition	Beckett	9781405139496
Chocolate Science and Technology	Afoakwa	9781405199063
Essentials of Thermal Processing	Tucker	9781405190589
Calorimetry in Food Processing: Analysis and Design of Food Systems	Kaletunç	9780813814834
Fruit and Vegetable Phytochemicals	de la Rosa	9780813803203
Water Properties in Food, Health, Pharma and Biological Systems	Reid	9780813812731
Food Science and Technology (textbook)	Campbell-Platt	9780632064212
IFIS Dictionary of Food Science and Technology, 2nd edition	IFIS	9781405187404
Drying Technologies in Food Processing	Chen	9781405157636
Biotechnology in Flavor Production	Havkin-Frenkel	9781405156493
Frozen Food Science and Technology	Evans	9781405154789
Sustainability in the Food Industry	Baldwin	9780813808468
Kosher Food Production, 2nd edition	Blech	9780813820934

FUNCTIONAL FOODS, NUTRACEUTICALS & HEALTH

Title	Author	ISBN
Functional Foods, Nutraceuticals and Degenerative Disease Prevention	Paliyath	9780813824536
Nondigestible Carbohydrates and Digestive Health	Paeschke	9780813817620
Bioactive Proteins and Peptides as Functional Foods and Nutraceuticals	Mine	9780813813110
Probiotics and Health Claims	Kneifel	9781405194914
Functional Food Product Development	Smith	9781405178761
Nutraceuticals, Glycemic Health and Type 2 Diabetes	Pasupuleti	9780813829333
Nutrigenomics and Proteomics in Health and Disease	Mine	9780813800332
Prebiotics and Probiotics Handbook, 2nd edition	Jardine	9781905224524
Whey Processing, Functionality and Health Benefits	Onwulata	9780813809038
Weight Control and Slimming Ingredients in Food Technology	Cho	9780813813233

INGREDIENTS

Title	Author	ISBN
Hydrocolloids in Food Processing	Laaman	9780813820767
Natural Food Flavors and Colorants	Attokaran	9780813821108
Handbook of Vanilla Science and Technology	Havkin-Frenkel	9781405193252
Enzymes in Food Technology, 2nd edition	Whitehurst	9781405183666
Food Stabilisers, Thickeners and Gelling Agents	Imeson	9781405132671
Glucose Syrups – Technology and Applications	Hull	9781405175562
Dictionary of Flavors, 2nd edition	De Rovira	9780813821351
Vegetable Oils in Food Technology, 2nd edition	Gunstone	9781444332681
Oils and Fats in the Food Industry	Gunstone	9781405171212
Fish Oils	Rossell	9781905224630
Food Colours Handbook	Emerton	9781905224449
Sweeteners Handbook	Wilson	9781905224425
Sweeteners and Sugar Alternatives in Food Technology	Mitchell	9781405134347

FOOD SAFETY, QUALITY AND MICROBIOLOGY

Title	Author	ISBN
Food Safety for the 21st Century	Wallace	9781405189118
The Microbiology of Safe Food, 2nd edition	Forsythe	9781405140058
Analysis of Endocrine Disrupting Compounds in Food	Nollet	9780813818160
Microbial Safety of Fresh Produce	Fan	9780813804163
Biotechnology of Lactic Acid Bacteria: Novel Applications	Mozzi	9780813815831
HACCP and ISO 22000 – Application to Foods of Animal Origin	Arvanitoyannis	9781405153669
Food Microbiology: An Introduction, 2nd edition	Montville	9781405189132
Management of Food Allergens	Coutts	9781405167581
Campylobacter	Bell	9781405156288
Bioactive Compounds in Foods	Gilbert	9781405158756
Color Atlas of Postharvest Quality of Fruits and Vegetables	Nunes	9780813817521
Microbiological Safety of Food in Health Care Settings	Lund	9781405122207
Food Biodeterioration and Preservation	Tucker	9781405154178
Phycotoxins	Botana	9780813827001
Advances in Food Diagnostics	Nollet	9780813822211
Advances in Thermal and Non-Thermal Food Preservation	Tewari	9780813829685

For further details and ordering information, please visit www.wiley.com/go/food

Food Science and Technology from Wiley-Blackwell

SENSORY SCIENCE, CONSUMER RESEARCH & NEW PRODUCT DEVELOPMENT

Sensory Evaluation: A Practical Handbook	Kemp	9781405162104
Statistical Methods for Food Science	Bower	9781405167642
Concept Research in Food Product Design and Development	Moskowitz	9780813824246
Sensory and Consumer Research in Food Product Design and Development	Moskowitz	9780813816326
Sensory Discrimination Tests and Measurements	Bi	9780813811116
Accelerating New Food Product Design and Development	Beckley	9780813808093
Handbook of Organic and Fair Trade Food Marketing	Wright	9781405150583
Multivariate and Probabilistic Analyses of Sensory Science Problems	Meullenet	9780813801780

FOOD LAWS & REGULATIONS

The BRC Global Standard for Food Safety: A Guide to a Successful Audit	Kill	9781405157964
Food Labeling Compliance Review, 4th edition	Summers	9780813821818
Guide to Food Laws and Regulations	Curtis	9780813819464
Regulation of Functional Foods and Nutraceuticals	Hasler	9780813811772

DAIRY FOODS

Dairy Ingredients for Food Processing	Chandan	9780813817460
Processed Cheeses and Analogues	Tamime	9781405186421
Technology of Cheesemaking, 2nd edition	Law	9781405182980
Dairy Fats and Related Products	Tamime	9781405150903
Bioactive Components in Milk and Dairy Products	Park	9780813819822
Milk Processing and Quality Management	Tamime	9781405145305
Dairy Powders and Concentrated Products	Tamime	9781405157643
Cleaning-in-Place: Dairy, Food and Beverage Operations	Tamime	9781405155038
Advanced Dairy Science and Technology	Britz	9781405136181
Dairy Processing and Quality Assurance	Chandan	9780813827568
Structure of Dairy Products	Tamime	9781405129756
Brined Cheeses	Tamime	9781405124607
Fermented Milks	Tamime	9780632064588
Manufacturing Yogurt and Fermented Milks	Chandan	9780813823041
Handbook of Milk of Non-Bovine Mammals	Park	9780813820514
Probiotic Dairy Products	Tamime	9781405121248

SEAFOOD, MEAT AND POULTRY

Handbook of Seafood Quality, Safety and Health Applications	Alasalvar	9781405180702
Fish Canning Handbook	Bratt	9781405180993
Fish Processing – Sustainability and New Opportunities	Hall	9781405190473
Fishery Products: Quality, safety and authenticity	Rehbein	9781405141628
Thermal Processing for Ready-to-Eat Meat Products	Knipe	9780813801483
Handbook of Meat Processing	Toldra	9780813821825
Handbook of Meat, Poultry and Seafood Quality	Nollet	9780813824468

BAKERY & CEREALS

Whole Grains and Health	Marquart	9780813807775
Gluten-Free Food Science and Technology	Gallagher	9781405159159
Baked Products – Science, Technology and Practice	Cauvain	9781405127028
Bakery Products: Science and Technology	Hui	9780813801872
Bakery Food Manufacture and Quality, 2nd edition	Cauvain	9781405176132

BEVERAGES & FERMENTED FOODS/BEVERAGES

Technology of Bottled Water, 3rd edition	Dege	9781405199322
Wine Flavour Chemistry, 2nd edition	Bakker	9781444330427
Wine Quality: Tasting and Selection	Grainger	9781405113663
Beverage Industry Microfiltration	Starbard	9780813812717
Handbook of Fermented Meat and Poultry	Toldra	9780813814773
Microbiology and Technology of Fermented Foods	Hutkins	9780813800189
Carbonated Soft Drinks	Steen	9781405134354
Brewing Yeast and Fermentation	Boulton	9781405152686
Food, Fermentation and Micro-organisms	Bamforth	9780632059874
Wine Production	Grainger	9781405113656
Chemistry and Technology of Soft Drinks and Fruit Juices, 2nd edition	Ashurst	9781405122863

PACKAGING

Food and Beverage Packaging Technology, 2nd edition	Coles	9781405189101
Food Packaging Engineering	Morris	9780813814797
Modified Atmosphere Packaging for Fresh-Cut Fruits and Vegetables	Brody	9780813812748
Packaging Research in Food Product Design and Development	Moskowitz	9780813812229
Packaging for Nonthermal Processing of Food	Han	9780813819440
Packaging Closures and Sealing Systems	Theobald	9781841273372
Modified Atmospheric Processing and Packaging of Fish	Otwell	9780813807683
Paper and Paperboard Packaging Technology	Kirwan	9781405125031

For further details and ordering information, please visit www.wiley.com/go/food